The Origins of Geology in Italy

edited by

Gian Battista Vai
Dipartimento di Scienze della Terra e Geologico-Ambientali
Università di Bologna Alma Mater Studiorum
Via Zamboni 67
I-40127 Bologna
Italy

and

W. Glen E. Caldwell
Department of Earth Sciences
The University of Western Ontario
London, Ontario, N6A 5B7
Canada

THE
GEOLOGICAL
SOCIETY
OF AMERICA

Special Paper 411

3300 Penrose Place, P.O. Box 9140 ▪ Boulder, Colorado 80301-9140, USA

2006

Copyright © 2006, The Geological Society of America, Inc. (GSA). All rights reserved. GSA grants permission to individual scientists to make unlimited photocopies of one or more items from this volume for noncommercial purposes advancing science or education, including classroom use. For permission to make photocopies of any item in this volume for other noncommercial, nonprofit purposes, contact the Geological Society of America. Written permission is required from GSA for all other forms of capture or reproduction of any item in the volume including, but not limited to, all types of electronic or digital scanning or other digital or manual transformation of articles or any portion thereof, such as abstracts, into computer-readable and/or transmittable form for personal or corporate use, either noncommercial or commercial, for-profit or otherwise. Send permission requests to GSA Copyright Permissions, 3300 Penrose Place, P.O. Box 9140, Boulder, Colorado 80301-9140, USA.

Copyright is not claimed on any material prepared wholly by government employees within the scope of their employment.

Published by The Geological Society of America, Inc.
3300 Penrose Place, P.O. Box 9140, Boulder, Colorado 80301-9140, USA
www.geosociety.org

Printed in U.S.A.

GSA Books Science Editors: Marion E. Bickford and Abhijit Basu

Library of Congress Cataloging-in-Publication Data

The origins of geology in Italy / edited by Gian Battista Vai and W. Glen E. Caldwell.
    p. cm.—(Special paper ; 411)
  Includes bibliographical references.
  ISBN-13 978-0-8137-2411-9 (pbk.)
  ISBN-10 0-8137-2411-2 (pbk.)
    1. Geology--Italy--History. I. Vai, Gian Battista. II. Caldwell, W. G. E., 1932-.
  Special papers (Geological Society of America) ; 411.

QE13.I8 O75 2006
551.09--dc22

2006043464

Cover: Luigi Ferdinando Marsili's Orographic Map of Major Mountain Trends in Europe (unpublished and unfinished manuscript map drawn before 1728; see chapter by Vai (chapter 7) for details; from Marsili, Fondo Marsili, ms. 90, A, 21, c. 148/1555; courtesy of Biblioteca Universitaria di Bologna).

10 9 8 7 6 5 4 3 2 1

In memory of
Nicoletta Morello
1946–2006

Contents

Preface vii

1. *Italian gemology during the Renaissance: A step toward modern mineralogy* 1
   Annibale Mottana

2. *Agricola and the birth of the mineralogical sciences in Italy in the sixteenth century* 23
   Nicoletta Morello

3. *Geology and the artists of the fifteenth and sixteenth centuries, mainly Florentine* 31
   David Branagan

4. *Ulisse Aldrovandi and the origin of geology and science* 43
   Gian Battista Vai and William Cavazza

5. *Kircher and Steno on the "geocosm," with a reassessment of the role of
   Gassendi's works* 65
   Toshihiro Yamada

6. *Steno, the fossils, the rocks, and the calendar of the Earth* 81
   Nicoletta Morello

7. *Isostasy in Luigi Ferdinando Marsili's manuscripts* 95
   Gian Battista Vai

8. *Luigi Ferdinando Marsili (1658–1730): A pioneer in geomorphological and
   archaeological surveying* 129
   Carlotta Franceschelli and Stefano Marabini

9. *Mattia Damiani (1705–1776), poet and scientist in eighteenth century Tuscany* 141
   Giancarlo Scalera

10. *The "classification" of mountains in eighteenth century Italy and the lithostratigraphic
    theory of Giovanni Arduino (1714–1795)* 157
    Ezio Vaccari

11. *The geological work of Gregory Watt, his travels with William Maclure in Italy (1801–1802),
    and Watt's "proto-geological" map of Italy (1804)* 179
    Hugh S. Torrens

12. *Giovan Battista Brocchi's Rome: A pioneering study in urban geology* 199
    Renato Funiciello and Claudio Caputo

13. *Leopoldo Pilla (1805–1848): A young combatant who lived for geology
    and died for his country* 211
    Bruno D'Argenio

Preface

Despite the term *geology* being coined in Bologna in 1603 by Ulisse Aldrovandi (1522–1605) and other Italian naturalists holding positions of prominence in the growth of this most historical of the natural sciences during the sixteenth to eighteenth centuries, the magnitude of the Italian contribution to the founding of this science has never been as widely recognized as it should be.

No one held this view with more conviction than Sir Charles Lyell (1797–1875), who in the first edition of his *Principles of Geology* (1830–1833; first volume, chapter III) said:

> It was not till the earlier part of the sixteenth century that geological phenomena began to attract the attention of the Christian nations. At that period a very animated controversy sprung up in Italy, concerning the true nature and origin of marine shells, and other organized fossils, found abundantly in the strata of the peninsula. (p. 23);

> [after referring to Steno's *De Solido*] This work attests the priority of the Italian school in geological research (p. 27–28); [and]

> We return with pleasure to the geologists of Italy, who preceded, as we before saw, the naturalists of other countries in their investigations into the ancient history of the Earth, and who still maintained a decided pre-eminence. (p. 41)

That Lyell should have come to appreciate the early Italian contributions is not altogether surprising. Born into a well-to-do Scottish family and nurtured in a nascent geology by no less an authority than William Buckland, professor of geology in the University of Oxford, Lyell had the means to travel widely in Europe and even farther afield, learning about geology and developing contacts with eminent natural scientists beyond his native Britain. He made several visits to Italy. The first of these took him, with Sir Roderick Murchison, to the Apennines (1828) and later, alone, south through the peninsula to Sicily (1829). This particular excursion was a powerful stimulant to his writing and publishing the first volume of the *Principles* in about one year. His second visit took place in 1832, when he happily combined honeymoon and exploratory geology. Another prolonged geological excursion took place in 1857–1858, when Lyell again traveled the length of Italy, reaching Sicily by way of Liguria and Tuscany. On this journey, he met the young Giovanni Capellini (1833–1922), still an undergraduate student, at his home in La Spezia, and discussed with him the geology of the Gulf of Spezia. In the autumn of 1857, both visited the classical localities of Vallisneri and Spallanzani. A consequence of that visit was that Lyell then wrote to Bartolomeo Gastaldi (1818–1879) to plan joint research involving Italian localities. Capellini and Lyell met again in Pisa, where Capellini introduced him to his teachers, Giuseppe Meneghini (1811–1889) and Paolo Savi (1798–1871), following which Capellini and Lyell spent some days examining the Paleozoic and Mesozoic sections in the Monti Pisani and Leghorn areas. It was not until December of 1857 that Lyell left Pisa for Sicily.

These few details illustrate that Lyell's concern for Italian geology and geologists was anything but superficial. He made geological observations and interpretations at a number of Italian localities, discussed them with Italian authorities, and then used the evidence they provided to support his arguments both in discussion and as set forth in the *Principles*. Among these localities, the Temple of Serapis at Puzzuoli in the Bay of Naples is doubtless best known, but that is probably only because Lyell selected the classical ruin to serve as frontispiece to the *Principles*. Nor was Lyell's recognition of, and reliance upon, the pioneering Italian studies by any means short lived; it persisted into much later editions of the *Principles*—witness, for example, the number of Italian naturalists cited in the third chapter ("Historical Progress of Geology") of the much-revised ninth edition of 1853.

Why was Lyell such an exception, and why did not the nineteenth- and twentieth-century Italian natural scientists take up the challenge of exposing, and seeking more appropriate recognition for, the prowess of their forebears?

The first question is easily answered. Lyell had an exceptional gift for observing, interpreting, and synthesizing geological information, and he had the ability to set forth his arguments in clear, simple prose. He could draw many pieces of seemingly disparate evidence under a single umbrella, integrate them, and marshal arguments for their unified interpretation with such cogency that his explanations generally won quick and unconditional acceptance. Thus, it should detract nothing from his abilities and accomplishments to point out that he was, nonetheless, greatly helped in his synthesis by the work of others, as he was by living in the generational shadow of such luminaries as James Hutton and Georges Cuvier. So far as his European colleagues were concerned, his acquaintance with them, gleaned through his travels, and his knowledge of the French and Italian languages gave him a special familiarity with much of their work. Among the Italian contributions, Lyell was impressed by Giovan Battista Brocchi's *Conchiologia Fossile Subapennina* (1814), which carries an 80-page history of the Earth sciences in Italy and other parts of Europe and speaks eloquently to the geological relevance of much Italian research conducted from the fourteenth to the late eighteenth centuries. This historical account was translated and summarized by Lyell and used by him as a basis for much of the third chapter of the *Principles*.

The second question is much more controversial and less easily explained, involving as it does a somewhat problematical change in attitude on the part of Italian natural scientists at the beginning of the nineteenth century, whereby they substantially surrendered to students of the humane disciplines their interests in matters of epistemology, natural philosophy, and the history of science—matters for which in prior times they had shown a more than healthy concern. The Italian humanists, in turn, by and large took an idealistic approach to these matters, inspired by the arts—one in which neither the cultural relevance of science nor the country's scientific heritage had any particular place. A contributing factor too was that this took place at a time when, in Italy, there was no national identity and pride, when parochial and even regional infighting was commonplace, and when the country lost its competitive edge over many other European nations.

Be that as it may, Italian prominence in the youthful science of geology, which lasted into the early nineteenth century, was revived during the late nineteenth century, when the distinction of organizing the second of the quadrennial International Geological Congresses was awarded to Italy. In this respect, it was not fortuity that led Thomas Sterry Hunt—prominent Canadian member of the "Philadelphia Committee" that founded these congresses—to propose at the conclusion of the inaugural event held in Paris in 1878 that a second congress be held, after a foreshortened interval of only three years, in Bologna under the presidency of Giovanni Capellini. The proposal won the unanimous support of the Paris delegates. Nor was the proposal merely a courtesy by Hunt, acknowledging that prior to the deliberations of the Philadelphia Committee, Capellini actually had proposed holding an international conference of this type in Italy. It seems quite clear that awarding the second congress to Italy was as much as anything a tribute also to the stature of Italian geology and to its history, testimony to which is borne by the publications distributed to all registered members of the Bolognese congress. These publications included, *inter alia*, geological maps of Italy and selected regions of the country and a bibliography of Italian geology and paleontology that ran to a substantial 630 pages! The congress in Bologna accomplished much in terms of resolving geological issues of international concern and in establishing and formalizing organizational procedures for future congresses. Many of these procedures remain in place to the present day. That the success of the Bolognese event was due in fair measure to the skillful guidance of Capellini, with the assistance of the distinguished Quintino Sella, president of the Italian Academy of Sciences, as honorary president of the meeting, is not in question. Nonetheless, the records of that congress also clearly imply that the convention's accomplishments owed much also to the scientific distinction that attached to Bologna, to the history of geology in Italy, to the stature of Italian geologists, and to public awareness of these distinctions.

It is not surprising, therefore, that when, after one and a quarter centuries, the decision was taken to hold the 32nd International Geological Congress in 2004 in Italy—with the city of Florence chosen this time as the central locale—the occasion should be viewed as an opportune one to rekindle a wider interest in the history of geology in that country and to launch initiatives that might expose, at least to some degree, the original contributions made to that science by Italian naturalists before, and immediately following, the time of Lyell's *Principles*. These initiatives took two forms: first, preparation of a commemorative volume to celebrate the Florentine congress, and, second, convening of a special session of the congress at which presentations on the "Origins of Geology in Italy" might be made. The commemorative volume, which appeared in advance of the congress, more than met its share of the challenge, offering original contributions on the historical interpretation of fossils, the birth of stratigraphy, and the founding of oceanography, to quote only a few illustrative examples (see Gian Battista Vai and William Cavazza, eds., 2003, *Four Centuries of the Word* Geology: *Ulisse Aldrovandi 1603 in Bologna,* 325 p., Bologna, Minerva Edizioni). The proceedings of the second promotional event—the special session of the congress—forms the basis of the present publication, which has an added objective of contributing further to the celebrations of the fourth centennial of Aldrovandi's death in 1605.

The content of this book is an eclectic group of papers by authors of varied background and nationality who describe contributions to philosophy and natural science by Italian naturalists or foreign naturalists working in Italy between the fourteenth and nineteenth centuries. These contributions have the added distinction of having a role in the emergence of modern geology as a distinct scientific discipline—a step that, despite Aldrovandi's discrimination of the botanical, geological, and zoological disciplines in 1603 when he coined the term *giologia* (geology), did not gain wide acceptance until the late eighteenth and early nineteenth centuries, when the Huttonian doctrine had been set forth and Lyell had espoused it so enthusiastically in the *Principles*. For convenience, the papers in this book are chronologically ordered in terms of the historical periods they address, which, if arbitrary to some degree, has the advantage of exposing that some of the subdisciplines of geology are of great antiquity, long predating *geology* itself—subdisciplines that many modern geologists, unprepared to accept or unaware of the breadth attaching to Aldrovandi's *geology* of 1603, now insist on recognizing as distinct components of the "Earth sciences," "geological sciences," or "geosciences."

Thus, for example, Mottana discusses the rise of gemology as a branch of mineralogy from the Renaissance until it developed its distinct identity in the later nineteenth and earlier twentieth centuries. In the same way, Franceschelli and Marabini and, separately, Vai devote their contributions to the remarkable Luigi Ferdinando Marsili (1658–1730), who played a prominent role in European science. Geologically speaking, Marsili was much concerned with global structure but is perhaps best known as the founder of marine geology and oceanography. In his paper, Vai asserts that Marsili also was the first to recognize the concept of isostasy. Yet another quite different example of the often-surprising antiquity of many modern geological fields—although one of much younger vintage—is offered by Funiciello and Caputo in their account of the geology of Rome, which illustrates that urban geology is a much less modern branch of inquiry than many might suppose. Interestingly enough, various aspects of Marsili's work on marine geology and tectonics were recognized by Lyell in the *Principles*, and as Funiciello and Caputo emphasize, the key element in the evolution of the geology of the Eternal City was a remarkable map by the same Giovan Battista Brocchi who inspired Lyell's account of the history of geology in the *Principles*.

Setting the contributions to this book in a chronological framework should not, however, blind the reader to the recognition of other common threads that not only link at least some of them but also more readily promote the drawing of useful comparisons and contrasts. Thus, for example, more than one author touches upon the same historical figure, if generally examining different aspects, or taking different approaches to the interpretation, of his work. In the case of Agricola, the aspects of his studies that had a bearing on the development of gemology are discussed by Mottana, whereas Morello, in her paper on the birth of the mineralogical fields in Italy, draws upon Agricola's mining background and shows how his fourfold classification of inorganic materials (*fossilia*) improved Aristotle's twofold division, and how this, in turn, promoted the somewhat segregated development of the mineralogical and the geological fields in Italy during the mid-sixteenth century. On a related theme, the influence of one historical figure on another and the impact of their relationship on their respective contributions to science are addressed by more than one author. Thus Morello, in her paper on Nicolas Steno, and Yamada in his paper on Athanasius Kircher and Steno follow different analytical pathways. Morello seeks to show that the study of fossils in Italy generated distinct views on their origin in each of Descartes, Kircher, and Steno, yet provided a means of developing a calendar of the Earth independent of (although not notably contrasted with) that based on biblical chronology. Yamada, on the other hand, claims that close examination of the works of Kircher and Steno clarifies differences and similarities between their theories of the Earth, which, in turn, argues for reducing the assumed Cartesian influence on Steno and enhancing that of Kircher and, through Kircher, that of the French atomist, Pierre Gassendi.

A number of authors, in their preoccupation with Nicolas Steno and his principles of original horizontality, continuity, and superposition, have unintentionally demonstrated that the international community may have failed to appreciate the breadth of Italian stratigraphical thinking in the sixteenth and seventeenth centuries. In this connection, Vai and Cavazza have contributed further to exposing the stature of the remarkable Ulisse Aldrovandi, who first proposed the term *geology* in 1603. Among his accomplishments, which included notable contributions to the understanding of such sedimentary processes as fossilization and lithification, was the assembly of the largest collections of animals, plants, and their fossilized remains since Aristotle, and he used these to propose a new classification. He theorized about a "new science" of taxonomy applicable to all naturally occurring materials and objects, and he made the first attempt to establish a typologically based system of binomial classification for living and fossil species. Similarly, Vaccari has focused renewed attention on Giovanni Arduino, less for his well-known primary divisions of the Phanerozoic time-scale than for his development of the "lithostratigraphic" and "chronostratigraphic" basis for drawing geological cross sections some decades before stratigraphy became a recognized subdiscipline of geology.

Yet another common thread links the papers by D'Argenio and by Torrens, who outline the early works of two young scientists whose potential to become outstanding figures in the nineteenth-century blossoming of geology was terminated by premature death. D'Argenio addresses the life of Leopoldo Pilla (1805–1843), who contributed to the acceptance in Italy of the general principles of the rapidly developing geological discipline (including Lyellian uniformitarianism); to the understanding of internal geological processes; and, based in Naples as he was for much of his young adult life, to interpreting the volcanology of the Roman province. His rate of scholarly productivity was prolific—about 120 scientific papers, many in publications of the stature of the *Bulletin de la Société Géologique de France*, between 1830 and 1847. Pilla died in 1848, marching with his students, in the battle of Curtatone during the Italian insurgency ("Risorgimento"). Torrens addresses the equally tragic figure of Gregory Watt (1777–1804), son of James Watt, the eminent engineer. The younger Watt died of consumption at 27 years, but only after traveling through (a hostile) France, and Italy as far south as Naples, with Scottish-American geologist William Maclure; visiting Vesuvius and seeing other evidence of recent volcanism; drafting a proto-geological map of Italy that attempts to record no less than 47 distinct lithotypes; and later, on return to England, conducting some of the earliest experiments in theoretical petrology—those on the melting of basalt.

Finally, the relationships between geology and the arts and letters, long justifiably upheld in Italy, is augmented by Branagan's account of geology and artists (mainly Florentine) in the fifteenth and sixteenth centuries. Not only does Branagan deal with the geology of landscapes by such painters as Leonardo and Francesco Botticini, he extends the field of research to illuminated manuscripts. Furthermore, he makes the case that the recorded interest in geological features of the Arno valley shown by so many Florentine artists clearly anticipates the thoughts of Nicolas Steno in the same general locale a century later that led to the formulation of his stratigraphic principles. Scalera, in contrast, considers the accomplishments of one, Mattia Damiani (1705–1776), as both poet and scientist, with implications for several geological fields arising through his work on climatology and meteorology. The principal scientific interest focuses on the newly discovered manuscript of Damiani's dissertation at Pisa in 1776, involving a comparison of the Cartesian and Newtonian hypotheses in providing a more satisfactory explanation for meteorological phenomena measured using laboratory techniques.

It is remarkable that, in the 175 years since Sir Charles Lyell drew attention to the inadequate recognition of Italian contributions to the founding of geology as a science—and this in no less a work than the *Principles of Geology* (1830–1833)—historians of a science have not sought to evaluate Lyell's claim more fully and thoroughly. Indeed, the failing is all the more remarkable, given Lyell's own assiduous referencing in that work to so many Italian geological localities of note and to the Italian naturalists who observed and interpreted them. Recently, however, there has been a more substantial effort to evaluate further, and, as appropriate, to recognize, the validity of Lyell's claim as part of a wider undertaking to make the historical foundations of the science more secure. There are, moreover, supplementary objectives to such an undertaking, viz., to broaden the cultural aspects of the science—an effort that is in many ways appropriately focused in Italy—and to enhance the already rich legacy of scientific accomplishment to be bequeathed to the younger generations of international geologists. Familiarity with that legacy should be part of determining the level of the foundational platform on which they begin to build, as it should enrich the building experience itself. It has been a traditional concern of the International Geological Congresses and the International Union of Geological Sciences to enhance and secure that legacy. This volume—an outgrowth of the most recent of these congresses—is offered, therefore, as a modest contribution to helping achieve these ends.

Gian Battista Vai
W. Glen E. Caldwell

Italian gemology during the Renaissance: A step toward modern mineralogy

Annibale Mottana*

Dipartimento di Scienze Geologiche, Università degli Studi Roma Tre, Largo S. Leonardo Murialdo, I-00146 Roma, Italy

ABSTRACT

Under the pressure of industrial demands following the discovery of South African diamonds, gemology became a science during the late nineteenth century by combining morphological mineralogy with mineral physics and chemistry. However, it underwent an empirical, pre- to semiscientific period during the Renaissance, when market novelties required development in gemological knowledge. Pliny's *Naturalis Historia* (1469) was the reference treatise on gemstones among scholars, but it was the Italian translation of this work by Landino in 1476 that made gem studies grow. Indeed, while scholarly mineralogy developed through Latin texts, practical arts related to minerals developed through light handbooks in the new European languages. In Italy, the most active trading center at that time, where luxury goods were brought to be set in gold and distributed to all of Europe, most gem traders possibly understood some Latin, but certainly their providers did not, nor their customers. This is why the first original Renaissance book on gems, *Speculum lapidum*, by Leonardi (1502), did not enjoy popularity until it was translated into Italian by Dolce in 1565. Similarly, Barbosa's accounts of travel to gem-producing India (1516) became known only after Ramusio translated them in 1554. Among gemological contributions in Italian, the most farsighted ones are Mattioli's translation of Dioscorides' *De materia medica* (1544) and Cellini's *Dell'oreficeria* (1568). Moreover, three manuscripts did not reach the stage of being printed: Vasolo's *Le miracolose virtù delle pietre pretiose* (1577), Costanti's *Questo è 'l libro lapidario* (1587), and del Riccio's *Istoria delle pietre* (1597). They survived, however, to help clarify gem interests and activities by the merchant class in the transitional time from the Renaissance to the Baroque. Then, Italy lost its top position in culture and trade, and a Fleming, A.B. de Boot, wrote the treatise that summed up the available knowledge on gems at that time (1609).

Keywords: lapidary, gem, cut, trade, use, Pliny, Leonardi, Biringuccio, von Calw, Mattioli, Cellini, Cardano, del Riccio, Bacci, de Boot, sixteenth century.

INTRODUCTION

It is widely accepted that gemology, the branch of mineralogy that studies precious stones, did not reach its full scientific development until the second half of the nineteenth century or even the early twentieth century, i.e., some 50–70 years later than the mother discipline developed into a true, modern science. The development of gemology took place by combining a long-time wealth of information on natural occurrences and physical properties of solid materials (minerals, gemstones) with data on the stoichiometric ratio of their constituent atoms (determined by chemistry) and hypotheses on the three-dimensional symmetric

*mottana@uniroma3.it

arrangement of these (suggested by theoretical crystallography, itself a branch of mathematics). The event that prompted the conversion of gemology from an applied *à-peu-près* practice to a discipline meeting precision standards (Koyré, 1948) was the discovery of the South African diamond (1866), with all its related economic, industrial, and scientific demands.

The added monetary value of gemstones, as opposed to minerals, lies primarily in a combination of economics (i.e., worth) and aesthetics (i.e., beauty), two hard-to-define aspects of the gems that involve subjective evaluations (on prices and jewels, respectively) lying well beyond the usual boundaries of science. Such an added value, however, is also intimately tied to the need for having some kind of evaluative criteria that may be considered irrefutable. In order to cope with these demands, gemology needs the support of universally reproducible results, and these can be obtained only by scientific methods. In turn, these methods should be upgradeable whenever the opportunity arises.

Just as mineralogy had a long prescientific stage before reaching its modern, sound scientific state at the beginning of nineteenth century so also did gemology. Actually, research on precious and semiprecious stones progressed stepwise, each phase triggered by the introduction of novelties to the gem market, such as those by the Portuguese when first importing corundum directly from the East Indies (ca. 1500–15), by the Spaniards for Colombian emerald (ca. 1530–50), by Jean-Baptiste Tavernier's travels to the Golconda diamond fields (1632–68), and by the Brazilian diamond rush (ca. 1723–30). All these events provoked market frenzy, and compelled the operators to deepen their knowledge. Taken together, this growth pattern of gemology, albeit unplanned and extemporary, accompanied the prescientific, steady evolution of the science of minerals and rocks and eventually contributed to the development of scientific mineralogy.

The transitional period from ancient to modern scientific appraisal of the natural world is commonly positioned during the Renaissance. However, the meaning of this term is unclear, as it spans from the twelfth to the sixteenth centuries according to different historians (cf. Burckhardt, 1855; Haskins, 1927). By contrast, historians are unanimous in considering Italy to be the best-qualified candidate for the country that functioned as the center, engine, and pivot of the Renaissance period directly resulting into the present-day epoch of human evolution. Most, if not all, innovation in economics, aesthetics, arts, letters, etc., were conceived and unfolded in Italy by the then emerging merchant class, even though some of them actually attained their practical application elsewhere in Europe. For example, double entry in accounting is documented for the first time in Flanders and Champagne, but it was invented most probably in Venice (*alla veneziana*: Pacioli, 1494) by the "Lombards," the bankers who were actually not from Lombardy, but from Tuscany, Venice, and other areas of Italy. As for precious stones, although first documented in France in the 1413 jewel inventory of the Duke of Berry, the cut and polished diamonds of which he boasted possession (e.g., *gros dyament poinctu taillé à plusieurs lozenges,* Guiffrey, 1894–96) had been imported from Florence and Genoa, and their craft possibly conceived and first executed in Venice (Tillander, 1995).

MEDIEVAL LAPIDARY PROGRESS IS THE FORERUNNER OF THE RENAISSANCE APPRAISAL FOR GEMS

Technical aspects, such as gem cutting, are deeply interwoven with the development of gemology into a scientific discipline, for they not only influenced the customers' taste but compelled the dealers to approach gems with a different viewpoint. During classical antiquity and the Middle Ages, the greatest value of gems lay in their color, combined with the "signs" they carried. The "signes" were the images believed to be recognizable in the stones. These signs were either natural stains or growth figures (e.g., Pyrrhus' agate with Apollo and the nine muses: Pliny, XXXVII.5), or even *intaglios* carved by skilled craftsmen (e.g., magical jewels, Evans, 1922). With the early Renaissance, other properties of gemstones began to be appreciated, among them hardness, especially for diamond. Possibly to bring out the best of this property, just as color is enhanced by polishing, craftsmen began manufacturing stones to a regular contour (*cabochon* cut). They discovered that certain gems, like diamond and spinel, could not be brought to a round shape because of their easy fracture along flat surfaces (Pazaurek, 1930; Falk, 1975), but by skillfully orienting the stones on the newly obtained flat directions, they could be made to exhibit dazzling plays of light ("fire"), which added much to their brightness and, consequently, to their appreciation and value.

Although at that time diamond was not the most important stone, it certainly was widely appreciated for its rarity and hardness. Jewelers first attempted to improve it by polishing the four octahedral upper faces (*poinctu, taglio a punta*). Then, around 1380, they began adding new diagonal facets (*lozenges*), probably using a lapidary wheel like the Dutchman Henricus Arnoldus (or Henri Arnault *viz.* Heinrich Arnold) drew on his sketchbook in 1439 (von Stromer, 1992, p. 120–121). Some years later, possibly in an attempt at getting the best from stones that were too thin or had been halved by chance, unknown craftsmen began cutting a series of triangular facets that gradually sloped down from the pointed top to a large flat base (*taglio piatto*); this made the diamond contour as round and smooth as possible (rose cut). The real breakthrough, however, occurred some time between 1530 and 1538 (Tillander, 1995; Wild, 1997). In 1538, a 8/8 type of diamond cut consisting of an octagonal section with sixteen lower facets and a flat top, the *tavola*, was practiced in Augsburg, and nearly at the same time, the "Antwerp rose"—hexagonal, with six nearly flat top facets (*di stella*) and twelve more inclined facets next to the table (*di traverso*)—was developed (Kockelbergh et al., 1992). All of these innovations modified the shape of the raw stone and involved a change in taste. They implied a new, intimate understanding of diamond crystal morphology (cleavage); these cuts not only required changing the slope of the octahedral faces from 54°44' (their natural value) to measured values ranging from 45° to 60° (Tillander, 1995), so as to maximize reflection, but also how to operate the coping saw by using oil-bound diamond powder as abrasive. The new fashionable outlines were no longer made only as squares, but as rectangles (*émeraude* cut), triangles, and even lozenges (*baguette*, *briolette*), all these being forms that rarely occur in natural diamonds; they are cut on purpose. The introduction of oil-dispersed diamond powder as the appropriate medium to cut stones into shape is credited to the Indians. Its presence in Europe was testified to in Venice in 1530, thus giving a *terminus post quem* the rose cut of diamond would start spreading across the world.

The Fleming Louis de Berquem (or Lodewijck van Berckem) is credited with inventing the "dop," i.e., the device for holding the stone while it is being cut, as well as the method of melting tin around the stone to fasten it and avoid casual breaking. He is also believed to have been the first, in 1476—a date much too early to be true, as this statement was made in 1661 by Robert de Berckem, who claimed to be his grandson—to introduce "absolute symmetry in the disposition of facets" (Tolkowsky, 1919, p. 12). This statement implies scientific understanding of crystal morphology, a fundamental aspect of the science of minerals, and this upgrade was reached by gemology, or—better said—owing to the use of certain minerals to make gems. Undoubtedly, most work on gemstones at that time was done outside Italy, following a tradition that probably originated much earlier than ca. 1110–40, as described by Theophilus Presbyter (Dodwell, 1961, p. 189–191; Bänsch and Linscheid-Burdich, 1985, v. 1, p. 374), and continues to date. Indeed, Italy, the world's current largest jewelry producing country, only sets imported cut stones in gold and exports the finished jewels. However, some technical innovation was certainly made in Italy, as tradition holds that a Venetian goldsmith conceived of, and first performed, the "brilliant" cut of diamond (see below).

AIM AND STRUCTURE OF THIS STUDY

The intent of this paper is to show, first, that a number of the basic steps by which gemology advanced from an ancient trade practice of procuring rare, prized objects carrying alleged magic and medical properties to a modern science firmly established on validated physical characteristics—albeit maintaining most of the aesthetic and commercial traits that make it distinctive from mineralogy—took place in Italy during the sixteenth century, i.e., during the same time span that, in the arts, encompasses trends that are called Renaissance (senso stricto) and "Mannerism."

Unfortunately, the role of gem studies in upbringing mineralogy to a scientific stage is hardly appreciated, overshadowed, as it was, by the impetuous growth of mining at that time and of crystallography in the following centuries. Regarding the history of Italian science in particular, not even one expert of precious stones is taken into consideration among those Renaissance thinkers who are considered to be worth mentioning as precursors of mineralogy and geology (see Gortani, 1963).

Second, I will try showing that, during the Renaissance, a gemology such as that now considered modern, or at least such as to be trending toward modern, was devised and described essentially in Italian, and progressed for a fairly long time in this language despite the fact that the two books that bracket this time span were written in Latin, as were most scientific books at that time. These books are *Speculum lapidum*, by Camillus Leonardi (1502), and *Gemmarum et lapidum historia*, by Anselmus Boetius de Boot (1609).

The first book simply demonstrates that the need to consider all evidence on gems gathered during the Middle Ages throughout Europe (and in the Arabic world) was first felt in Italy, and therefore stimulated Italian thinking. By contrast, the second book, the re-edition of which Adriaan Toll (1636[2]) reworked, is possibly the most complete book on gems written before Bauer's *Edelsteinkunde* (1896). It discloses that, within one century, the leadership in gem science had been lost by Italy and had shifted toward the impetuously growing countries to the northwest. Indeed, despite Galileo's outstanding achievements (Westfall, 1971), by ca. 1650 the new, modern science migrated from the Mediterranean countries to those surrounding the North Sea (Rossi, 1997), and Newton was on the verge of becoming the pole star (Koyré, 1965).

For a scientist, the study of an event or phenomenon implies following a path through several stages: first, to detect it, then to describe and interpret it, and finally to verify it in relation to some working hypothesis and to create a model that allows reproducibility. Now, the first and possibly foremost step along this path is the correct recognition of the case study, as all other steps are heavily dependent upon selecting the best way to validate the assumed model, i.e., to demonstrate that it will suitably explain the observed event or phenomenon.

In the case of the development of modern gemology, it may be inferred to have had an incubation stage in Italy and in the Italian language during the Renaissance, before growing and reaching full maturity elsewhere. The reason for this inference is easily found in the overall state of European affairs at that time:

> *La bilancia commerciale resta per tutto il Cinquecento favorevole all'Italia rispetto alla Francia, alla Spagna, anche a grandissimi centri commerciali come Anversa. Le città italiane rimangono al vertice dei traffici europei soprattutto per quel che riguarda i beni di lusso e il commercio di denaro* (Ruffolo, 2004, p. 270).

In other words, ever since 1204, when most eastern trade was snatched away from Constantinople, and all through the entire sixteenth century, Italy, despite being a secondary country in politics, under the domination of France at first, and then of Spain, was the center of the European luxury trade (and big expenditures). In particular, Italy could overshadow busy commercial centers, like those in the Low Countries, because it was on the main route through which luxurious commodities from gem-producing India entered wealthy, gem-thirsty Europe. Once the market for oriental gems and jewels had been established (e.g., at Venice and Genoa), Italy could absorb the market novelties coming from the West Indies too, although there were short slumps during which it looked as if Lisbon or Seville would take over (Falk, 1975). Actually, it took at least one century before the market for luxury commodities could be taken away from Italy and transferred to the countries around the North Sea. The Italian situation during the sixteenth century, therefore, was much the same as the Swiss one during the twentieth century: the country enjoyed high standing for being both small and politically unimportant, yet appropriately located along the unofficial trade routes linking the fighting partners.

If, however, large sums of money were being made available from all over Europe for buying jewels in Italy, then, for the customer's safety and satisfaction, the capacity had to exist there for the gems, which were to be set in these jewels, to be evaluated using appropriate and reliable methods. This is why gemology flourished in Italy during the Renaissance, albeit within the narrow limits of the science of the time.

A second argument is added here to support the starting assumption: new knowledge, if not widely communicated, dies rapidly, and therein lies a further reason way Italy was the world center of gemology during the Renaissance. Indeed, from 1350 to at least 1550, Italy hosted the superior universities in Europe. Thus, it enjoyed a key role in spreading scientific novelty to other countries. The outstanding minds that became the fathers of some branches of modern science, e.g., Copernicus (Mikolaj Kopernik viz. Nicolaus Koppernigk) and Agricola (Georg Pawer viz. Bauer), received their education in, and degrees from, the Italian universities at Ferrara and Padua, respectively (Schmitt, 1975, p. 62; Westfall, 1984, p. 132). At the time these institutions were in a state of turmoil and they struggled to renovate their curricula, seeking freedom from the narrow-minded limitations of scholasticism. Their clever foreign students, too, after graduation, began to disseminate at home the new knowledge with all the intrinsic anti-Aristotelian feelings they had acquired from their humanist Italian teachers (see Hall, 1954; Wieland, 1962). They did so having a full consciousness of the bright future their new ideas would produce (Hannaway, 1992), and indeed they succeeded.

As for gemology, Agricola's responsibility is particularly important. In building up mineralogy and steering its development into a science, he blended the ancient Greek and Roman theories he had learned during his studies in Italy with the mining practices he had acquired in his native Saxony (Suhling, 1983; Schneer, 1995, p. 722). In this context, he did not neglect gemology. Indeed, fifty-seven of the ~300 stones he deals with in his *De natura fossilium*, published in Basle in 1546, are listed as gems by his Italian editor, Michele Tramezzino (*De la natura de le cose fossili*; Venice, 1550). This, too, is a small number, which is the result of a bizarre means of selecting what constituted a gem and what did not. "*Zaffiro*," "*topatio*," "*smaraldo*," and "*diamante*" were not classified as gems, but "*rubino*" was! Why? Even admitting that this strange selection was not Agricola's, but Tramezzino's, there is little doubt that Agricola derived most of his information from Pliny (1469), the natural science textbook at Padua in his time (Suhling, 1983, p. 157). Thus, possibly, the weird assortment is rooted in some kind of misunderstanding of the content of that classical book, which at that time was already largely obsolete for both information and taste, as the appreciation for gems had progressed during the Middle Ages (see below). By contrast, Agricola accepted Pliny as all true, rebutted most medieval findings (by the Arabs, mainly), and added the few new data gathered while he was making his own training as a medical doctor and practicing medicine in a mining district. He did so in a conscious attempt at reestablishing the continuity of the ancient mining world with what he knew directly from contemporaneous experience, but inevitably this diminished the role of gems. For example, when describing the sparkling, white, hard stone associated with gold specks that was dug out at Cotteneheida, Saxony, he used the local name "*quertz*" (in *Bermannus*, 1530, Appendix: *Rerum metallicarum appellationes iuxta vernaculam Germanorum linguam, autore Plateano*; note that this addendum is not in the 1550 Italian translation), which he could also find in the *Bergbüchlein* (attributed to Rülein von Calw, 1505), apparently not realizing that it was the same material as the rock crystal (*cristallo di rocca*) he knew well because he mentions its use to cut not only beautiful vases but the common *paternostri* too i.e., rosary beads (Agricola, 1550, p. 285). Alternatively, and mor e probably, he was aware of their common identity and wanted to stress the difference existing between a mining object and a decorative stone, the more so because *cristallus* had been misused for centuries as being a medical ailment, mainly under the persistent fame of Marbode's *Liber lapidum* (Riddle, 1977, p. 77). Quartz, from *quarzo*, the Italian transliteration of the Saxon word (Agricola, 1550, p. 449), now identifies all varieties of such a mineral, the most widespread individual mineral on Earth, at least in the present paradigm of species understanding. Of his native mining information, Agricola had a further chance of applying his practical knowledge during his 1522–26 period of apprenticeship in Italy; he stated this openly, when acknowledging Biringuccio's work in *De re metallica* (1556, p. 4) and in *De l'arte de metalli* (1563, p. 4): "*quando io lessi queste sue cose, mi tornaron a mente quei che gia vidi fare in Italia.*"

Indeed, it is proof of Agricola's farsightedness that, when at home again (1526), he reconsidered all of what he knew by practice on the basis of the new awareness he had acquired in Italy. He did not back away from rebutting classical authors when their theories were at odds with his own experience, but always referred to them when he looked for a general explanation. Consequently, he brought about substantial innovation in the mining field, which for thousands of years had suffered a complete separation between learned and practical men (Suhling, 1983, p. 151–153). He recreated mineralogy on new grounds: he adopted most names and ideas from the ancient word and blended it with the new ones developed by miners (i.e., technicians) who had been cast aside from the learned men during the previous centuries. Mineralogy could bypass mining, always implicated in business problems, and look for the scientific core of solid matter. The outcome was so good that his textbooks were almost immediately translated into Italian and provided increased information on, and heightened interest in, minerals and ores in the learned world, as well as the treasure inherent in precious stones.

THE EARLY RENAISSANCE: REVITALIZING THE ANCIENT RECORDS OPENED THE WAY TO MODERN STUDIES

Some time before 18 September 1469, the *editio princeps* of the elder Pliny's *Naturalis historia* was printed in Venice by Johannes de Spira. The first book ever published on natural science, it became "the most popular natural history book ever published [...] almost until the dawn of the nineteenth century" (Gudger, 1924, p. 270; cf. Serbat, 1984). Indeed, by the end of the fifteenth century, it had no less than fifteen Latin editions, which rapidly became forty-seven by 1550 and increased again slightly by the end of the sixteenth century when some fifteen more editions had appeared (Schiavone, 1982). Altogether, more than 190 full or partial separate editions in several languages were printed before the year 1600 (Sinkankas, 1993).

Almost immediately afterwards, in 1471 (or slightly later), Berthold Ruppel in Basle and William Caxton in Cologne independently printed Bartholomaeus Anglicus' *De proprietatibus rerum*, a medieval encyclopedia written ca. 1200–40, which also was reprinted several times at various places. Although a minor work when compared to Pliny's and considered per se, this book enjoyed long-time popularity during the Middle Ages because it appealed to common people who liked to get simple information on the natural world around them, such as its numerous references to the supposed medical properties of gems. Apparently, the simple approach of *De proprietatibus rerum* appealed more to readers than the rigorous one of Albertus Magnus's *De mineralibus*—"an impressive attempt to organize a science of mineralogy" (Wyckoff, 1967, p. xxxv)—which, although written ca. 1248–62, was not printed until 1476 and re-edited twice before the end of that century.

Another technical book of great impact was *Lapidarius* by the so-called "pseudo-Aristotle," translated into Latin by Alfredus Anglicus ca. 1200. It was printed at Merseburg in 1473 and reissued a few times, either alone under the alternative titles *Mineralia* or *De mineralibus* or together with *Secreta secretorum*. Although a fake, probably by some Syrian monk of the eighth century, later reworked many times by Arab, Hebrew, and even Latin translators up to the thirteenth century (Ruska, 1912; Thorndike, 1960, p. 21), *Lapidarius* contributed enormously to the transfer of information on precious stones from the then highly developed Muslim countries to Christian Europe, and to the creation of a new taste for gems as carriers of intrinsic magic and medical properties besides being objects of outstanding prestige. These considerations added more than a little value to a natural material, the primary price of which had been perfunctorily determined by its being a *lapis parvus, rarus, durus, and pulcher à natura procreatus*—a natural stone that is small, rare, hard and beautiful (de Boot, 1636^2, p. 13)! Furthermore, the attribution of *Lapidarius* to Aristotle (*summo in omni doctrina viro*, i.e., a great man in all fields of learning: Pliny, 1469, VIII.17) allowed it to share in the enormous prestige that classical Greek and Roman authors had among humanists (Hall, 1954; cf. also Funkenstein, 1986), if with due recognition of the decreasing credit that accrued to Aristotle because of the way the book had been interpreted by supporters of the scholasticism. (Aristotle was respected from the thirteenth to the fifteenth centuries A.D., then was cast aside by true scientists, owing to the misuse of his work by the church [i.e., philosophers of scholasticism], who tried to convert his thinking into a dogmatic imposition. Modern science was born as a reaction to the bible and Aristotle.) Nonetheless, the overall credit easily superseded the gemological tradition originating from Marbode of Rennes' (Marbodus Redonensis) *Liber lapidum seu de gemmis* (written ca. 1061–81), a poem that had dominated during the late Middle Ages (Riddle, 1977), but which faded during the Renaissance, with the result that the *editio princeps* of Marbode's work was delayed until 1511 (cf. Wiemann, 1983).

Additionally, in some year between 1473 and 1478, Vincent of Beauvais' (Vincentius Bellovacensis) *Speculum naturale* went into print, i.e., another bulky encyclopedia written just before 1244 that had been influential during the late Middle Ages, but which had lost its impact with new discoveries, and in 1497 Theophrastus' *De lapidibus* was published in Greek (the first book on gems and minerals in that language). These works complete the writings on stones available to scholars before the 1502 appearance in Venice of Camillus Leonardi's *Speculum lapidum* (Fig. 1). Indeed, it was the first original work on stones, gems, and minerals conceived after more than two centuries of silence by the scientists, but certainly not of neglect by the wealthy ones.

Leonardi's work, composed of three books written sometime between 1480 and 1500, describes all the stones he could find reference to, mostly drawing information from Pliny, but also from many other sources, especially the Arab ones mediated by the late medieval authors. His first book details how and where gems are generated, how they develop their forms, colors, and properties, and even how natural stones can be distinguished from artificial ones. Leonardi's second book presents an impressive list of 279 gem names and descriptions in alphabetical order, the longest list compiled during the Middle Ages and early Renaissance. The impact of Leonardi's work, however, was somewhat lessened by the inclusion of much false information he felt a duty to report upon. Afterwards, he let his mind wander free, so that the third book turned out to be a lengthy account of the magical and medical properties of stones, including the signs to be engraved on them to enhance their powers (De Bellis, 1985). All this does not contribute to our appreciation of gemstones.

Figure 1. Title page of the second Latin edition (Venice, 1516) of Camillus Leonardi's *Speculum lapidum*. This book, the printing of which occurred in 1502, closes the Middle Age speculations on gemstones and opens the door for Renaissance studies.

Leonardi's treatise enjoyed only a moderate positive reception. However, it was reprinted twice, in Venice (1516) and Augsburg (1533), before falling into oblivion (partially due to its translation into Italian by Lodovico Dolce; see below) from which it did not emerge until the following century, when it was reedited by Petrus Arlensis de Scudalupis (Paris, 1610) with many changes and even more fantastic additions.

ITALIAN CONTRIBUTIONS TO GEMOLOGY DURING THE HIGH RENAISSANCE

The previous review sets the stage for the following evaluation of the works printed during the high Renaissance and Mannerism, as well as those not printed by then but discovered later, which represent the bulk of Italian studies on gemstones at that time.

The most influential treatise considered during the entire time span was, beyond a doubt, Pliny's *Naturalis historia*, his thirty-seventh and last book. There was nothing in it that was against the Christian religion and required censorship—as a kind of debasement (*defloratio*)—and yet even this book underwent numerous cuts and amendments (*castigationes*) to free it of all the contamination suffered during fifteen centuries of handwritten transmission. Indeed, *Castigationes plinianae* (Rome, 1492–93), by Hermolaus Barbarus (Ermolao Barbaro), boasts the correction and removal of some 5,000 mistakes so as to reconstruct philologically a text more closely adhering to the author's original draft. It is another, much stricter type of *castigatio*, however, that better shows the Renaissance scholars' renewed attention to Pliny's textbook. This began with Nicolaus Leonicenus (Niccolò da Lonigo), who in 1492 wrote a *De Plinii et pluriorum medicorum in medicina erroribus*. Leonicenus's work criticized the *medicina Plinii*, a collection of healing recipes based mostly on herbs, compiled in the fourth century A.D., which had become almost dogma in medical teaching throughout the Middle Ages. The work found a printer in 1509 (Martini, 1977) and Pliny's entire voluminous encyclopedia was increasingly submitted to several careful scrutinies. The cornerstones of this activity are the works by Jakob Ziegler (1536), who criticized the astronomical and geographical books on the basis of the constantly increasing new discoveries of the time, and by Otho Brunfels (1530–36), Adam Lonicerus (Lonitzer, 1551), Conradus Gesnerus (Konrad Gesner, 1551–58), and Guillaume Rondelet (1554–55), who systematically checked and improved Pliny's botanical, zoological, and ichthyological data. It is often said that through these amendments (*castigatione*), modern natural science was built because research progressively became more independent of tradition and reliant on findings and experience. However, most, if not all, of these critics concerned themselves with books other than the thirty-seventh one. Therefore, Pliny's reputation and esteem as the top reference on stones and gems appears to have remained undiminished for some time, all through the Renaissance and early Baroque, until it was found out that this book, too, contained a number of medieval interpolations (Hardouin, 1685).

As for mineral studies, however, more than Pliny's original *Naturalis historia*, the Italian translation of which was done by Cristoforo Landino, published in 1476 and reprinted six times before 1543, won recognition from all those who were involved with precious stones (and in general for decorative stones as well). Pliny's encyclopedia was translated anew into Italian twice more during the sixteenth century—by Alessandro Brucioli in1548 and Lodovico Domenichi in 1561. The former translation was close to a failure from the editorial point of view, but the latter, which imitates Landino's but is written in even plainer Italian, enjoyed great success and was reprinted dozens of times. It actually remained the standard Pliny translation in Italian for centuries despite not always being scientifically reliable; it was finally superseded a few years ago (Conte et al., 1982–88). Note that all these Italian translations occurred before the first French translation by Antoine du Pinet went into print in 1562, and much earlier than the first English translation was made by Philemon Holland in 1601.

While the main stream of mineral studies in the sixteenth century developed steadily through treatises in Latin (e.g., Georgius Agricola's *Bermannus* [Basle, 1530] and *De natura fossilium* [Basle, 1546]; Hieronymus Cardanus' [Girolamo Cardano] *De subtilitate* [Nuremberg, 1550; Basle, 1554²; Basle, 1560³]; Christophorus Encelius' [Christoff Entzel] *De re metallica* [Frankfort, 1551]; Johannes Centmannus' [Johann Kentmann] *Nomenclaturae rerum fossilium* [Zurich, 1565]; Conradus Gesnerus' *De omni rerum fossilium genere* [Zurich, 1565]; etc.)—Latin remaining the European language of scholars for two more centuries—the use of Latin did have the effect of keeping open the gap that separated the mineralogists from the miners (cf. Suhling, 1983). Subordinate streams of mineralogical development, such as those dealing with the applied arts that made use of minerals, relied on handbooks in the new European languages for their advancement (e.g., mining geology using Ulrich Rülein von Calw's *Nützlich Bergbüchlein* [Augsburg, 1505] and the anonymous *Probierbüchlein* [Magdeburg, 1524], as well as—and best of all—Vannoccio Biringuccio's *De la pirotechnia* [Venice, 1540]). All these handbooks underwent several reprintings in different places and by different, often unknown, publishers before merging into the more mainstream Latin, as represented by Agricola's *De re metallica* (Basle, 1556; cf. *Praefatio*), in which the handbooks are acknowledged, as is another lost book by Paldulfus Anglus not otherwise known. It was at this last stage that mining finally came to be acknowledged as deserving of a prescientific status.

As for gemology, most gem traders could grasp some Latin, but certainly their providers could not, nor could their customers, who were high-class, yet poorly educated, noblemen, warriors, or wealthy merchants (with some rare exceptions among clergymen, a social group always greedy for gemstones). This is possibly why the first original book on gems written during the Renaissance, Leonardi's *Speculum lapidum*, although reprinted twice (see above), did not enjoy popularity until Lodovico Dolce translated it (Venice, 1565) verbatim, but without reporting the name of the author! This translation (Fig. 2) influenced the gem trade deeply, not only because it introduced quite a number of vernacular technical expressions, but also because it appeared in Venice, at that time the best selling center for gem trade.

The same fate happened to Duarte Barbosa's accounts of his travels to southern India. They had been written in 1516, but they became known only after Giovanni Battista Ramusio summarized and translated them into Italian (Venice, 1554). Despite the inevitable reduction of information caused by the translations, the few chapters concerning gemstones are still important for our evaluation of the state of the art not only because they give a *prima manu* description of the Portuguese trade (Vassallo e Silva, 1993), but also because they devoted considerable attention to the weight vs. price relationships of ten important gems, including diamond. This book, indeed, offers the first notification of this evaluative method and of several other marketing practices still fundamental in gemology.

One should not undervalue the contribution of travel accounts to the development of the Renaissance's new feeling for gems. Gold and spices had been the actual reasons behind geographical discoveries, and greed for gemstones developed just a little later. The Spaniards proved to be less interested in gemstones than in precious metals, and the Portuguese, having found most gem mining activities already operated by strong, advanced states, turned to trade instead. However, both the *conquistadores* (and the Italian merchants behind them, mainly from Genoa: cf. Cipolla, 1996; Ruffolo, 2004), and the missionary settlers who accompanied them, immediately began to assemble information on all the riches available in the new, fabulous lands. They flooded Europe with travel accounts full of data, and yet most of these manuscripts kept lying in the archives for years, or they were printed in secluded sites and remained unknown until G.B. Ramusio and Sebastianus Munsterus (Sebastian Münster), both around 1550, organized them and made them available to

Figure 2. Title page of the Italian translation of Camillo Leonardi's *Speculum lapidum*, translated by Lodovico Dolce (Venice, 1568) (cf. Fig. 1): an unscrupulous, Renaissance-style plagiarism, and yet an effective way of spreading knowledge on gemstones.

the public. The most interesting gem-related accounts from the West Indies were Petrus Martyr ab Angleria's (Pietro Martire d'Anghiera) *Decades etc.*, first published at Alcalá (1516), then shortened into Italian (Venice, 1534), and eventually republished in short form by Ramusio together with Barbosa's travels (Venice, 1554), and Gonzalo Fernández de Oviedo *Historia natural y general de las Indias* (first published in Toledo in 1526), the summary of which was translated into Italian and published in Venice as early as 1534. The most influential accounts from the East Indies, reached either through the land route or by circumnavigating Africa, were, indeed, Duarte Barbosa's and Lodovico de Varthema's (Vartomanus) works, the latter of which was translated first from vernacular Italian into Latin (Milan, 1511). Antonio Pigafetta's log of Magellan's (Fernando Magalhães) travels around the world in 1519–22, which he wrote in a queer mix of Venetian Italian and Castellano Spanish, was also a strongly influential account. It was first published as a pirate French translation by Jacques-Antoine Fabre (Parisii, 1526) before entering Ramusio's revised Italian collection (Venice, 1550).

Vannoccio Biringuccio's *De la pirotechnia* (1540) was the first extensive and completely original treatise on stones as sources of useful metals written in Italy during the Renaissance. It was printed in Venice in 1540, two years after the author's death (Fig. 3). Although it intended to deal primarily with metallurgy (i.e., the art of freeing metals from their gangue by the use of fire, in which both craftsmanship and science must go hand in hand to achieve good results), this book covered several other applied aspects of ore mining, exploitation, and dressing, as well as more remote matters related to mining, such as those bearing on precious stones. Biringuccio first stated that gems are stones (*son pietre, & han natura di pietre*) with special color effects (*respetto a colori*). Then he defined rock crystal to be a typical gem, a natural stone that is lustrous, transparent, and limpid (*trasparente, lucida & chiara*). He added that rock crystal is composed predominantly of a watery substance mixed with subtle earthiness, much air, and little fire (*di sustantia aquea con terrestita suttile con molto aere & pocho fuoco*): this involves nothing more than Aristotle's theory of the four elements applied to stones by Theophrastus (*De lapidibus*, I.2; see Mottana and Napolitano, 1997, p. 158). Then, leaving out the humble stones of the mountains having greatest earthiness (which he promised to deal with later), Biringuccio divided precious stones into two groups: the transparent and lustrous type (*lucide*: diamond, ruby, emerald, sapphire) and the opaque type (*opache*: sardonyx, nicolo, agate, and others), and stated that both groups assume various colors depending upon where they are found and their proximity to certain ore minerals. He went further with a description of the four main gems, but here he failed, because he drew his information from Pliny, adding only some traditional data on their supposed medical properties. The sole new information of interest concerns diamond (*pietra picchola & lucentissima, anzi fulgente & di durezza incomparable*): the largest gemstone of this kind known at the time was a little less than half the size of a walnut (*pocho men de una mezza noce*) and was in the possession of Suleiman, the magnificent Emperor of the Turks. Finally, Biringuccio abruptly ended this chapter with another promise: he would write a special treatise (*farvene un di un particular trattato*) on gems, "for it is a very useful and honorable thing for a man to have knowledge of these things and to know how to talk of them" (*in* Smith and Gnudi's translation, 1990, p. 125). Thus, he was well aware that gem science was of great interest, and had to be made open, to those who had not mastered Latin, yet he never did fulfill this promise. More than likely, he did not know about gem matters as deeply as he knew about metallurgy and mining, and thus could not follow through on his promise. Alternatively, the possibility exists that his unpublished manuscript has not yet been identified and retrieved from the archives of some Italian library (see below).

Consequently, the first really important contribution to gemology, which appeared in Italian during the sixteenth century, is contained in Pietro Andrea Mattioli's *I discorsi* (Venice, 1544)

Figure 3. Title page of the first edition (Venice, 1540) of Vannoccio Biringuccio's posthumous treatise *De la pirotechnia*, which deals with gemstones as minor materials, but promises to also address them elsewhere.

(Fig. 4)—the translation of and commentary on Dioscorides' *De materia medica*—the fifth book of which deals with minerals, gems, and inorganic artificial products, mainly for medical use. Dioscorides' original text, however, almost disappears in the pages of Mattioli's voluminous book, even in the comparatively short first edition, because Mattioli has included so many italicized comments (reminiscent of the medieval *glossae*) before and after each translated paragraph. Mattioli brought to print seven editions during his lifetime (the last and best being that at Venice, in 1568[7], which is quoted in the following) and constantly added extensive, new, original, and updated information. Each addition aimed not only at better illustrating the pharmaceutical properties of the individual stones (following his Greek originals, he never drew a line between these stones and other inorganic products), but also at providing more complete coverage by including new findings, about both natural stones and the artificial products of the then rapidly expanding medical-chemical research. Mattioli began by dealing with artificial products (*pietre metalliche*) and progressively mixed them with natural earths and salts (*terre*, *sali*) before moving on to colored stones (*pietre*) and gems (*gioie*), the last of which included sapphire (*saphiro*), ruby (*rubino*), lapis lazuli (*lapis lazuli*), and little else.

Rather than individual descriptions of the gems, which are scanty, what makes Mattioli's contribution significant is the extensive introduction (*discorso*) he wrote to Dioscorides' fifth book. Indeed, it is a short but complete theoretical treatise in itself (1568[7], p. 1369–1375 in folio); it begins by stating the problem of mineral formation, then goes on with summarizing the ancient authors' opinions and criticizing them (in particular Pliny, for example, for believing that rock crystal was deeply frozen ice), and ends by describing the properties and inferred origins of a number of gem materials. There, Mattioli stated (p. 1373):

bisogna che le gemme bianche si generino d'un succo simile all'acqua, & che però ci si dimostrino piu lucide, & piu chiare di tutte le altre, come è il cristallo, & parimenti l'iride; [...] le altre gemme lucide di qual si voglia colore, ò sieno fatte di succhi verdi, come sono gli smeraldi, & le prasme: ò di cerulei, come sono i sapphiri, i ciani, & alcune spetie di diaspri: ò di rossi, come sono i carbonchi: ò di porporei, come sono i giacinthi, & gli amethisti: overo di color d'oro, come sono i chrisoliti, & i chrisopatij: ò di misti, come gli opali.

In other words, he believed that all gems were derived from liquids by condensation, and that they inherited the colors of these original liquids: clear, colorless gems like rock crystal came from water; green gems like emerald from a green "sap" (*succo*); sapphires from a blue "sap," rubies from a red one, and so on, to opals, which, because of their sparkle in several colors, must form from mixed "saps." When Mattioli claimed, therefore, that diamond derived from a "sap" that is more brownish than the "saps" forming rock crystal and iris, it may confidently be inferred from this gemologically significant information that, in his time, the diamonds arriving in Europe certainly did not meet the highest levels of the 4 C scale, for at least their color would fall into the M grade, i.e., the commercial type formerly known as "Cape" type.

Furthermore, another statement by Mattioli is worth mentioning (1568[7], p. 1373):

Generasi l'ombra nelle gemme, ogni volta che la materia succosa loro è in qualche parte piu scura: & le nuvole vi si fanno, per esservi alcuna parte piu bianca: & i peli, da cui sono offesi spetialmente i sapphiri, il sale, che offusca particolarmente gli opali. & la piombaggine, che occupa gli smeraldi, sono veramente tutti impedimenti di altri colori differenti dal proprio di quelle gioie, in cui si ritrovano.

In other words, the flaws that decrease clarity, another valuable property of gemstones, already had been noticed by him, especially silk and clouds in sapphires and emeralds, as well as the small white inclusions (*piqué*) blurring the flicker of opals,

Figure 4. Etching of Pietro Andrea Mattioli (1500–77) taken from the inside front page of the seventh edition of his treatise *I discorsi nelli sei libri di Pedacio Dioscoride Anazarbeo della materia medicinale* (Venice, 1568).

DI GIORGIO
AGRICOLA.

DE LA NATVRA DE LE CO=
se foßili, e che da la terra si cauano.
LIBRI X.

A L'ILLVSTRE S. MAVRI=
tio Duca di Sassonia, e Principe
di Turingia, e di Misena.
Giorgio Agricola.

LA Philosophia; che tratta de la generatione de le cose, e de le cause e nature loro; contiene in se, Illustre Principe, difficilißime materie; e si diuide in molte parti: per che ella disputa di Iddio e de le cose diuine: del cielo, e de le stelle: de gli elementi, de le cause, de le cose con queste anneße; de gli intorbidamenti de l'aere, de gli animali; de le piante; de le cose sotterranee: Tutte le nationi si sono merauigliosamente trauagliate ne la cognitione d'Iddio: ma gli Hebrei principalmente, gli Egittij, e i Greci hanno questa cognitione acquistata de la natura diuina: Gli Caldei con lunghe oßeruationi, e i Greci con molto studio hanno conseguito la scientia de le stelle, e del potere andare misurando il cielo: I Greci ancho hanno sommamente disputato de gli elementi, de le cause, e de le cose con queste anneße: Tra li quali Aristotele ha ancho parlato

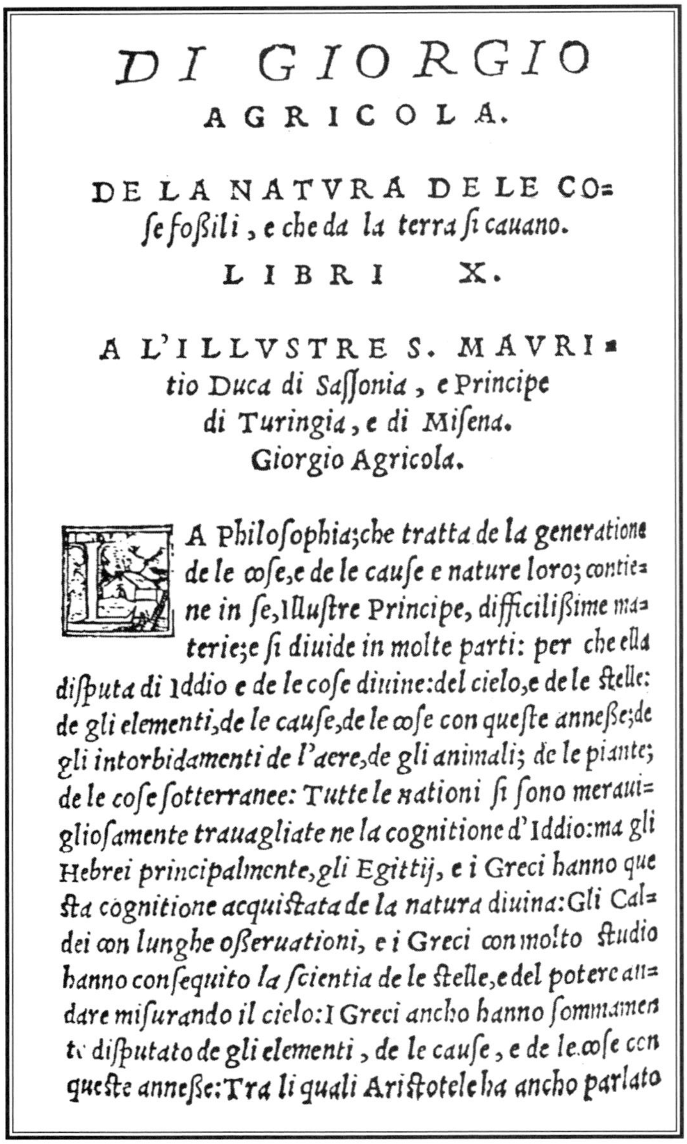

Figure 5. The beginning of the Italian translation of Georgius Agricola's *De natura fossilium*. It is leaf 163v of the miscellaneous book printed in Venice (1550) by Michele Tramezzino following the Latin first edition by Hieronymus Frobenius and Nicolaus Episcopius (Basle, 1546).

and the obscure feathers dimming the light passing through emeralds. All these observations, if not made with the naked eye, were made under a plain magnifying lens, because the microscope was not invented until some 40 years later, the first published account of its use on minerals being by Robert Hooke (1665) (see also Vai and Cavazza, this volume, Chapter 5).

In the time sequence of books on gemstones that appeared in Italian, the translation of Agricola's five Latin texts, jointly printed at Basle in 1546, appeared next. They were all published jointly in Venice (*Della natura delle cose fossili,* 1550) (Fig. 5), having been translated by someone unknown (possibly the same Michele Tramezzino who edited the volume) into such an appropriate and fluent Italian as to immediately become the standard for spelling and structure of mineral-related works in a scientific field (for which the Italian vernacular, and all the other European ones for the matter, was still rather backward). As already mentioned, the content of these texts is seminal for the development of mineralogy. Their gemological content, however, is not especially valuable. Obviously, Agricola had a penchant for a different kind of mineral study, in which he could provide his best knowledge (*De re metallica*, Basle, 1556, also translated into Italian by Michelangelo Florio as *De l'arte de metalli* and printed again in Basle, 1563). Nevertheless, he could not—and indeed he did not—forget gems when organizing the then-scanty information about minerals. Thus, he classified gems among stones, i.e., among the natural materials defined as those that do not get soft when impregnated with water, or melt when fired, but rather break up into small pieces. He then used other criteria, such as color, (specific) weight, transparency, luster, hardness, and the like to tell the various gems apart, mostly drawing his information from Pliny, but also from Leonardi, whom he respectfully acknowledged (Agricola, 1550, p. 306v: *Camillo da Pesaro*). What Agricola achieved on his own was to bring about a renewed attention to the visual characteristics of the stones and to add scattered data on new findings and uses, mostly from the German lands he knew so well. Thus, he listed the dig localities of the agate, cornelian, onyx, and zircon incorporated into the Bohemian crown jewels, and wrote that the best ruby came from *Zaylon* (Sri Lanka). Unfortunately, Agricola did not discriminate among the items of information he received and transmitted; thus, many traditional rumors are also recorded in his work. At the end (Agricola, 1550, p. 312v), he also listed the gems he knew in their decreasing order of value: diamond, pearl (*Unio dell'India*: curiously enough, the shell name and origin are given, but the name of the precious object, *perla*, is not), emerald, opal, ruby, jasper, sapphire, asteria, topaz, chrysolith, and so on—a rather different evaluation scale from that accepted today!

In the same year that Agricola's composite book was translated into Italian, another book which won much greater popularity, came into print: this was Giorgio Vasari's *Vite dei più eccellenti pittori, scultori e architettori* (Florence, 1550; 1568^2). It may seem odd to find information on gemstones in a series of biographical sketches of artists, but it is there nonetheless. First of all, Vasari, himself a master artist and equally proficient in architecture and painting, knew the constraints that technical aspects could put on an artist's work. Consequently, although most modern editions omit them, he thought it best to begin each one of the three parts of his *Vite* with an introduction in which all technical peculiarities of the works of art to be described were carefully detailed (Fig. 6). This included some aspects which may now seem insignificant, but at that time were an essential part of the artist's training and life, such as choosing and milling the earths to make fast colors and dyes (e.g., Giotto's tempera on wood) and how to select the marble block that would best fit the

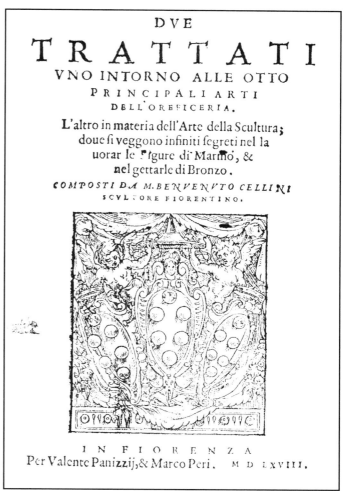

Figure 6. Title page of Giorgio Vasari's first edition (Florence, 1550) of *Le vite de piu eccellenti architetti, pittori, et scultori italiani*. Even here, Vasari stresses his point that the book includes "a useful and necessary introduction to their arts."

Figure 7. Title page of Benvenuto Cellini's *Due trattati: uno intorno alle otto principali parti dell'Oreficeria*, printed in Florence (1568). The craftsman's viewpoint on how jewels should be made The book details the craftman's viewpoint on how jewelry should be made and includes original data on how gems should be set into jewelry.

sculpture to be carved (e.g., Michelangelo on the Apuan Alps). Such introductory sections are now valuable because they are a first-hand account of how Renaissance artists operated (Rodolico, 1952, 1976). Secondly, following again the Renaissance way of thinking, Vasari did not diminish certain aspects of art that would now be considered trivial but at that time were considered at the same rank as masterpieces. Indeed, during the Renaissance, artists were considered—and believed themselves to be—no more than craftsmen (Hall, 1962), and were equally content to carve mantelpieces as statues, or to paint chests (*cassoni*) as they did crucifixes (*crocefissi*). They worked on those simple items with the same ingenuity and dedication we now assume they devoted to their masterpieces. Therefore, Vasari did not refrain from also listing the gems some of them had used to cut intaglios and cameos, thus providing a list of precious informational items on types and appraisal methods.

Benvenuto Cellini's accomplishments demonstrate the validity of this statement. He is now celebrated for his bronze statues, like *Ninfa* or *Perseo*, and for his gold *Saliera*, and indeed he made a great fuss about them in his *Vita* (dictated 1558–66, printed in Naples, 1728). Nevertheless, he actually made his reputation by casting and chiseling small objects, such as coins, buttons, and medals, as well as jewels, in which gemstones had to be set in such a way as to be ostentatious. He described his achievements and techniques in a book on jewelry (*Dell'oreficeria*), which was published only in part during his lifetime (Florence, 1568) (Fig. 7). Fortunately, the manuscript survived and could later be edited and printed in full (Florence, 1857; 2002), providing more recent scholars with much data recorded from the viewpoint of the worker, rather than the amateur and customer.

Cellini was blunt when stating his views (2002, p. 43): *Le qual gioie non sono altre che quattro, le quali son fatte per i quattro elementi, cioè il rubino è fatto per il fuoco, il zaffiro si vede veramente esser fatto per l'aria, lo smeraldo per la terra, e il diamante per l'acqua.* In other words: there are only four true gems, one each for every element. Ruby is made of fire, sapphire of air, emerald of earth, and diamond of water. Could anything be said against the extant Aristotelian book tradition that would make all gems mixtures of all four elements in different proportions? Cellini cared for nothing but practical experience, and the evidence deduced from it became for him the only scientific rule (p. 49: *io credo che fussi prima la pratica che la teorica di tutte le scienzie, e che alla pratica se le ponesse di poi regola, a tale che la si venissi a fare con quella virtuosa ragione*). Indeed, he was the first modern scientist, at his time the only comparable mind being Bernard Palissy in France, another craftsman turned to writing. The opinion Cellini had of contemporaneous scholars was very low: they were ignorant and presumptuous (*ignoranti [...] arroganti*), knew nothing of gems, and, although they may have called to mind the names of many of them, they were so stupid they did not even know that pearl is *osso di pesce* (fish bone)! Another mistaken statement, but how effective in a dispute! After these polemical outbursts, Cellini entered into a description of the four main gems, giving them in the order of their value at his time: ruby, emerald, diamond, and sapphire—further confirming the poor quality of diamonds then reaching Europe from India. Then, he listed another seventeen less-precious stones, some of them with accompanying expressions of supercilious disdain. This is not the case with topaz, however, which he appreciated and, although the name is ancient, correctly identified for the first time with the stone almost as hard as sapphire that shines as if it were sun, i.e., [yellow] topaz. He carefully described how he saw cheap doublets made so cleverly that they would fool even experienced people (p. 52: *scaglietta di rubini indiani [...] e quel resto che va nascosto nel castone [...] fatto di cristallo; di poi gli hanno tinti et appiccicati insieme, et appresso gli hanno fatti legare in oro*), and how he himself could prepare metal sheets (*foglia*) and set them into the cast behind the gem (*specchietto*) so as to enhance the stone's outward appearance. He provided details on a black glue (*tinta*) he made to improve the flash of a diamond so that it would sell for 20,000 *scudi* rather than 12,000 (p. 77). As a whole, his treatise—had it not been deprived of some interesting parts by the 1568 editor—would certainly have made gemology grow much faster than it did because it contains information on gemstones and their manipulation that, although difficult to retrieve because of a confusing organization of the text, is not to be found in later books until the *Encyclopédie* by Denis Diderot and Jean Baptiste d'Alembert, published in 1751!

Most (but not all) new information on gems published between 1502 and 1550–60 was reflected in Girolamo Cardano's *De subtilitate*, a comprehensive treatise on natural philosophy, which he started around 1546–47 (2004, p. 14), unfolded in its bulk during 1547–49 (p. 83), and progressively implemented into a final edition (1560³) that he carefully revised after a harsh controversy with Iulius Caesar Scaliger (1557), another top Italian scholar of that time. The seventh book of *De subtilitate*, titled *De lapidibus*, which immediately followed *De metallis* and contained new data on mining, was dedicated to stones in general and gemstones in particular. It contains an enormous amount of information drawn from classical, medieval, Arab, and contemporaneous sources. Indeed, Cardano's brain was a real sponge that could absorb all the writings he came across, and organize even minute pieces of information: a true *Plinius redivivus* (Morhof, 1673). Elio Nenci, the editor of *De subtilitate*'s critical edition (Milan, 2004), lists dozens of passages, which demonstrate that Cardano, although always writing in an effective and fairly elegant Renaissance Latin, drew his new information mainly from contemporary sources in Italian, Spanish, and Portuguese and actually translated them back into Latin. There are even passages from Agricola's *De natura fossilium* that Tramezzino had rendered into Italian and Cardano retranslated back into Latin. He defined gems to be *lapidem omnem nitentem, natura rarum atque parvum* (2004, p. 585). Thus, while confirming the small size and rarity of gemstones, he pinpoints the fact that gems must be bright to attract amateurs. And while still recalling color, he emphasized the new value given to hardness, claiming that *verae* [gemmae] *dicuntur hae solum, quae eam* [limam] *penitus non sentient* (p. 586), i.e., those unaffected by filing. Indeed, when suggesting the order of appreciation, he stated that *nitor, primum, inde durities, tertio opacitas, quarto color,* i.e., brightness, hardness, murkiness, and, last (!), color are to be accounted for, and he even returned to the subject of the added value given to them by *figurae*, i.e., the signs allegedly imprinted by heavenly powers. He then described thirty-one gems (among which, three to be found in living beings, e.g., in the stomach of a hen or in plants), plus a number of false ones. He did not refrain from reporting curiosities and prejudices, but he also described experiments he claimed he himself made or saw, and he ended by suggesting a visual method to distinguish a true stone from one made of glass, *dum in annulis,* i.e., even when it is already set into a ring (p. 695). He does not avoid listing gems in their order of commercial value (p. 595), as any Renaissance merchant would do: emerald, opal, ruby, diamond, pearl, sapphire, chrysolite, hyacinth, and *prannium* (smoky quartz), all identified by their own appropriate color. However, right after, he also did not refrain from asserting that gems get sick, old, and die (*lapidibus morbos, senium ac mortem advenire*: p. 595)!

As a whole, Cardano's treatise is astonishing for its wealth of learning, but also for its lore of credulity. It's no wonder that it became the reference book for generations of naïve scholars who avoided learning from experience (Accordi et al., 1981)! And it was a great misfortune for him that his attempt at changing the Aristotelian theory of elements (he denied that fire was the fourth element, and maintained that it was just burning air) made his entire thinking suspicious to the Catholic church; certain horoscopes he dared to make (e.g., Jesus Christ, Edward VI of England) almost dragged him to the scaffold!

Cardano wrote a second treatise on the same subject, *De rerum varietate* (Basle, 1557), where he himself claimed to have assembled all the information that he could not appropriately organize in *De subtilitate*. Indeed, the fourth and fifth books are once again titled *De metallis* and *De lapidibus*, the latter of which even contains a section on *Gemmae*. Nevertheless, what he had left out in the second edition of *De subtilitate* (1554), he introduced into the third one (1560), so that this book, in fact, sums up all of his thoughts on stones (and, more generally, on natural philosophy, the end result of which meant that *De rerum varietate* practically disappeared from printing).

The height of the merchant class concern for gems during the sixteenth century was reached around 1575 and is proved by Paolo Veronese's portrait of Jakob König of Füssen (*Jacobus Kinig Germanus Fiessensis*) (Fig. 8). This German jeweler and merchant was proudly portrayed holding an emerald in his right hand and showing it to the viewer. Indeed, the gem deserves it: a magnificent, nearly perfect, pinacoid-terminated hexagonal prism ~3 cm across and 7–8 cm long, of a good green color—actually, a little too dark, but this may be due to color decay—and with just a few marginal cracks! Obviously, this was not the first picture of a raw gem in the arts, but it is probably the first one ever made to pinpoint the holder so as to advertise his business. Indeed, König was a well known jeweler and gem provider who had left Bavaria to settle in Venice in order to be able to more efficiently carry out his trade all over Europe. It is well documented that he supplied diamonds to not only the Gonzaga and Medici dukes, but even Emperor Rudolph II himself, who was so pleased with him that he purchased König's portrait after König's death (ca. 1605) and stored it in his own Prague castle, where it still remains (Martin, 1999).

The incomplete publication of Cellini's treatise on jewels was an alarming indication of the overall decrease in interest in gemology that was going to pervade Italy during the final quarter of the sixteenth century. Indeed, none of the three other gemological treatises written in Italian at that time were ever printed.

The first of these treatises, *Le miracolose virtù delle pietre pretiose*, written by Scipione Vasolo in Rome (1577), probably did not deserve to be printed, as it is a mere compilation of the alleged medical properties of twenty-eight precious stones, some of them fabulous. It belongs to the field of lithotherapy more than gemology, never having contributed to the scientific growth of the latter (Mottana, 2005).

By contrast, *Questo è 'l libro lapidario* (Fig. 9), a manuscript compiled by Niccolò Costanti (a merchant at Siena, 1587), recently published by Paganini and Poli (Sesto San Giovanni, 1987), is full of interesting data. It is a composite text consisting of two markedly different sections. The first, allegedly transcribed from a book written by a Grey friar named Tommaso, who in turn claimed to have translated it from an earlier work in an Indian language (*Abessinia*), revisited—and described in, if possible, even more imaginative terms—the ancient lapidary by Damigeron and Evax (Halleux and Schamp, 1985). This lapidary was transposed in Latin verses by Marbode (see above), and, in turn, translated into Italian verses by an anonymous poet at the end of the thirteenth century, thus becoming common reading among the merchants. In general, the first part is of little interest as it draws information mostly from ancient sources. However, in places, it also contains new data, probably taken from personal experience if not plucked from unpublished accounts of travels to oriental gem markets. For example, *cetrino* is said to be identical to ruby and sapphire, i.e., it is yellow corundum coming from *Schilano* (Sri Lanka) and is without a doubt to be differentiated from *topazio*. Red stones were distinguished by their external characteristics, and *rubino* was stated to be the heaviest (densest) of them. More precisely, ruby was said to be *più pesante che tutte l'altre pietre*, i.e., heavier than all other gemstones, as indeed corundum is. There is also new information about amethyst, turquoise, lapis lazuli, and others.

In marked contrast with this section, the second part (which is written by the same hand, but stated to have been copied from an older, rotten manuscript by a *messer* [sir] Alessandro Vanocci) is of great interest and is totally Renaissance in its approach. It deals with the properties and values of the most important gems, starting with some general data that helps to qualify their quality and, consequently, to establish their price. For example, a good

Figure 8. Jakob König's portrait by Paolo Caliari, named Veronese (ca. 1575). A magnificent raw emerald is proudly exhibited by this merchant and jeweler as a display of his trade.

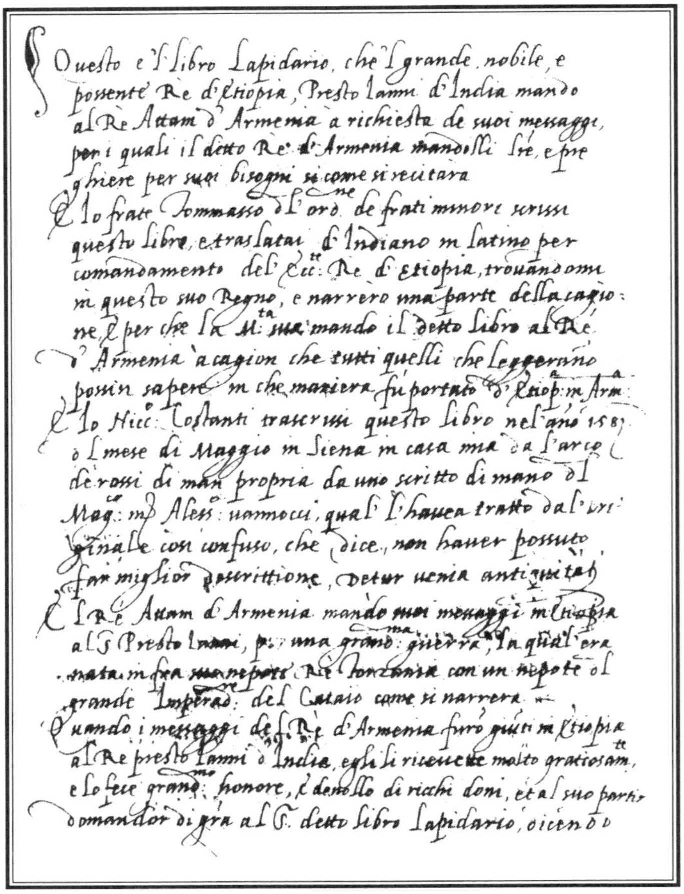
Questo e 'l libro Lapidario, che 'l grande, nobile, e
possente Re d'Etiopia, Presto Ianni d'India mando
al Re Attam d'Armenia a richiesta de suoi messaggi,
per i quali il detto Re d'Armenia mandolli lre, e pre
ghiere per suoi bisogni si come si recitara
Io frate Tommasso dell'ordine de frati minori scrissi
questo libro, e traslatai d'Indiano in latino per
comandamento del Ecc.mo Re d'Etiopia, trouandomi
in questo suo Regno, e narrero una parte della cagione, e per che la M.ta sua mando il detto libro al Re
d'Armenia a cagion che tutti quelli che leggeranno
possin sapere in che maniera fu portato d'Etiop. in Arm.
Io Nicc. Costanti trascrissi questo libro nel anno 1587
del mese di Maggio in Siena in casa mia da l'arco
de rossi di man propria da uno scritto di mano del
Mag.co m. Aless. Uannocci, qual l'haueua tratto dal originale con confuso, che dice, non hauer possuto
far miglior descrittione, detur uenia antiquitati
Il Re Attam d'Armenia mando suoi messaggi in Etiopia
al S. Presto Ianni, per una grande guerra, la qual era
nata infra suo nepote Re Tonzania con un nepote del
grande Imperad. del Cataio come si narrera.
Quando i messaggi del Re d'Armenia furo giunti in Etiopia
al Re presto Ianni d'India, egli li riceuette molto gratiosam.
e lo fece grande honore, e donolli di ricchi doni, et al suo partir
domandor di gra al S. detto libro Lapidario, dicendo

Figure 9. The first leaf of the manuscript *Questo è 'l libro lapidario*, transcribed (1587) by Niccolò Costanti from texts by friars Tommaso and Alessandro Vannocci. This book represents the most complete example of how Renaissance merchants dealt with gemstones.

diamond should have five properties: (1) it must be square in outline; (2) its faces should be equal; (3) it must be well pointed with sharp edges; (4) the faces should be flat; and (5) it must show good, pretty water (*bone e belle acque*). Apparently, color did not matter: diamond may be colorless, or turn to yellow or blue shades, but the best type is said to be the one similar to a mirror made of steel (*somiglianti al colore dello spechio d'acciaro*), i.e., highly reflective. If these conditions were met, then the diamond price increased quadratically with weight. Indeed, the book reports detailed tables of prices (in *ducati*) vs. weight (in *grani*) for diamond, ruby, emerald, and spinel. The data on emeralds are quite puzzling: the oriental ones, or *vecchi*, believed to have come from Egypt, are stated to be twice as expensive as the new ones, i.e., those coming from *Indie di Spagna* (Colombia). This is in marked contrast with our present evaluation, and with the bare fact that the Muzo and Chivor mines in *Colombia* were already well known sources. Such a statement indicates that, at least in Italy, the taste for clear, transparent, grass-green emeralds had not yet developed; rather, size, combined with a darker shade of green such as oriental emeralds usually show, was still preferred. Perhaps, therefore, there is no good reason to argue for color decay in the recently restored König's portrait by Veronese! The above four gems are the most prized ones. Other stones are also priced, but in *reali*, i.e. on a much lower money scale (~1:12; Cipolla, 1987, p. 88).

Clearly, the second part of this manuscript reflects the prevailing interests of the merchant class and fits the then economic significance of precious stones much more than any other contemporaneous book. The language not only changes somewhat from the first part, becoming more colloquial, but the extensive reliance on information from travels to the eastern countries shows that the Italian merchants were still trying to find out the places of gem production, where they could buy the stones at the lowest possible prices, thus increasing their final profit, if also their risk. Note, also, that these merchants were concentrating their attention upon the eastern source, either because they still gave little consideration to gem potential of the West Indies, or because they were trying to revitalize the eastern trade route as an alternative to the western one dominated by the Spaniards.

To complete the survey of Italian original studies of gemstones during the late sixteenth century, it is a great pity that *Istoria delle pietre*, written by Agostino del Riccio in Florence (1597), could not be published at that time. It appeared only in 1979 as a photostatic reprint of the original manuscript (Fig. 10) (an edited print with comment followed in 1996). It is masterly and, unfortunately, the last evidence of the interests of the merchant class in Italy during the final period of the Renaissance, i.e., Mannerism. Later on, during the Baroque, interests and tastes changed as a result of the Catholic Counter Reformation, which at that time began to establish its definitive, firm grasp on the best Italian minds.

Agostino del Riccio was a Black friar, disliked by his brothers for the great fervor he used to speak about the numerous interests (*propter magnam prolixitatem qua uti consuevit in prolatione*) he had cultivated thanks to his many travels throughout central Italy. He claimed he did not bother reading too much, preferring to rely on direct observation. However, this is not true: his book contains collected facts together with a number of pieces of hearsay information, but it also has pages copied from previous Italian books, such as those by Mattioli (1544, 1568), Dolce (1568), and Vasari (1550). It even has pages that closely follow certain Latin texts, both ancient and near contemporary to him (cf. the 1996 edition by Gnoli and Sironi, p. 229–230). The first part of the treatise describes some 100 decorative stones, both brought by the Romans to Italy from elsewhere in the Mediterranean lands (*le pietre pellegrine che i Romani hanno condotto in Italia*) and extracted anew from local quarries (*quelle che si cavano nella bella Italia*). This was appropriate enough because Renaissance architects, no longer able to use ancient salvage (*materiali di spoglio*) for their building, had to open new quarries to cope with the increased demand for structural and decorative stones. The second part of the book begins with a general survey of precious

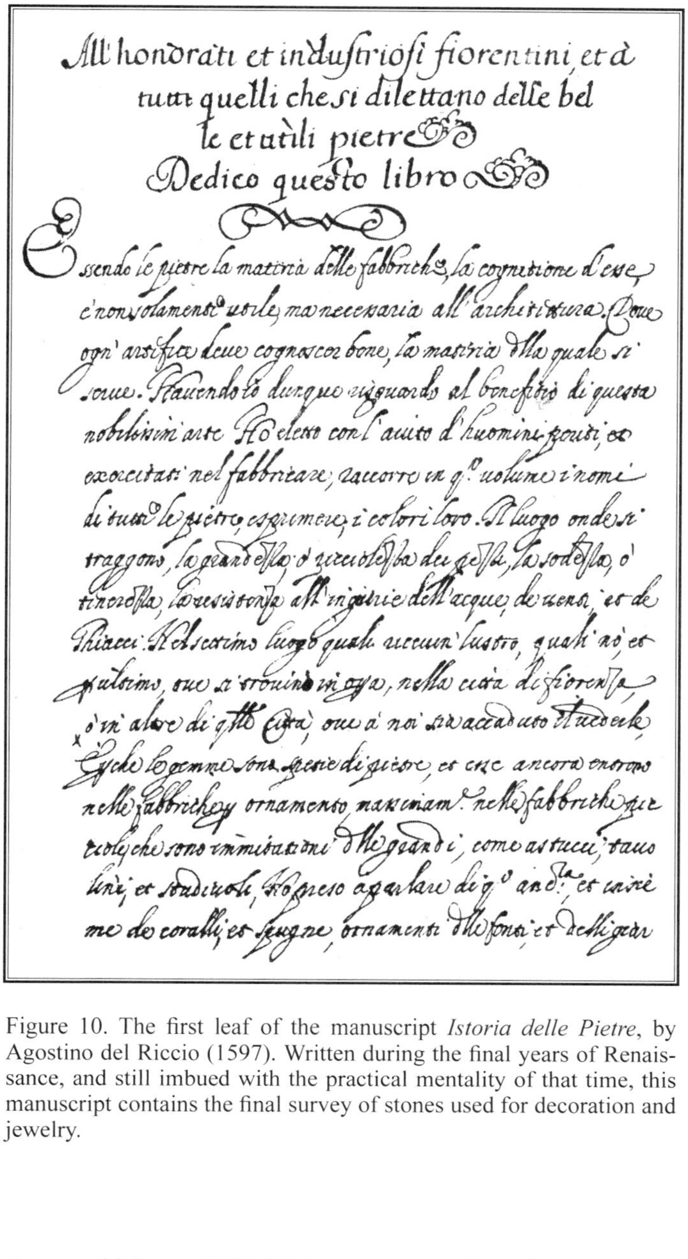
All'honorati et industriosi fiorentini, et à
tutti quelli che si dilettano delle bel
le et utili pietre
Dedico questo libro

Figure 10. The first leaf of the manuscript *Istoria delle Pietre*, by Agostino del Riccio (1597). Written during the final years of Renaissance, and still imbued with the practical mentality of that time, this manuscript contains the final survey of stones used for decoration and jewelry.

stones, which are divided into two groups (namely transparent and nontransparent), and continues with enthusiastic praise for stones in general as carriers of numerous properties conferred on them by God in order to favor humankind. Then follows a description (*si ragiona delle pietre preziose e perle*) of twenty-four precious and semiprecious stones, one by one, and of some other gemstones of organic derivation, such as pearl and coral. This second part is, on the one hand, not as interesting as the first, and yet, on the other hand, it is intriguing because it contains information not only on the gems used at that time for exhibition (size, worth, aspect, etc.), but also on new uses for them other than for jewels. For example, it contains the first description of *commesso* work, i.e., the Florentine method of setting semiprecious hard stones in a naturalistic mosaic that promoted a luxurious artisan craft highly appreciated by all the courts of Europe for more than three centuries (Giusti, 1992). However, intermixed with original data, this second part also contains long passages copied from Dolce's (1568) and Mattioli's (1568) books, as well as from Cesare Federici's account (Venice, 1587) of his travels to southern India and *Pegù* (Burma) that had appeared just a few years before. Although a compilation, this section is again useful because, if compared to Barbosa's (1554) report, it illustrates how the eastern gemstone trade changed within the sixteenth century. Surprisingly, Agostino del Riccio neither mentions Gasparo Balbi's book (Venice, 1590) nor makes use of his travel descriptions, although these, too, contain sections that are highly interesting from a jeweler's viewpoint. The book ends abruptly, announcing a forthcoming, comprehensive table of all the stones (*tavola universale*), but because of del Riccio's death this was never written.

The fact that such a well-informed and well-conceived text did not go to press during its time does not indicate that the author could not bring it to completion (indeed, the manuscript was copied several times; see the edition by Gnoli and Sironi, 1996, p. 86 [del Riccio, 1597]), but rather that the social class, whose interests and concerns had pushed Italy ahead of all the other European nations from 1350 onward, was losing influence and determination. This occurred, unfortunately, at the same time that all of Europe was entering an economic stagnation, presumably in part because of the implicit abolition of intellectual freedom imposed by the decisions of the Trento Council in the Catholic lands and of the harsh reaction to these in the Protestant ones. Giordano Bruno and Miguel Servet are just two of the numerous unfortunate victims of such a trend.

Thus, interest in gems rapidly moved toward another form of appreciation, in which gemstones mattered less for their intrinsic value or as carriers of mysterious, magical, and medicinal properties, or as luxury commodities, than for the prestige they would convey to the owner. This change of attitude also shows that interest was moving toward other forms of understanding natural materials. They were no longer sought for their applications, but rather as tools to illustrate God's action in the natural world, this being often extended to include odd, deformed animals, vegetables, and fossils: all materials that would be able to stir up emotions as well as to frighten people not strictly adhering to God's words (whatever their faith was).

Andrea Bacci's *Le XII pietre pretiose* (Rome, 1587) was another book published in Italian during the very last quarter of the sixteenth century. This treatise revealed the meaning of the twelve stones set in the breastplate of the Jewish high priest according to *Exodus* 29, as well as those mentioned by Saint John in *Apocalypsis* 22, plus other miscellaneous stones. It contributed nothing to gem appraisal and to the development of gemology, yet this treatise was highly considered by Catholics and Protestants alike. It was vastly implemented by both and finally re-edited in Latin (Frankfort, 1603) after the author's death because it was a testimony of the uninterrupted Christian tradition of interpreting gemstones allegorically.

Moreover, the second half of sixteenth century saw the formation of several museums (Olmi, 1982; Findlen, 1989, 1994; Vai, 2003), such as those organized in Bologna by Ulisse Aldrovandi (De Bellis, 2001), in Verona by Francesco Calzolari (Accordi, 1977), in Rome by Michele Mercati (Accordi, 1980), and in Naples by Ferrante Imperato (Stendardo, 2001). All of these museums started on the right track, i.e., as collections of materials that could properly represent the natural world (*teatrum naturae*), but they rapidly developed into *Wunderkammern*, such as the museum emperor Rudolph II assembled in Prague (von Schlosser, 1974), where weird and wonderful things were sought after and bought for a high price, the stranger the better, and used to amaze people. Minerals and gems were certainly present in those museums, but most often their role was minor, and they were confused among other odd, shocking things.

Inevitably, with such a change in customer interests, demand for gems and minerals among merchants changed, too. Thus, merchants stopped looking for gems for the purpose of selling them and craftsmen stopped studying their intrinsic properties with the purpose of enhancing their outward appearance, and, accordingly, their value. Consequently, the study of gems fell back again to scholars writing in Latin or, even if writing in Italian, wishing to interpret gems allegorically as a part of God's gifts to human beings. Around the turn of the sixteenth century there were several books printed in Latin that deal with gems and minerals (e.g., Andrea Caesalpinus's *De metallicis*, Rome, 1596), but most of them were just erudite classifications containing little or nothing new. Strangely enough, even *Musaeum metallicum*, the book written by Aldrovandi to describe the huge collection of natural history he had assembled from 1547 to 1603 (Olmi, 2001; Sarti, 2003; Vai, 2003), which contained numerous original data especially on Apennine localities, was not published until after his lifetime (Bologna, 1648). The same unlucky fate occurred to Mercati's *Metallotheca*, written between 1576 and 1589 to illustrate the collection of stones and fossils he had kept in the Vatican. It did not reach the printer in his lifetime, although he had taken care to have 127 superb prints prepared to illustrate it, probably because his premature death left the commentary of the book unfinished. Eventually, one century later, *Metallotheca* found an editor and publisher (Rome, 1717, 1719^2), but by that time the impact of its scientific content had fallen close to nil (Accordi, 1980). By contrast, Ferrante Imperato finished and published his book, *Dell'historia naturale*, in Italian (Naples, 1599), although the entire book is somewhat of a compilation, with just a minor amount of gemological content.

It is to be concluded—sadly—that at the end of the sixteenth century mineralogy and gemology had been deserted by most active, practical workers and returned to the hands of scholars— good, proficient scholars, as a matter of record, who could accumulate all of the previous data and arrange them in a logical way (Accordi, 1981), but inadequate and possibly incapable of further innovation in the field. Indeed Italy, as a minor part of Europe struggling for political survival, was falling into backwardness under the pressure of conformity. It was a slow downgrading, however, because for a while Italy continued attracting foreign students eager to learn from teachers of very high scientific fame, such as Ulisse Aldrovandi, Galileo Galilei, Evangelista Torricelli, and Marcello Malpighi, to quote just a few. William Harvey, for example, took his doctor's degree at Padua in 1602 (Rossi, 1997) and Niels Stensen, better known as Nicolaus Steno, left Leiden, where he had been granted a doctorate in 1664, to continue his studies at Pisa (Mottana, 1995). It was there that he found the scientific atmosphere conducive to conceiving the *Prodromus* (published in Florence, 1669, when he had already left for his religious mission), by which he brought radical new thinking to both stratigraphy and crystallography.

Nevertheless, the scientific climate went from bad to worse: modern science needed a more stimulating atmosphere and more consideration than was allowed by religious and political conformity. The downgrading was slow, but continuous. In terms of gem studies, gem cutting as a practical art kept being practiced in Italy with some success. For example, a Venetian craftsman, Vincenzo Peruzzi, is credited with introducing a fourfold-symmetry cut for diamond resulting in fifty-eight facets, which became the precursor to today's brilliant cut. This happened some years before the end of the seventeenth century; the first mention of two "brilliants" (those in the casket of Amalia of Orange) was in 1690 (Tillander, 1995). Nevertheless, this was just innovation in the cutting process, implying no better understanding of the crystallographic reasons behind it or why the stone acquired such an intense play of colors (fire), i.e., of the physical and geometrical reasons by which the back-reflection path imposed on light passing through the crystal enhances dispersion. Indeed, no new scientific work on gems appeared in Italy before Giacinto Gimma's *Sulla storia naturale delle gemme* (Naples, 1730). Elsewhere in Europe, Anselm Boetius de Boot and his followers Adriaan Toll and Jan de Laet were busy establishing the fundamentals of gemology as a science, describing newly discovered, bright, high-priced stones, as well as upgrading knowledge on traditional ones. Unfortunately, however, they failed to link their gemology to mineralogy, which was then subject to a continuous supply of new practical and theoretical considerations by Johannes Kepler, Pierre Gassendi, Robert Hooke, Rasmus Barthelsen (Bartholin), Robert Boyle, Domenico Guglielmini, Christiaan Huygens, and others. The need to establish that link was not felt until much later.

CONCLUDING REMARKS

When de Boot acknowledged the authors he had studied to compose his *Gemmarum et lapidum historia* (Hanau, 1609; Leiden, 1636^2, p. 10), thus also summarizing the state of the art of stones and gems in his time, he ignored Camillus Leonardi completely and quoted, besides obvious ancient and medieval writers, twelve Renaissance scholars. Among them, four were Italians (*Prosper Alpinus, Andreas Baccius, Hieronymus Cardanus,* and *Ludovicus Dulcis*), the last of which was the only one writing in Italian, but also the least reliable, for he made his reputation by plagiarism!

The reason for such a choice, which seems to contradict the goal of the present study, lies in de Boot's approach to gemology. Indeed, when evaluating the contribution of his sources, he correctly pointed out (de Boot, 1636[2], p. 1) that Konrad Gesner had been the first to classify gems by a method based on their similarity, names, and costs, making the value an important parameter for gem evaluation. He asserted, nevertheless, his dislike for such an approach and was sufficiently strong in his view that he decided, in order to restore the matter to a sound basis, to develop a method of his own based on size, rarity, transparency, hardness, and color. Clearly, de Boot wanted to show his contempt for the merchant approach to gemstones in an attempt to base gemology on more durable and reliable scientific criteria. Cardano's essentially theoretical essays were certainly up to his goal. As for Bacci's treatises on the allegorical significance of the biblical gems, it was probably appropriate for a Fleming to seek support from an eminent man whose work was equally appreciated by both Catholics and Protestants, and Dolce's book contained too many traditional data to be dismissed altogether (if one did not know Leonardi's). There is poor reason for de Boot's mention of Alpinus, who is better known as a botanist and as the first European who measured the Egyptian pyramids than for his contributions to gemology. The criteria de Boot suggested at the end might seem the most appropriate to the Renaissance scientific context and to the following development of mineralogy as the general study of inorganic materials, but by no means were they adequate to meet a major aim of gem studies, which was, and is, to support the potential customer with accurate information on the quality of the stones he was about to buy, this quality being one of the factors determining the price—a consideration as important then as it is now.

Unfortunately, de Boot's opposition of the practical trend in gem appraisal that had developed in Italy during the Renaissance triumphed at the time and remained dominant for a long time thereafter because it best fit the scholarly mind. Even more unfortunately, it did no service to gemology, which began to lag behind mineralogy, while two other related branches of this science, mining and crystallography—the first, a totally practical field run by men scorning theory when not fully ignorant of it, the second, struggling for wide-ranging concepts and a global theoretical foundation—grew and flourished, having respectively Agricola and Steno as their founding fathers. Anyway, despite de Boot's prevailing approach of the time, the Italian contributions to gemology during the Renaissance, and their value as carriers of new ideas and methods, remain fully demonstrated: they did not vanish into the dark eddy (*oscuro gorgo*) where all bad sciences sink, but they kept swimming underwater. Eventually, these contributions reappeared back on the surface during the late nineteenth century, becoming the informing principles that made modern gemology finally take off.

Strangely enough, while practically oriented gemology was dying in Italy and elsewhere in Europe because of the attack by de Boot's influential book, it continued to be practiced in Antwerp, a town some 80 km east of Bruges, de Boot's hometown. Most of the technical innovations developed during the Renaissance, such as gem cutting and polishing, continued to be practiced proficiently there (Kockelbergh et al., 1992). As a result, the trade pattern changed: gems no longer moved initially from Venice to the rest of Europe, but from Antwerp to Venice and only then to elsewhere. During the seventeenth century, raw gems were imported into Europe from both Indies through Lisbon and Seville, mostly, then shipped to Antwerp to be cut and ameliorated. Only then, in this new form, were they transferred from there to Venice, which remained the European center of luxury spending for the entire seventeenth and most the eighteenth centuries. It was in Venice that most gems were set in fashionable jewels, thus occasionally needing some additional cut or polishing, and from Venice they continued their trip into the world market. Only some top quality raw gems would be disposed directly to the royal courts (Falk, 1975; Tillander, 1995) and processed according to princely wishes. Occasionally, certain gems would travel from Venice to as far back as India, where it even happened that a Venetian lapidary, Ortensio Borgis, was entrusted (ca. 1665) to cut the Grand Mogol diamond—at that time the world's largest—into an unusual rose.

ACKNOWLEDGMENTS

Gian Battista Vai encouraged me to write this paper for the 32nd International Geological Congress, the second to be held in Italy in more than a century. This congress has signaled international acknowledgment of a new Renaissance of geological studies in Italy after a period of scattered, although impressive, contributions by individual, dedicated researchers poorly supported by national funding agencies. My interest in gems dates back thirty years, to the stimulating discussions I had with Isidoro Zaro and Guido Devoto, and has received much stimulus from the study I made five years ago on the mineralogical content of the first scientific book ever written in an Italian vernacular, by Restoro d'Arezzo (1282). The financial support by Accademia Nazionale dei Lincei, Commissione per i musei naturalistici e i musei della Scienza, and the advice of Marco Guardo, Director of Biblioteca Accademica e Corsiniana, are gratefully acknowledged. The manuscript has been reviewed by Chiara Boroli and Michael Carroll. Francesco Paolo Sassi, an unknown French referee, and the editors contributed greatly to improve it by their many suggestions and thoughtful editing.

REFERENCES CITED

(The primary references include information concerning different editions, marked with superscript numbers, as well as translations which are referred to in text by the name of the tranlator(s) and the date of publication.)

Accordi, B., 1977, Contributions to the history of geological sciences: the Musaeum Calceolarium (XVIth century) of Verona illustrated in 1622 by Ceruti and Chiocco: Geologica Romana, v. 16, p. 21–54.

Accordi, B., 1980, Michele Mercati (1541–1593) e la Metallotheca: Geologica Romana, v. 19, p. 1–50.

Accordi, B., 1981, Tentativi di classificazione delle pietre e delle gemme nei secoli XVI e XVII: Physis, v. 23, no. 3, 32 p.

Accordi, B., Stacchiotti, L., and Tagliaferro, C., 1981, Traduzione e commento del *De lapidibus* di Gerolamo Cardano (dal libro VII del *De subtilitate*, edizione del 1560): Geologica Romana, v. 20, p. 125–169.

Agricola, 1530, Georgii Agricolae Medici, Bermannus sive De re Metallica dialogus. Basileae, in aedibus Frobenianis, 184 p. (quoted according to the Italian translation, 1550, p. 419–467, cared by Michiele Tramezzino *in* Di Giorgio Agricola De la generatione de le cose, che sotto la terra sono, e de le cause de' loro effetti e nature. Lib. 5. De la natura di quelle cose, che da la terra scorrono. Lib. 4. De la natura de le cose fossili, e che sotto la terra si cauano. Lib 10. De le minere antiche e moderne. Lib. 2. Il Bermanno, o de le cose metallice, dialogo, recato tutto hora dal latino in buona lingua volgare, *in* Vinegia, Tramezzino, M., XXVIII + 467 p.).

Agricola, 1546, Georgii Agricolae De ortu & causis subterraneorum Lib. V. De natura eorum quae effluunt ex terra Lib. IIII. De natura fossilium Lib. X. De ueteribus & nouis metallis, Lib. II. Bermannus, siue De re Metallica dialogus. Interpretatio Germanica vocum rei metallicae addito Indice faecundissimo: Basileae, apud Hieronimymum Frobenium et Nic. Episcopium, 487 + 53 p. (quoted according to the Italian translation, 1550, p. 163–390, cared by Michiele Tramezzino *in* Di Giorgio Agricola De la generatione de le cose, che sotto la terra sono, e de le cause de' loro effetti e nature. Lib. 5. De la natura di quelle cose, che da la terra scorrono. Lib. 4. De la natura de le cose fossili, e che sotto la terra si cauano. Lib 10. De le minere antiche e moderne. Lib. 2. Il Bermanno, o de le cose metallice, dialogo, recato tutto hora dal latino in buona lingua volgare, *in* Tramezzino, M., Vinegia, XXVIII + 467 p.).

Agricola, 1556, Georgii Agricolae De Re Metallica Libri XII. Quibus Officia, Instrumenta, Machinae, ac omnia denique ad Metallicam spectantia, non modo loculentissimè describuntur, sed & per effigies, suis locis insertas, adiunctis Latinis, Germanicisque appellationibus ita ob oculos ponuntur, ut clarius tradi non possint. Eiusdem de Animantibus subterraneis Liber, ab Autore recognitus; cum Indicibus diversis, quidquid in opere tractatum est, pulchrè demonstrantibus: Basileae, [H Froben & N Episcopius], 502 p. + 86 p. of glossary and index (quoted according to the Italian translation by Michelangelo Florio, 1563, titled "Opera di Giorgio Agricola de l'arte de metalli, etc.": Basilea, per Hieronimo Frobenio et Nicolao Episcopio, 542 p. + 6 p. of index; anastatically reprinted, 1994, with introduction by Macini, P., and Mesini, E.: Segrate, ANIM, XXXII + 542 p.).

Albertus Magnus, 1476, De mineralibus et rebus metallicis libri quinque, correxit Nicolaus de Pigaciis: Patavii, Petrus Maufer per Antonium de Albricis, n.n.

Aldrovandi, U., 1648, Ulyssis Aldrovandi Patricii Bononiensis Musaeum Metallicum in Libros IIII distributum Bartholomaeus Ambrosinus labore et studio composuit, Marcus Antonius Bernia edidit: Bononiae, typis Jo. Baptistae Ferronij, 992 p.

Anonymous, 1524, Probirbüchleyn tzu Gotteslob: Magdeburg, n.n.

Bacci, A., 1587, Le XII pietre pretiose, le quali per ordine di Dio nella santa legge, adornavano I vestimenti del sommo sacerdote. Aggiuntevi il diamante, la margarie, e l'oro, poste da S. Giovanni nell'apocalisse, in figura della celeste Gierusalemme: con un sommario dell'altre pietre pretiose. Discorso dell'alicorno, et delle sue singolarissime virtù. Et della gran bestia detta alce da gli antichi: Roma, appresso Giouanni Martinelli, nella stamparia di Vincenzo Accolti, in Borgo novo, 130 p. (reprinted, 1992, under the title: Le 12 pietre preziose: le quali per ordine di Dio nella santa legge adornavano il manto del gran sacerdote: secondo la interpretazion di S. Ieronimo e S. Epifanio arcivescovo di Cipri: Milano, Philobyblon, 53 p.).

Baccius, A., 1603, De gemmis et lapidibus pretiosis, eorumq; viribus & usu tractatus, italica lingua conscriptus; nunc vero non solum in latinum sermonem conversus verum etiam utilissimis annotationibus & observationibus auctior redditus. A Wolfango Gabel, chovero, medicinae doctore & phisico Calvvensi ordinario. Cui accessit disputatio, de generatione auri in locis subterraneis, illiusq. Temperamento: Francofurti, ex officina Matthiae Beckeri, impensis Nicolai Steinii, 231 p.

Bänsch, B., and Linscheid-Burdich, S., 1985, Theophilus, Schedula diversarum artium: Textauszüge, *in* Legner, A., ed., Ornamenta Ecclesiae, Kunst und Künstler der Romanik in Köln (2 vols.): Köln, Greven & Bechtold, v. 1, p. 363–375.

Balbi, G., 1590, Viaggio dell'Indie orientali, di Gasparo Balbi, gioielliero venetiano. Nel quale si contiene quanto egli in detto viaggio ha veduto per lo spatio di 9. anni consumati in esso dal 1579. fino al 1588. Con la relatione de i datij, pesi, & misure di tutte le citta di tal viaggio, & del gouerno del re del Pegu, & delle guerre fatte da lui con altri re d'Auua & di Sion. Con la tauola delle cose piu notabili: In Venetia, appresso Camillo Borgominieri, XVI + 149 + XXIII p. (re-edited, 1962, at p. 71–233, *in* Viaggi di C. Federici e G. Balbi alle Indie Orientali, care of Pinto, O.: Roma, Istituto poligrafico dello Stato—Libreria dello Stato, XLIV + 438 p.).

Barbarus, H., 1492–93, Castigationes plinianae et in Pomponium Melam. In Plinium Glossemata (2 vols.): Romae, impressit Eucharius Argenteus Germanus, 348 p. (re-edited, 1973, by Pozzi, G.: Padova, Antenore, CLXVIII + 318 p.)

de Barbosa, O., 1554, Libro dei viaggi, p. 319–358, *in* Ramusio, G.B., ed., Delle nauigationi et viaggi raccolto gia da messer Giovanni Battista Ramusio, & con molti & vaghi discorsi, da lui in molti luoghi dichiarato & illustrato (vol. I): Venetia, nella stamperia de Giunti, XXXIV + 394 (re-edited as Libro di Odoardo Barbosa *in* Ramusio, G.B., Navigazioni e viaggi, v. II, care of M. Milanesi, Torino, Einaudi, 1979, XI + 1160 p.; full translation, 1918–21, from the original Portuguese manuscript, first published 1812, by Dames, M.L., The Book of Duarte Barbosa. An account of the countries bordering on the Indian Ocean and their inhabitants, written by Duarte Barbosa, and completed about the year 1518 A.D. (2 vols.): London, Hakluyt Society, LXXXV + 238 p., XXXIX + 2 maps. Re-edited, 1989: ASEA, Indian reprint).

Bartholomaeus Anglicus, 1471, Liber de Proprietatibus Rerum: Basileae, Berthold Ruppel.

Bartholomaeus Anglicus, 1471–72?, De proprietatibus rerum: Coloniae, Johannes Schilling per William Caxton.

Bauer, M., 1896, Edelsteinkunde. Eine allgemein verständliche Darstellung der Eigenschaften, des Vorkommens und der Verwendung der Edelsteine, nebst einer Anleitung zur Bestimmung derselben, für Mineralogen, Edelsteinlieber, Steinschleifer, Juweliere. Leipzig, Chr. Herm. Tauchnitz, XVI + 766 p.

Biringuccio, V., 1540, De la Pirotechnia. Libri X dove ampiamente si tratta non solo di ogni sorte & diuersita di Miniere, ma anchora quanto si ricerca intorno à la prattica di quelle cose che s'appartiene a l'arte de la fusione ouer gitto de metalli come d'ogni altra cosa simile à questa. [Composti per il Signor Vannoccio Biringuccio Sennese]: Venetia, per Venturino Roffinello, ad instantia di Curtio Nauo & fratelli, VIII + 168 p. (reprinted, 1977, with an introduction in Italian and English and a glossary care of Carugo A.: Milano, Il Polifilo, LXXXVIII + 336 p.; English translation, 1990, by Smith, C.S., and Gnudi, M.T.: "The Pirotechnia of Vannoccio Biringuccio. The classic sixteenth-century treatise on metals and metallurgy": New York, Dover, XXVII + 477 p.).

Boot, A.B. de, 1609, Gemmarum et lapidum historia, qua non solus ortus, natura, vis & precium, sed etiam modus quo ex iis olea, salia, tincturae, essentiae, arcana & magisteria arte chymica confici possint, ostenditur: Hanouiae, typis Wechelianis apud Claudium Marnium & heredes Ioannis Aubrii, VIII + 288 p. (new edition, 1636², improved and added with indices by Adrianus Toll: Lugduni Batavorum, ex officina Ioannis Maire, 576 p.).

Brucioli, A., 1548, Historia naturale di C. Plinio Secondo. Nuouamente tradotta di latino in vulgare toscano per Antonio Brucioli: Venetia, per Alessandro Brucioli, & i fregatli, XL + 1068 p.

Brunfels, O., 1530–36, Herbarum viuae eicones ad naturae imitationem, summa cum diligentia & artificio effigiatae, una cum effectibus earundem, in gratiam ueteris illius, & iamiam renascentis herbariae medicinae per Othonem Brunfels recens editae. Quibus adiecta ad calcem, Appendix isagogica de usu & administratione simplicium. Item index contentorum singulorum (2 vols. with illustrations by Hans Weiditz): Argentorati, apud Ioannem Schottum, 166 + 201 p.

Burckhardt, J., 1855, Die Cicerone: eine Anleitung zum Genuss der Kunstwerke Italiens: Basel, Schweighauser'sche Verlagsbuchhanglung, 1287 p. (new critical edition, 2001, München, Beck).

Caesalpinus, A., 1596, De Metallicis Libri Tres, Andraea Caesalpino auctore. Ad Sanctissumum Dominum Nostrum Clementem VIII, Pont. Max: Romae, ex typographia Aloysij Zannetti, XIV + 222 p.

Cardanus, H., 1550, De subtilitate libri XXI: Norimbergae, apud Ioh. Petreium, XXXVI + 371 p. (second edition, 1554: Basileae per Ludovicum Lucium, XXXV + 561 p; third, final edition, 1560, De subtilitate libri XXI. Ab authore plusquam mille locis illustrati, nonnullis etiam cum additionibus. Addita insuper Apologia aduersus calumniatorem, qua uis horum librorum aperitur: Basileae, ex officina Petrina, LXXXVIII + 1426 p.; quoted according to the critical edition, 2004, of the first seven books edited by Nenci, E.:

Milano, Angeli, 720 p.).

Cardanus, H., 1557, De rerum varietate libri XVII. Adiectus est capitum, rerum & sententiarum notatu dignissimarum index: Basileae, per Henricum Petri, 12 + 707 + 33 p. (quoted according to the references contained in "H.C. Mediolanensis medici Liber de libris propriis": Lugduni apud Gulielmum Rouillium, sub scuto Veneto, 1557, 192 p.; edited, 2004, with introduction and chronology by Maclean, I.: Milano, Angeli, 400 p.)

Cellini, B., 1568, Due trattati: uno intorno alle otto principali parti dell'Oreficeria. L'altro in materia dell'Arte della Scultura; dove si veggono infiniti segreti nel lavorare le Figure di Marmo, et nel gettarle in Bronzo: Firenze, per Valente Panizzij e Marco Peri (quoted according to the improved partial edition, 2002, which follows the critical edition "I trattati dell'oreficeria e della scultura di Benvenuto Cellini novamente messi alle stampe secondo la originale dettatura del codice marciano" care of Milanesi, C.: Firenze, Sansoni, 1857, re-edited under the title "Dell'oreficeria" by Capitanio, A.: Torino, Aragno, XXVI + 208 p.).

Cellini, B., 1728, Vita di Benvenuto Cellini orefice e scultore fiorentino, da lui medesimo scritta, nella quale molte curiose particolarita si toccano appartenenti alle arti ed all'istoria del suo tempo, tratta da un'ottimo manoscritto, e dedicata all'eccellenza di Mylord Riccardo Boyle, a cura di Antonio Cocchi: Colonia [but Naples], Pietro Martello [Pierre Marteau] x + 318 p. (quoted according to a new partial edition, 1988, which follows the critical edition titled "Vita di Benvenuto Cellini / testo critico con introduzione e note storiche" care of Bacci, O.: Firenze, Sansoni, 1901, LXXXVIII + 451 p., as re-edited under the title "Vita" by Argan, P. and Lazotti, L.: Milano, SEDES, 128 p.).

Centmannus, J., 1565, Nomenclaturae rerum fossilium quae in Misnia precipue et in aliis quoque regionibus inveniuntur (95 p. together with seven other works by other authors in the miscellaneous book "De omni Rerum Fossilium Genere, Gemmis, Lapidibus, Metallis, et huiusmodi, libri aliquot, plerique nunc primum editi, opera Conradi Gesneri: quorum catalogus sequens folium continet": Tiguri, excudebat Iacobus Gesnerus, 549 p.).

Cipolla, C.M., 1987, La moneta a Firenze nel Cinquecento: Bologna, Il Mulino, 112 p.

Cipolla, C.M., 1996, Conquistadores, pirati, mercatanti. La saga dell'argento spagnolo: Bologna, Il Mulino, 84 p.

Conte, G.B., Barchiesi, A., and Ranucci, G., 1982–88, Gaio Plinio Secondo. Storia naturale (5 vols. in 6 parts): Torino, Einaudi, LXXIV + 844 + 706 + 996 + 907 + 642 + 966 p.

Costanti, N., 1587, Questo è 'l libro lapidario (manuscript of 42 leaves, transcribed by Paganini, C., and Poli, G., with an introduction by de Michele, V.): Sesto S. Giovanni, Istituto Gemmologico Italiano, 1987, n.n.

De Bellis, C., 1985, Astri, gemme e arti medico-magiche nello "Speculum lapidum" di Camillo Leonardi, in Formichetti, G., ed., Il mago, il cosmo, il teatro degli astri. Saggi sulla letteratura esoterica del Rinascimento: Roma, Bulzoni, p. 67–114.

De Bellis, N., 2001, La mineralogia, in Simili, R. ed., Il teatro della natura di Ulisse Aldrovandi: Bologna, Editrice Compositori, p. 88–89.

Del Riccio, A., 1597, Istoria delle pietre (riproduzione in facsimile del cod. 230 della Biblioteca Riccardiana, a cura di Paola Barocchi): Firenze, SPES—Studio per edizioni scelte, 1979, XXXIX + 224 + 68 p.

Del Riccio, A., 1597, Istoria delle pietre (a portion of ms. Magl. II, I, 13: "Arte della Memoria", Biblioteca Nazionale di Firenze, collated with cod. misc. 230, Biblioteca Riccardiana di Firenze, and with col ms. V E, IO, Biblioteca Nazionale di Napoli) Gnoli, R., and Sironi, A., eds., with an introduction by Gnoli, R.: Torino, Allemandi, 1996, 253 p.

Dodwell, C.R., 1961, Theophilus Presbyter. De diversis artibus: London and Edinburgh, Thomas Nelson, LXIII + 171 p. (new edition, 1979, by Hawthorne, J.G., and Smith, C.S., titled: Theophilus: On diverse arts. The foremost medieval treatise on painting, glassmaking and metalworks, translated with introduction and notes: New York, Dover, xxxv + 216 p.).

Dolce, L., 1568, Libri tre di M. Lodovico Dolce; ne i quali si tratta delle diuerse sorti delle Gemme, che produce la Natura, della qualità, grandezza, bellezza & virtù loro: Venetia appresso Gio. Battista, Marchio Sessa, et Fratelli, 100 p.

Domenichi, L., 1561, Historia naturale di G. Plinio Secondo, tradotta per M. Lodouico Domenichi con le postille in margine, nelle quali, o vengono segnate le cose notabili, o citati altri autori, che della stessa materia habbiano scritto, o dichiarati i luoghi difficili, o posti i nomi di geografia moderni: Vinegia, appresso Gabriel Giolito de' Ferrari, LXVIII + 1188 p.

Encelius, C., 1551, De Re Metallica, hoc est, de Origine, Varietate, & Natura Corporum Metallicorum, Lapidum, Gemmarum, at; aliarum, quae et fodinis eruuntur, rerum, ad Medicinae usum deseruientium, libri III: Franc.[ofurti], apud Chr. Egenolphum, 271 p. + Index.

Evans, J., 1922, Magical jewels of the Middle Ages and the Renaissance, particularly in England: Oxford, Clarendon, 264 p. (anastatic reprint, 1976: New York, Dover).

Falk, F., 1975, Edelsteinschliff und Fassungsformen im späten Mittelalter und im 16. Jh. Studien zur Geschichte der Edelsteine und des Schmuckes: Ulm, Kempter, 150 p.

Federici, C., 1587, Viaggio di messer Cesare de' Federici nell'India orientale e oltra l'India per via di Soria: Venetia, tipografia A. Muschio, XVI + 186 p. (re-edited, 1962, p. 234–428 in "Viaggi di C. Federici e G. Balbi alle Indie Orientali", care of Pinto, O.: Roma, Istituto poligrafico dello stato—Libreria dello stato, XLIV + 438 p.).

Fernández de Oviedo y Valdes, G., 1526, Primera parte de la historia natural y general de las Indias, isles y tierra firme del mar Oceano: Toledo 1526. (translated into Italian, 1534, as "Summario de la naturale et general historia de l'Indie occidentali. 2. Libro secondo delle Indie occidentali: Vinegia, s.i.p., 64 + 2 p.; vol. 3 in G.B. Ramusio, 1550, "Delle nauigationi et viaggi nel quale si contengono le nauigationi al mondo nuouo": Venetia, nella stamperia de Giunti, 6 + 34 + 456 p., 5 plates; new edition, 1992, as "Sommario della storia naturale delle Indie"; care of Giletti Benso, S.: Palermo, Sellerio, 246 p.).

Findlen, P., 1989, The museum: its classical etymology and Renaissance genealogy: Journal of the History of Collections, v. 1, p. 59–78.

Findlen, P., 1994, Possessing nature: museums, collecting, and scientific culture in early modern Italy: Berkeley, University of California Press, 449 p.

Funkenstein, A., 1986, Theology and the scientific imagination from the Middle Ages to the seventeenth century: Princeton, New Jersey, Princeton University Press, 440 p.

Gesnerus, C., 1551–58, Conradi Gesneri medici Tigurini Historiae animalium libri (5 vols.): Tigurii, apud Cristoph. Froschouerum, 1104 + 110 + 779 + 1297 + 779 p.

Gesnerus, C., 1565, De Rerum Fossilium, Lapidum et Gemmarum maxime, Figuris et Similitudinibus Liber (169 p. together with seven other works by other authors in the miscellaneous book "De omni Rerum Fossilium Genere, Gemmis, Lapidibus, Metallis, et huiusmodi, libri aliquot, plerique nunc primum editi, opera Conradi Gesneri: quorum catalogus sequens folium continet": Tiguri, excudebat Iacobus Gesnerus, 549 p.).

Gimma, G., 1730, Sulla Storia naturale delle gemme, delle pietre e di tutti i minerali, overo della fisica sotterranea in cui delle gemme, e delle pietre stesse si spiegano la nobilta, i nomi, i colori, le spezie, i luoghi, la figura, la generazione, la grandezza, la durezza, la madrice, l'uso, le virtu, le favole; divisa in libri 6 o tomi 2 colle tavole de' capitoli: Napoli, a spese di Gennaro Muzio e di Felice Mosca, 551 + 603 p.

Giusti, A.M., 1992, Pietre dure. L'arte europea del mosaico negli arredi e nelle decorazioni dal 1500 al 1800: Torino, Allemandi, 311 p.

Gortani, M., 1963, Italian pioneers in geology and mineralogy: Journal of World History, v. 7, p. 503–519.

Gudger, E.W., 1924, Pliny's Historia naturalis, the most popular natural history ever published: Isis, v. 6, p. 269–281, doi: 10.1086/358236.

Guiffrey, J., 1894–96, Les inventaires de Jean duc de Berry (1401–1416), (2 vols.): Paris, Leroux, 480 + 274 p.

Hall, A.R., 1954, The scientific revolution, 1500–1800: The formation of the modern scientific attitude: London, Longmans Green and Co., III + 394 p.

Hall, A.R., 1962, The scholar and the craftsman in the scientific revolution, in Clagett, M. ed., Critical problems in the history of science (Proceedings of the Institute for the History of Science at the University of Wisconsin): Madison, University of Wisconsin Press, p. 3–24.

Halleux, R., and Schamp, J., 1985, Les lapidaires grecs. Lapidaire orphique. Kérygmes. Lapidaires d'Orphée, Socrate et Denis. Lapidaire nautique. Damigéron-Evax (traduction latine). (Text établi et traduit par R. H. et J. S.): Paris, Les Belles Lettres, XXXIV + 349 p.

Hannaway, O., 1992, Georgius Agricola as humanist: Journal of the History of Ideas, v. 53, p. 553–560.

Hardouin, J., 1685, Caii Plinii Secundi Naturalis historiæ libri XXXVII (5 vols.): Parisiis, apud Franciscum Muguet.

Haskins, C.H., 1927, The Renaissance of the twelfth century: Cambridge, Mass., Harvard University Press, VIII + 437 p. (reprinted, 1960: New York, Meridian).

Hooke, R., 1665, Micrographia: or some physiological descriptions of minute bodies made by magnifying glasses with observations and inquiries thereupon: London, printed by Jo. Martin & Ja. Allestry, printers to the Royal Society, 36 + 246 + 10 p. + 38 plates.

Imperato, F., 1599, Dell'historia naturale di Ferrante Imperato Napolitano.

Libri XXVIII. Nella quale ordinatamente si tratta della diversa condition di miniere e pietre. E con alcune historie di piante et animali fin'hora non date alla luce: Napoli, nella Stamperia di Porta Reale per Costantino Vitale, 728 p.

Kockelbergh, I., Vleeschdrager, E., and Welgrave, J., 1992, The brillant story of Antwerp diamonds: Antwerpen, MIM Publishing Corporation, 302 p.

Koyré, A., 1948, Du monde de l'à-peu-près à l'univers de la précision. Critique, no. 28 p. (reprinted in Ètudes d'histoire de la pensée philosophique: Paris, Leclerc, v. 1981, p. 87–111.

Koyré, A., 1965, Newtonian studies: London, Chapman & Hall, VIII + 288 p. (second edition, 1972: Cambridge Mass., Harvard University Press).

Landino, C., 1476, Historia naturale di C. Plinio secondo tradocta di lingua latina in fiorentino per Christoforo Landino fiorentino: Venetiis, opus Nicolai Ioansonis Gallici, 414 p.

Leonardi, C., 1502, Speculum lapidum clarissimi artium et medicinae doctoris Camilli Leonardi Pisaurensis: Venetiis, per Ioannem Baptistam Sessa, LXVI p. (second, unaltered edition, 1516: impressum Venetiis, per Melchiorem Sessam & Petrum de Rauanis sociis, 66 + 40 p.; third edition, 1533: Augustae Vindelicorum, apud Henricum Siliceum, 66 p.).

Leonardi, C., 1610, Speculum lapidum Camilli Leonardi. Cui accessit Sympathia septem metallorum ac septem selectorum Lapidum ad planetas. D. Petri Arlensis de Scudalupis prespyteri Hierosolimitani: Parisiis, apud Carlum Seuestre, & Dauidem Gillium via jacobaea e regione Mathurinorum, et Joannem Petitpas, 44 + 244 + 39 p. (reprinted, 1717, as "Speculum lapidum, et D. Petri Arlensis de Scudalupis, Sympathia septem metallorum ac septem selectorum lapidum ad planetas. Accedit Magia astrologica, hoc est: Petri Constantii Albini Villanovensis, Clavis Sympathiae septem Metallorum et septem selectorum Lapidum et Planetas. Pro maiori illius elucidatione. Opus tam Astrologicis, quam Chymicis perutile et iucundum. Liber olim impressus Parisiis 1611 apud Carol. Sevestre et David Gillium, iam propter eius raritate recursus": Hamburgi, apud Christianum Liebzeit. Anno 1716. 183 p.).

Leonicenus, N., 1492, De Plinii et pluriorum medicorum in medicina erroribus: Ferrariae, Laurentius de Rubeis et Andreas de Grassis.

Lonicerus, A., 1551, Naturalis Historiae Opus Novum, in quo tractatur de Natura et Viribus Arborum, Fructium, Herbarum, Animantiumque Terrestrium, Volatilium et Aquatilium; item, Gemmarum, Metallorum, Succorumque concretorum etc.: Francofurti, apud C. Egenolphum, XVIII + 353 p. + 631 tables.

Marbodus Redonensis, 1511, Libellus de Lapidibus Pretiosis nuper editus. Curis Joann. Cuspiniani: Viennae Pannoniae, per Hieronymum Victorem Philonallem, 22 p.

Martin, A.J., 1999, Scheda n. 192, in Aikema, B., Brown, B.L., and Nepi Scirè, G., eds., Il Rinascimento a Venezia e la pittura del Nord ai tempi di Bellini, Dürer, Tiziano: Milano, Bompiani, p. 622–623.

Martini, M.C., 1977, Piante medicamentose e rituali magico-religiosi in Plinio: Roma, Bulzoni, 190 p.

Mattioli, P.A., 1544, Di Pedacio Dioscoride, Anazarbeo libri cinque Della historia, et materia medicinale tradotti in volgare Italiana: Venetia, per Nicolò de Bascarini da Pavone di Brescia, 707 p.

Mattioli, P.A., 1568, I discorsi [di M. Pietro Andrea Matthioli sanese, medico cesareo, et del serenissimo principe Ferdinando archidvca d'Avstria &c] nei sei libri di Pedacio Dioscoride Anazarbeo della materia Medicinale. Hora di nvovo dal svo istesso avtore ricorretti, & in più di mille luoghi aumentati. Con le figure grandi tutte di nuouo rifatte, & tirate dalle naturali & uiue piante, & animali, & in numero molto maggiore che le altre per auanti stampate. Qvinto & Sesto Libro. Tavola di tvtte le cose. Tavola delli rimedi semplici: Venetia, appresso Vincenzo Valgrisi (quoted according to the reprint, 1970, edited by Peliti, R., with a note by Barberi, F.: Roma, Julia, 1527 p. + n.n.).

Mercati, M., 1717, Michaelis Mercati Samminiatensis Metallotheca. Opus posthumum. Auctoritate et Minificentia Clementis Undecimi P.M. e tenebris in lucem eductum; Opera autem et studio Ioannis Mariae Lancisii Archiatri Pontificii illustratum. Cui accessit appendix cum XIX recens inventis iconibus: Romae, apud Io. Mariam Salvioni Typographum Vaticanum in Archigymnasio Sapientiae, LXIV + 378 p. + index.

Mercati, M., 1719, Michaelis Mercati Samminiatensis Metallotheca. Opus posthumum. Auctoritate et Minificentia Clementis Undecimi P.M. e tenebris in lucem eductum; Opera autem et studio Ioannis Mariae Lancisii Archiatri Pontificii illustratum. Appendix ad Metallothecam Vaticanam Michaelis Mercati. In qua Lectoribus exhibentur XIX Icones ex Typis aeneis nuper Florentiae inventis, quorum XIV Pontificia liberalitate suppleti jam fuerant: Quinque vero penitus desiderabantur. Additis notis, et novis Iconibus Choclearum Cornu Ammonis forma: Romae, apud Io. Mariam Salvioni Typographum Vaticanum in Archigymnasio Sapientiae, LXIV + 378 + 53 p. + syllabus.

Morhof, D.G., 1673, De metallorum transmutatione ad Virum Nobilissimum & Amplissimum Joelem Langelottum, Serenissimi Principis Cimbrici Archatrum Celeberrimum epistola: Hamburgi, Schultze, 616 p.

Mottana, A., 1995, Niccolò Stenone—Su un corpo solido contenuto naturalmente entro un altro solido. Prodromo a una dissertazione (I Fondamenti della Scienza: collana a cura di Antonio di Meo, n. 8, Supplemento al n. 1 di Teknos): Roma, Edizioni Teknos, xxx + 65 p.

Mottana, A., 2005, "Le Miracolose Virtù Delle Pietre Pretiose Per salute Del Vivere Humano" di Scipione Vasolo: un trattatello rinascimentale sulle gemme come mezzi per mantenersi in salute senza ricorrere a medicine, Rendiconti Lincei Scienze Fisiche e Naturali, s. 9, v. 15, p. 19–73.

Mottana, A., and Napolitano, M., 1997, Il libro "Sulle pietre" di Teofrasto. Rendiconti Lincei Scienze Fisiche e Naturali, s. 9, v. 8, p. 151–234.

Munsterus, S., 1550, Cosmographiae uniuersalis lib. 6 in quibus, iuxta certioris fidei scriptorum traditionem describuntur, omnium habitabilis orbis partium situs, propriaeque dotes: Basileae, apud Henrichum Petri, 24 + 1162 p., 14 + 3 plates (first published in German, 1544, "Cosmographey oder Beschreibung aller Lander herrschafftenn und furnembsten Stetten des gantzen Erdbodens, sampt jhren Gelegenheiten, Eygenschafften, Religion, Gebreuchen, Geschichten unnd Handthierungen [et]c.": Basel, in der Officin Henricpetrina, 130 + 1414 p.).

Olmi, G., 1982, Ordine e fama: il museo naturalistico in Italia nei secoli XVI e XVII, Annali dell'Istituto Storico Italo-germanico di Trento, v. 8, p. 225–274.

Olmi, G., 2001, Il collezionismo scientifico, in Simili, R. ed., Il teatro della natura di Ulisse Aldrovandi: Bologna, Editrice Compositori, p. 20–50.

Pacioli, L., 1494, Trattato di partita doppia (a fragment, p. 198–210, of the book "Aritmetica, geometria, proporzioni e proporzionalita": Vinegia, Paganino de Paganini, 224 p., printed and critically edited, 1994, by Conterio, A., with an introduction and a comment by Yamey, B., and a philological note by Belloni, G.: Venezia, Albrizzi, 189 p.).

Pazaurek, G.E., 1930, Mittelalterlicher Edelsteinschliff: Belvedere, v. 9, p. 145–194.

Petrus Martyr ab Angleria, 1516, De orbe novo P. M. ab A. Mediolanensis protonotarij Cesaris senatoris decades: Alcalá de Henares, s.i.p., 119 p. (reprinted, 1533, Basileae, per Io. Bebelium, 12 + 92 p.; translated and edited in Italian, 1550, as "Sommario dell'istoria dell'Indie occidentali cavato dalli libri scritti dal signor don Pietro Martire milanese", pp. 25–205 vol. 3, in G.B. Ramusio, "Delle nauigationi et viaggi nel quale si contengono le nauigationi al mondo nuouo": Venetia, nella stamperia de Giunti, 6 + 34 + 456 p., 5 plates; re-translated and re-edited, 1990, as: "Il nuovo mondo: attraverso le pagine di un contemporaneo di Cristoforo Colombo l'incontro dell'Europa con un universo sconosciuto", care of Forni Bisogniero, P.: Roma, Logart Press, 119 p.).

Pigafetta, A., 1550, Viaggio atorno il mondo fatto e descritto per messer A.P. vicentino, cavalier di Rhodi, e da lui indrizzato al reverendissimo gran maestro di Rhodi messer Filippo di Villiers Lisleadam tradoto di lingua francesa nella italiana, p. 867–948, v. 1, in Ramusio, G.B., Primo volume delle nauigationi et viaggi nel qual si contiene la descrittione dell'Africa, et del paese del Prete Iannii: Venetia, appresso gli heredi di Lucantonio Giunti, 4 + 405 p.

Pliny, 1469, C. Plinii Secundi Historia Naturalis Libri XXXVII noctibus et horis successivis conscripti: Venetiis, Johannes de Spira, n.n.

Riddle, J.M., 1977, Marbode of Rennes' (1035–1123) De Lapidibus considered as a medical treatise, with text, commentary and C.W. King's translation together with text and translation of Marbode's minor works on stones (Sudhoffs Archiv, Beiheft 20): Wiesbaden, Steiner, XII + 144 p.

Rodolico, F., 1952, Il capitolo sulle pietre nelle arti di Giorgio Vasari: opportunità di un commento particolare; in AA.VV., Studi vasariani: atti del Convegno internazionale per il 4. centenario della prima edizione delle Vite del Vasari: Firenze, Palazzo Strozzi, 16–19 settembre 1950: Firenze, Sansoni, p. 129–133.

Rodolico, F., 1976, Lessico petrografico vasariano, in AA.VV., Il Vasari storiografo e artista: atti del Congresso internazionale nel 4. centenario della morte: Arezzo-Firenze, 2–8 settembre 1974: Firenze, Istituto Nazionale di Studi sul Rinascimento, p. 65–74.

Rondelet, G., 1554–55, Libri de piscibus marinis, in quibus verae piscium effigies expressae sunt. Quae in tota piscium historia contineatur indicat elenchus pagina nona et decima. Postrema accesserunt indices necessarii. Universae aquatilium historiae pars altera, cum veris ipsorum imaginibus

(2 vols.): Lugduni, apud M. Bonhomme, 483 + 242 p. + 470 plates.

Rossi, P., 1997, La nascita della scienza moderna in Europa: Roma-Bari, Laterza, XXI + 418 p.

[Rülein von Calw, U.], 1505, Eyn Nützlich Bergbüchlein: Augsburg, Erhard Ratdolt, 24 p.

Ruffolo, G., 2004, Quando l'Italia era una superpotenza. Il ferro di Roma e l'oro dei mercanti: Torino, Einaudi, VII + 314 p.

Ruska, J., 1912, Das Steinbuch des Aristoteles mit literaturgeschichtlichen Untersuchungen nach der arabischen Handschrift der Bibliothèque nationale herausgegeben und übersetzt: Heidelberg, Winter, VI + 208 p.

Sarti, C., 2003. The geology collections in Aldrovandi's Museum, in Vai, G.B., and Cavazza, W., eds., Four centuries of the wordgGeology: Ulisse Aldrovandi 1603 in Bologna: Bologna, Minerva, p. 153–167.

Scaliger, I.C., 1557, Exotericarum exercitationum liber quintus decimus, de subtilitate, ad Hieronymum Cardanum. In extremo duo sunt indices: prior breuiusculus, continens sententias nobiliores: alter opulentissimus, pene omnia complectens: Lutetiae, apud Federicum Morellum ex officina typographica Michaelis Vascosani, 4 + 476 + 32 p.

Schiavone, A., 1982, Dall'editio princeps della naturalis historia ad opera di Giovanni di Spira all'edizione di Lione del 1561, in AA. VV., Plinio e la Natura: Como, Camera di Commercio, p. 95–108.

von Schlosser, J., 1974, Raccolte d'arte e di meraviglie del tardo Rinascimento: Firenze, Sansoni, 188 p. (first German edition, 1908, titled: Die Kunst— und Wunderkammern der Spätrenaissance: ein Beitrag zur Geschichte des Sammelwesens: Leipzig, Klinkhardt & Biermann, 178 p.).

Schmitt, C.B., 1975, Science in the Italian universities, in Crossland, M.P., ed., The emergence of science in Western Europe: London and Basingstoke, MacMillan, p. 89–109.

Schneer, C.J., 1995, Origins of mineralogy: the age of Agricola: European Journal of Mineralogy, v. 7, p. 721–734.

Serbat, G., 1984, Pline l'Ancien. État présent des études sur sa vie, son oeuvre et son influence, in Temporini, H., and Haase, W., eds., Aufstieg und Niedergang der Römischen Welt: Geschichte und Kultur Roms im Spiegel der neueren Forschung (3 vols.): Berlin and New York, W. de Gruyter, v. II, section 32, subsection 4, p. 2069–2200.

Sinkankas, J., 1993, Gemology: an annotated bibliography (2 vols.): Metuchen, New Jersey, and London, Scarecrow Press, 1179 p.

Stendardo, E., 2001, Ferrante Imperato. Collezionismo e studio della natura a Napoli tra Cinque e Seicento (Quaderni della Accademia Pontaniana, 31): Napoli, Accademia Pontaniana, 120 p.

von Stromer, W., 1992, Modell der Edelstein-Schleifmaschine des Heinrich Arnold aus Zwolle von 1439, in Ebert-Schifferer, S., and Harms, M., eds., Faszination Edelstein. Aus den Schatzkammern der Welt. Mythos Kunst Wissenschaft (28. November 1992–25. April 1993, Hessisches Landesmuseum Darmstadt): Wabern, Benteli Verlag, p. 120–121.

Suhling, L., 1983, Georgius Agricola und der Bergbau. Zur Rolle der Antike im montanistischen Werk des Humanisten, in Buck, A., and Heitmann, K., eds., Die Antike-Rezeption in den Wissenschaften während der Renaissance (Deutsche Forschungsgemeinschaft: Mitteilungen der Kommission für Humanismus- forschung, 10). Winheim, Acta Humaniora, p. 149–165.

Theophrastus, 1497, Aristotelis et Theophrasti Opera, Volumen II. Eorum quae hoc uolumine continentur nomina & ordo. Aristotelis uita ex Laertio. Eiusdem uita per Ioannem Philoponum. Theophrasti vita ex Laertio. Galeni de philosopho historia. Aristotelis de physico auditu, libri octo. De coelo, libri quatuor. De generatione & corruptione, duo. Meteorologicorum, quatuor. De mundo ad Alexandrum, unus. Philonis Iudaei de mundus, liber unus. Theophrasti de igne, liber unus. Eiusdem de ventis liber unus. De signis aquarum & uentorum, incerti auctoris. Theophrasti de lapidibus, liber unus: Excriptum Venetiis manu stamnea ☐ domo Aldi manutii Romani, & graecorum studiosi, IV + 268 p. (partial edition, 1965, with introduction, translation, and commentary by Eichholz, D.E. under the title: "Theophrastus. De lapidibus": Oxford, Clarendon, VII + 141 p.).

Thorndike, L., 1960, De lapidibus: Ambix, v. 8, p. 6–23.

Tillander, H., 1995, Diamond cuts in historic jewelry, 1381–1910: London, Art Books International, 248 p.

Tolkowsky, M., 1919, Diamond design. A study of the reflection and refraction of light in a diamond: London and New York, Spon & Chamberlain, 104 p. (cited according to the Web edition as edited by J. Paulsen, Seattle, 2001).

Vai, G.B., 2003, Aldrovandi's Will: introducing the term 'geology' in 1603, in Vai, G.B., and Cavazza, W., eds., Four centuries of the word geology: Ulisse Aldrovandi 1603 in Bologna: Bologna, Minerva Edizioni, p. 64–111.

Vai, G.B., and Cavazza, W., 2006, this volume, Ulisse Aldrovandi and the origin of geology and science, in Vai, G.B., and Caldwell, W.G.E., The Origins of Geology in Italy: Geological Society of America Special Paper 411, doi: 10.1130/2006.2411(04).

Varthema, L., 1517, Itinerario de L. de V. bolognese nello Egypto nella Suria nella Arabia deserta & felice nella Persia nella India & nella Ethiopia la fede el viuere & costumi de tutte le prefate prouincie: Rome, per mastro Stephano Guillireti de Lorenno, 132 p. (Latin translation, 1511, by Madrignani, A. under title: "Ludouici Patritii Romani Nouum itinerarium Aethiopiae, Aegipti, vtriusque Arabiae, Persidis, Siriae, ac Indiae, intra et extra Gangem": Milano, Giovanni Giacomo Da Legnano & fratelli, 8 + 62 p.; original Italian reprinted, 1969, care of Bacchi Della Lega, A.: Bologna, Commissione per i testi di lingua, LI + 285 p. + 1 plate).

Vasari, G., 1550. Le vite de piu eccellenti architetti, pittori, et scultori italiani, da Cimabue insino a' tempi nostri: descritte in lingua Toscana, da G. V. Pittore Aretino. Con una sua utile & necessaria introduzione e le arti loro: in Firenze, [per I tipi di Lorenzo Torrentino], 537 p. (second edition, 1568, titled "Vite dei più eccellenti pittori, scultori e architettori. Introduzione di G.V. pittore aretino alle tre arti del disegno cioè Architettura, Scultura e Pittura": Fiorenza, appresso i Giunti, VII + 257 p.; quoted from the partial edition, 1996, titled "Le tecniche artistiche", which follows the text of the critical edition by Milanesi, G-: Firenze, Sansoni, 5 vols., 1878–85, as printed, 1906, under the title "Vasari on technique", with introduction and comment by Baldwin Brown, G.: Vicenza, Neri Pozza, XXXVIII + 313 p.).

Vasolo, S., 1577, Le miracolose virtù delle pietre preziose per salute del viuere humano col mantenersi allegro, e senza pericolo di sorte alcuna di male (manuscript of 14 ff. bound in the miscellaneous codex 901 titled Gioie e pietre: Roma, Biblioteca Corsiniana, 65 p.) (to be published in Mottana, A., 2005).

Vassallo e Silva, N., 1993, The Portuguese gem trade in the sixteenth century: Jewellery Studies, v. 6, p. 19–28.

Vincentius Bellovacensis, not later than 15 June 1476, Speculum naturale, libri XXXII: Strassburg: The R-Printer (Adolf Rusch).

Westfall, R.S., [1971], 1984, The construction of modern science. Mechanism and mechanics: New York, Wiley, XIII + 171 p. (second edition, 1984).

Wieland, W., 1962, Die aristotelische Physik—Untersuchungen über die Grundlegung der Naturwissenschaft und die sprachlichen Bedingungen der Prinzipienforschung bei Aristoteles (third edition, 1992): Göttingen, Vandenhoek & Ruprecht, 194 p. (Italian translation, 1993, by Gentili, C., La fisica di Aristotele: Bologna, Il Mulino, 480 p.).

Wiemann, W., 1983, Cuspinians Kommentar des liber lapidum Marbods: Kritischer Text und deutsche Übersetzung von Cod. 5195 (sowie der editio princeps von 1511): Dissertation Universität Heidelberg, XXVII + 467 p.

Wild, K.E., 1997, Zur Geschichte der Edelsteinverarbeitung, in AA.VV. Im Strome sein, heißt, in der Fülle des Lebens zu stehen. Festschrift für Adolf Grub: Birkenfeld, Gymnasium Birkenfeld, p. 331–352.

Wyckoff, D., 1967, Albertus Magnus book of minerals (translated by Dorothy Wyckoff): Oxford, Clarendon Press, XLI + 309 p.

Ziegler, J., 1536, De sphaerae solidae constructione, in Anonymous, Sphaerae atque astrorum coelestium ratio, natura, & motus: ad totius mundi fabricationis cognitionem fundamenta: Basileae, Johannes Walder.

MANUSCRIPT ACCEPTED BY THE SOCIETY 17 JANUARY 2006

Agricola and the birth of the mineralogical sciences in Italy in the sixteenth century

Nicoletta Morello*

Dipartimento di Storia Moderna e Contemporanea, Università degli Studi di Genova, via Balbi 6, I-16126 Genova, Italy

ABSTRACT

Agricola's *Bermannus* (1530) and his "minor" works describe his career as an expert in mining knowledge. If we examine the period after publication of his collected works in 1546, *Bermannus* and the other works provide additional keys for understanding the influence of Agricola on development of the mineralogical and geological sciences in Italy. These publications also offer a way of understanding the link between the culture of the humanists and that of the practitioners; this, after all, led to the birth of the empirical sciences.

Agricola's fourfold classification of fossil objects (earth, concretionary juice, stone, metal) improved considerably the twofold classification by Aristotle and became an influential paradigm for the scientists of the late sixteenth century that was further refined and developed as to the genetic environment of different types by Aldrovandi and Imperato.

Keywords: mining, mineralogy, geology, Aristotle, Agricola, Bermannus, De re metallica, De natura fossilium, classification of fossils, earth, juice, stone, metal, Dioscorides, Galenus, Avicenna, Gessner, Encelius, Kentmann, Fabricius, Cesalpino, Cardano, Falloppio, Aldrovandi, Imperato.

INTRODUCTION

In 1546, Froben and Episcopius in Basel published an edition of all works of Agricola on mineralogical topics (Agricola, 1546). In this edition are his writings on the genesis of inorganic bodies, *De ortu et causis subterraneorum libri V*; on the nature of the phenomena of the subterranean world rising to the surface (rivers, earthquakes, hot springs), *De natura eorum quae effluunt ex terra libri IIII*; on the nature of "fossils" (in the ancient meaning), *De natura fossilium libri X*; on ancient and modern quarries, *De veteribus et novis metallis libri II*; and, at the end, a short dialogue, the *Bermannus*, which was in fact the first of Agricola's works on mining, published previously in 1530 by Froben in Basel. The first Italian translation of the 1546 Agricola's edition appeared in 1550 (Fig. 1).

The 1546 edition was the most widespread of Agricola so called "minor works"—an unfortunate and inappropriate designation of these books used by some modern historians that stems from their non-historical, direct comparison with the *De re metallica* (Agricola, 1556). The latter—the largest book that concluded Agricola's career as an expert in mining—was not published until a full decade after the "minor works" and, for that matter, a few months after Agricola's death.

In reality, for his contemporaries and for several successive generations, the "minor" writings of Agricola represented the most fundamental and systematic source of a theoretical treatment of the *res metallica*. This topic was not treated in detail in printed works of the sixteenth century, for instance in works of Agricola's circle of acquaintances, such as Konrad Gessner (1516–1565) (Gessner, 1565) (Fig. 2), Christoph Enceliu (fl. sec.

---

*Nicoletta Morello died unexpectedly on Easter day, 2006, before publication of this volume.

Morello, N., 2006, Agricola and the birth of the mineralogical sciences in Italy in the sixteenth century, *in* Vai, G.B., and Caldwell, W.G.E., The Origins of Geology in Italy: Geological Society of America Special Paper 411, p. 23–30, doi: 10.1130/2006.2411(02). For permission to copy, contact editing@geosociety.org. ©2006 Geological Society of America. All rights reserved.

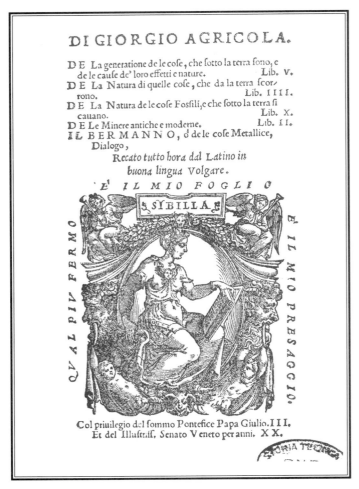

Figure 1. Title page of the Italian translation (1550) of the Agricola's book printed in Basel 1546

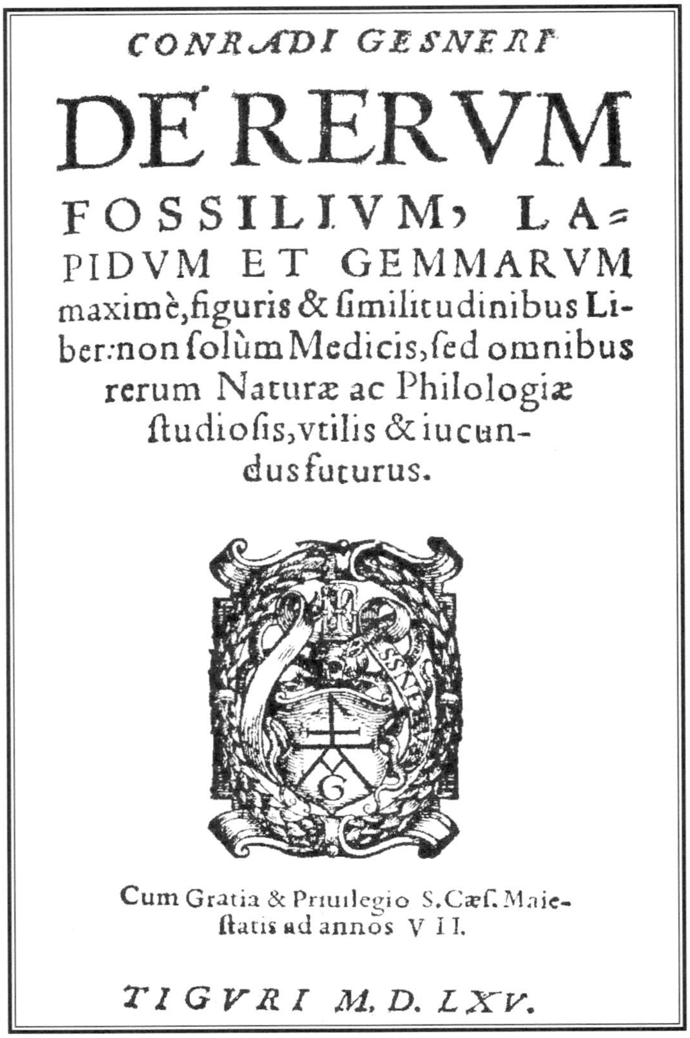

Figure 2. Gessner's book on inorganic bodies, Zurich 1565.

sixteenth century) (Encelius, 1551) (Fig. 3), Johann Kentmann of Torgau (1518–1574) (Kentmann, 1565) (Fig. 4), Georg Fabricius (1516–1571) (Fabricius, 1565) (Fig. 5), and finally Ulisse Aldrovandi (1522–1605) (Aldrovandi, 1648), who corresponded with these scholars. Their writings, and in general the contributions of the sixteenth century to the study of inorganic bodies, are organized into works of different kinds: comments on the *Meteorologica* by Aristotle, encyclopedias, monographs and descriptions of collections and of naturalistic museums. Each of these had its particular aim and its own method of investigation.

The writings on mineralogy, although structured differently, originated, as in the case of other natural sciences, from the confluence of various factors and strands of reasoning, from both outside and within naturalistic investigations. Between the end of the fourteenth and the beginning of the fifteenth century, the task of recovering scientific heritage of the classic antiquity, written in Greek, was on the shoulders of the humanists and was completed by the first half of the sixteenth century. In order that their exegetic effort would succeed, these humanists had to acquire specific knowledge of botany, zoology, mineralogy, astronomy, and such, which they generally did by practice through a kind of apprenticeship with the practitioners of these disciplines. Thus, the humanists assumed a double competence in, and a critical attitude toward, the ancient knowledge, of which they had become capable of recognizing the limits. Stimulated to correct the "errors" of the past, they began a process of revision, which led them to the elaboration of the first naturalistic books of the Modern Age.

Botany, zoology, anatomy, astronomy, mineralogy (to use the modern term), and the other observational sciences enriched their views within a relatively short time. In other words, the general orientation that the study of nature assumed in the sixteenth century focused the attention on daily and local experiences, as well as on those from the practice of professions. At the same time, new observations from exploration of new countries became available, and new discoveries were offered to the old

Figure 3. Encelius on metals, Frankfurt, probably 1551(?).

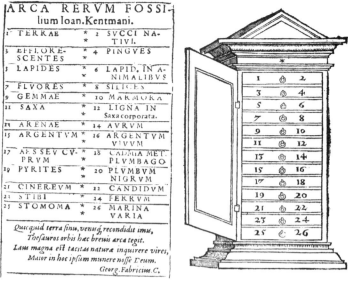

Figure 4. Johann Kentmann of Torgau: his cabinet of *fossilia*.

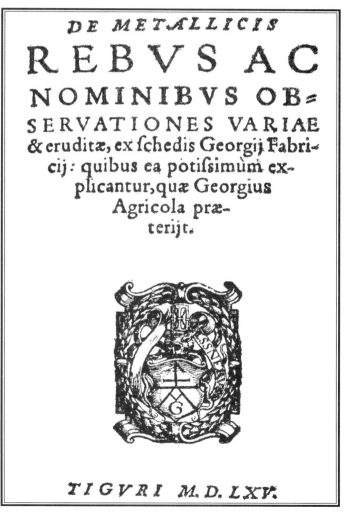

Figure 5. Fabricius' book, "Things, names and observations" on metals.

world. Together with the recovery of past knowledge, the new findings were of equal value for research. As a result of this direction, a gradual independence from the authority of the ancient scholars, first of all from Aristotle, developed.

The personality of Agricola conformed to this cultural background, and his work was perfectly aligned with the principal themes of naturalistic research of the sixteenth century. A brief outline is given here. In the years following the *Bermannus*, when Agricola dealt with the problems of the *bergmedicin*, he studied other more specific mineralogical topics. Among these was the classification of *fossilia*, which is of particular interest because of the influence his conclusions exercised in certain circles, including those in Italy.

Associated with the difficulty of determining individual *fossilia* due to the ambiguity of their morphological and organoleptic characters—already outlined in the *Bermannus*—was the problem of their classification. This subject is discussed by Agricola in *De natura fossilium*, in which he begins with a revision of the ancient classifications. Then he touches on an interpretation of the mode of formation of inorganic bodies and concludes with a classification that he himself elaborates and justifies with extensive documentation. His classification weighed heavily on the theoretical issues, such as the acceptance or rejection of the statements of Aristotle and other ancient scholars. It influenced also its immediate practical application, as in the division of the subject into books and distribution of the *specimina* within the rooms of naturalistic collections.

What could tradition offer to classification of inorganic bodies?

ARISTOTLE'S TWOFOLD CLASSIFICATION

According to Aristotle (382–322 B.C.), the inorganic bodies could be divided into *orycta* and *metalleuta*, formed respectively by dry (*cserà*) and humid (*atmidodes*) exhalations (*Meteorologica*, 378a 20). The former did not melt when subjected to the heat of fire and could be used without previous modification of their physical nature. The second had to be melted or worked in order to be used.

The twofold classification of Aristotle was taken up again by Theophrastos (372–288 B.C.) (Heinsius, 1593, p. 391; Hill, 1774, p. 3–4). In considering water and earth as basic component elements of the inorganic bodies, Theophrastos divided the latter into "*metalleuomena*" and "*lithoi*," corresponding to metals (gold, silver) and to both precious and common stones and "earths" of various kinds. This subdivision into only two large categories proved itself in time inadequate to embrace all the known natural bodies, which showed much greater variety, especially in their empirically recorded physicochemical behavior (reactions with fire and water).

Dioscorides (fl. 50 A.D.) in *Materia medica* (Ruellius, 1547, p. 433), while describing different pathologies and their relative therapy, often defined individual inorganic bodies, calling them *metallica*, *lapides*, *terrae* or *succi*. The source of Dioscoride's divisions is not known, and he himself did not propose a reasoned classification in his treatise. Divisions such as *terrae* or *succi* would seem to have a descriptive, and not a systematic, value.

Galenus (129–199 A.D.) distinguished three groups (Galenus, 1586, p. 64r, 64v, 71). In the ninth book of the *De simplicium medicamentorum facultatibus*, in dealing with the properties (*facultates*) of the single inorganic bodies, he divided them into *terrae*, *lapides*, and *metalla*, using a method of partitioning that Galenus himself stated to be already in use. This tri-partition was not used for systematic purposes. Galenus retained the two basic elements of Aristotle (water and earth) and the two exhalations (humid and dry) as characterizing the origin of all the inorganic bodies. Their consequent cold or hot nature allowed them to be used in the medical therapy based on the principle of "contraries" (for example, a "cold" illness could be healed by a "hot" medicine).

Later, during the stage of the Arabian domination, remarkable progress in the knowledge of individual inorganic bodies occurred under the influence of their practical applications in daily life (from cosmetics to art) and in alchemy. Ibn Sina (979–1037) (called Avicenna in Latin), in his book on *De congelatione et conglutinatione lapidum*, proposed a four-part classification into *lapides*, *sulphura*, *sales*, and *metalla*, which was favorably received by the alchemists.

The great diffusion achieved by Aristotles' *Meterologica* in the medieval age, however, determined the success over a long period of time for the twofold classification proposed by him, even if subject to some marginal revisions. An example of such revisions is given by Albertus Magnus (1193 or 1206–1280) who, in the *De mineralibus*, used the division of the Aristotelian matrix and added a third group to the *lapides* and *metalla* within which he assembled the bodies of an intermediate nature between the former two divisions: the *media mineralia*.

The sixteenth century was characterized by the general reconsideration of the Aristotelian works, and in this context, his classification of inorganic bodies also began to be revised by the humanists, who translated and commented upon the *Meterologica* and the work of Theophrastus. At that time, "orycta" was translated into Latin as *fossilia*, and "metalleuta" as *metalla*.

AGRICOLA'S FOURFOLD CLASSIFICATION

It was the distinction between *fossilia* and *metalla* that displeased Agricola, and from this came his revision of the inorganic bodies in the first book of the *De natura fossilium* (Agricola, 1550, p. 181v–185r, see also p. 185v for a synoptic table of the classification of *fossilia*) (Fig. 6). First he defined what was meant by fossils. Fossils were all those bodies, the parts of which were formed either of the same substance (for example, gold; simple fossils) or of different substances (composite fossils) which could be reduced, however, to earth, stone, or metal. It was evident, therefore, that a natural distinction between *fossilia* and *metalla* did not exist as Aristotle had maintained, and that the metals were a subgroup of the "fossils." It was from this moment that the term *fossilis* assumed the general meaning of an inorganic body. Agricola reused the fourfold division we already saw in Dioscorides' *Materia medica* without reference to the Greek physician and defined what was intended by "earth" (*terra*), "concretionary (or frozen) juices" (*succi concreti* or *succi congelati*), "stones"(*lapides*), and "metals" (*metalla*). The successive books of *De natura fossilium* deal specifically with these individual groups of *fossilia*.

The clarity Agricola introduced to such a complex and confused subject and the vast amount of empirical data by which he supported his assertions is one of the principal reasons for Agricola's dominance among naturalists interested in the inorganic world during the sixteenth century. This is not the place to look in detail at the position he assumed with regard to Aristotelianism and alchemy. Suffice it to recall that even though the origin of the inorganic bodies from the two exhalations was accepted, it was found that the empirical data

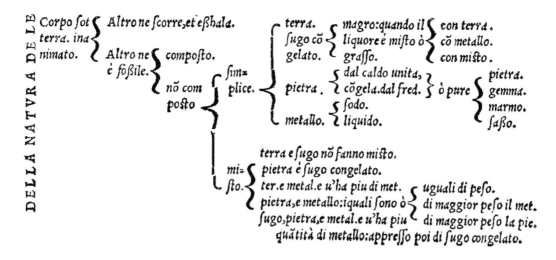

Figure 6. Agricola's classification of fossil bodies, Basel and Venice, 1546 (1550).

Aristotle used to support his ideas were very scarce: the Stagirite formulated a classification of the inorganic bodies that to Agricola seemed completely without foundation.

In addition, Agricola denied that a correlation existed between the formation of the subterranean bodies and the influence of the stars, as Aristotle had claimed (at least according to the Arabian commentators). Instead, Agricola promoted the theory of "petrifying juices" that are material substances inherent in nature, which brought about formation and growth of the *fossilia*. With respect to the alchemists, the position taken by Agricola was common among the mining experts of the early sixteenth century. The operations that the alchemists carried out were in vain: all mining practices showed that no metal, noble or otherwise, could originate if not from its metalliferous parent mineral according to the processes predetermined by its inherent physicochemical properties. This was in agreement with the general conceptions, widespread in the sixteenth century: that nature worked according to its own principles (*iuxta propria principia*), with a constant and regular manner that conserved the identity of the specific forms from one generation to another.

This brief synthesis allows us to appreciate the novelty of Agricola's arguments and may help us to understand the reason why the 1546 edition of the minor works was published in Italian within a short time. In fact, in 1550, the Venetian bibliophile, Michele Tramezzino, translated and published it at his own expense and with the *placet* of Pope Iulius III who gave him the copyright for a ten-year period (Fig. 1). It was not only the "sharpness of the language," together with the "excellence of the doctrines," and the remarkable level achieved in the art of mining that induced Tramezzino to translate the writings of Agricola. It was, above all, the desire to make known to the greater public, which did not speak Latin, the arguments that the author dealt with exhaustively and completely.

AGRICOLA'S INFLUENCE

Read in Italian or Latin, Agricola exerted a remarkable influence on mineralogy in the second half of the sixteenth century; his divisions were not necessarily adopted, but certainly they were kept into account. Andrea Cesalpino (1519–1603), for example, who in the *De metallicis* (p. 7: "At subterraneorum partitio, quae ab Aristotele traditur omnium exactissima, clarissimaque reperitur, quod hinc fiet manifestum…"), while recalling the many classifications proposed, agreed with that of Aristotle based on two essential basic elements, water and earth (Cesalpino, 1596). However, the division of the treatise into three books—*terrae*, *lapides*, and *metalla*—which appears to contradict what he had stated previously, tells us that frequently the *terrae* (as probably also the *succi concreti* in Dioscorides) did not always constitute a distinct category from the *lapides*, which are bodies prevalently formed of earthy elements.

Among those who instead adopted Agricola's classification and above all his ideas on the nature of the *fossilia* were Cardano and Falloppio.

Gerolamo Cardano (1501–1576), in his *De subtilitate* (1554), used Agricola's assertions, while not quoting him, to show that it was impossible, considering the passage of states, to have inorganic bodies of more than the four types: earths, juices, stones, and metals.

Gabriele Falloppio (1523–1562) accepted the opinion of Agricola that subterranean bodies could be defined as those which emerged naturally from Earth—a definition corresponding to the title of Agricola's *De natura eorum quae effluunt ex terra*—and those which required exploration by man. Falloppio (1564, p. 85v, 86v, 91v) willingly accepted the Agricola's classification including one unique category of *fossilia*, divided into four subgroups. It seemed to him the best, and he noted that it had been adopted by Dioscorides in the *Materia*

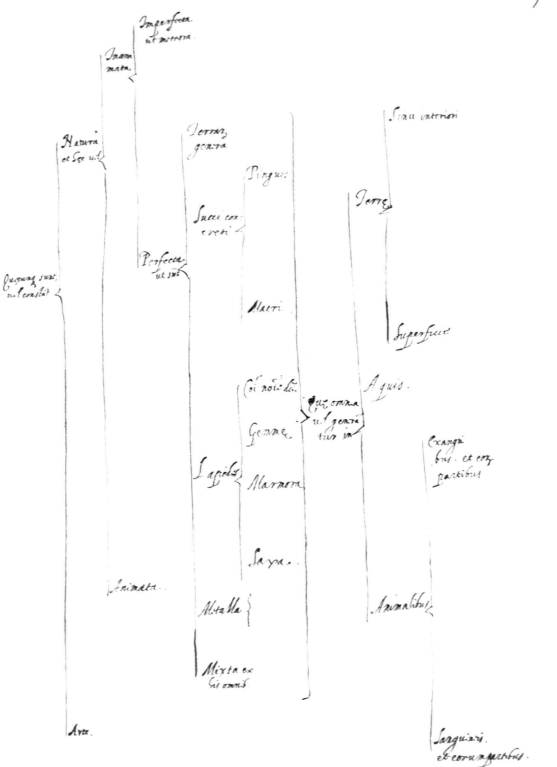

Figure 7. Aldrovandi's classification of fossil objects, Bologna, second half of the sixteenth century after 1571 and before 1594 (by permission of the BUB, FA, ms. 94, c. 92, photo Marabini and Vai; see Marabini et al., 2003).

medica. Nevertheless, Falloppio was not everywhere following the Agricola's classification. For instance, he criticized Agricola's category of *lapides*, finding the group was too difficult to define because of the great variety of bodies that could be included within it (Falloppio, 1564).

Near and at the end of the century, two contemporaries, Ulisse Aldrovandi (1522–1605) and Ferrante Imperato (1550–1625) (Fig. 7), again demonstrated the extent of Agricola's influence in mineralogical matters. While holding opposite positions, neither could afford to ignore the works of Agricola. Ulisse Aldrovandi dedicated a part of his encyclopedia to the inorganic world. This part was the *Musaeum metallicum*, published posthumously by Bartolomeo Ambrosini, the supervisor of the final drafting of the work (Aldrovandi, 1648).

In contrast to the monographs of Kentmann and Fabricius, for example, the encyclopedias assembled the greatest possible amount of information on each inorganic object, from past to present, showing for each how much had been preserved by tradition and how much generated by recent observations. In addition, the *Musaeum* by Aldrovandi is not only a virtual museum in print, it also reflects the reality of Aldrovandi's collection (only a minor part of the *specimina* of this collection still exists in Bologna) presumably arranged, as in the text, according to the four types established by Agricola (*metalla*, *lapides*, *terrae*, and *succi concreti*) but within those types in a sequence that instead corresponds to the Aristotelian method of dealing with natural things in the order of most well known to the less known. Each book (*liber*), therefore, deals with one type of *fossilia*, whereas the chapters (*capita*) are monographs dedicated to single natural objects presented, not in alphabetical order but according to their notoriety. For each of these, all the known information is outlined in paragraphs, starting with the name, etymon, synonyms, figures, and mode of generation, as well as the medicinal and magical properties, with their symbolic and significance in dreams, their use in cooking, in proverbs, and so on, until as many as twenty-eight types of diverse subjects (paragraphs) were reached.

While not wishing to abandon Aristotle, Aldrovandi, in fact, showed a remarkable independence from the Stagirite and also from the ancient literature by not only adopting Agricola's classificatory scheme but introducing a new definition and classification of *fossilis* (Marabini et al., 2003) (Fig. 7). He also had expressed the intention to entitle the volume *de fossilibus* preceded by the term *giologia*, which seems to have been introduced for the first time in scientific literature by him (Vai, 2003). Furthermore, by adopting the theories put forward by Agricola, which are dispersed throughout the text among the various descriptions of the inorganic bodies and are not elaborated in detail in the *Musaeum*, Aldrovandi made a modern choice which allowed him to organize a remarkable amount of heterogeneous information and ancient traditions within a conceptually new scenario.

At the end of the century, others who, like Ferrante Imperato (Fig. 8), did not fully accept Agricola, could not, however, ignore one of his most important lessons, which was to try to determine by "the evidence," that is, by empirical methods, the minerals

Figure 8. View of Imperato's museum in Naples 1599 (from *Dell'Historia Naturale*, Venezia, 1672).

and various types of *fossilia*. This was done not only for practical and therapeutic but also for systematic purposes. Both ends had dominated the research and the reflections of Agricola in order to identify those diagnostic features on which could be based the difficult, debated and ambiguous classification of *fossilia*.

REFERENCES CITED

Agricola (or Bauer), G., 1546, Georgii Agricolae De ortu et causis subterraneorum libri V. De natura eorum quae effluunt ex terra libri IIII. de natura fossilium libri X. De veteribus et novis metallis libri II. Bermannus, sive de re metallica dialogus. Interpretatio germanica vocum rei metallicae: Basileae, per Frobenium et Episcopium.

Agricola, G., 1550, Della natura delle cose fossili: Tramezzino, Venezia.

Agricola, G., 1551(?), De re metallica hoc est de origine, varietate et Natura corporum metallicorum, lapidum, gemmarum atque aliarum quae ex fodinis eruuntur, rerum ad medicinae usum deservientium libri III: Francofurti, apud Christianum Egenolphum, 1551(?).

Agricola, G., 1556, Georgii Agricolae Medici De re metallica libri XII: Basileae, Froben.

Aldrovandi, U., 1648, Ulissis Aldrovandi Patricii Bononiensis Musaeum metallicum in libros IIII distributum Bartholomaeus Ambrosinus in patrio Bononiensi Archigymnasio Simpl. Med. Professor ordinarius Musaei illustrissimi Senatus Bononiensis et Horti publici Praefectus labore et studio composuit [...]: Bononiae, Typis Joannis Baptistae Ferronii.

Cardano, G., 1554, Hieronymi Cardani mediolanensis medici De subtilitate libri XXI nunc demum recogniti atque perfecti, Basileae, per Ludovicum Lucium,. Cfr. Liber quintus, de mixtis, p. 141 e 1663 (I vol.), Hyeronimi Cardani Mediolanensis Philosophiae ac medici celeberrimi Opera omnia tam hactenus excusa [...] cura Caroli Sponii doctoris medici collegio med. Lugdunorum aggregati: Lugduni, sumptibus Ioannis Antonimi Huguetan et Marcii Antonimi Ravavd.

Cesalpino, A., 1596, De metallicis libri tres Andrea Caesalpino autore Ad sanctissimum dominum nostrum Clementem VIII pont. Max.: Romae, Aloysio Zannetti.

Fabricius, G., 1565, De metallicis rebus ac nominibus Observationes variae ex schedis Georgii Fabricii.

Falloppio, G., 1564, Gabrielis Falloppii Mutinensis medici ac philosophi praestantissimi, De medicatis Aquis atque de fossilibus Tractatus pulcherrimus ac maxime utilis ab Andrea Marcolino Fanestri medico ipsius discipulo amantissimo collectus. Accessit eiusdem Andreae duplex epistola: in quarum altera ad lectorem et huius libri inter reliqua utilitas, et docendi modus, ac totius rei, quae in hoc ipso opere continetur, summa breviter explicatur. Apud Ludovicum Avantium, Venetiis.

Encelius, C., 1551(?), De re metallica hoc est de origine [...] libri III, Francofurti, apud Christianum Egenolphum.

Ferrante, I., 1599, Dell'Historia naturale libri XXVIII, nella quale ordinatamente si tratta della diversa condition di miniere e pietre Con alcune historie di piante et animali fin'ora non date in luce: Napoli, per Costantino Vitale.

Galenus, C., 1586, Galeni Opera, ex sexta Iuntarum edizione: Venetiis, apud Iuntas.

Gessner, K., 1565, De omni rerum fossilium genere, gemmis, lapidibus, metallis et huiusmodi libri quinque plerique nunc primum editi opera Conradi Gesneri, Tiguri, Jacobus Gesnerus.

Heinsius, Daniel, 1593, Theophrasti Eresii Graece et Latine Opera omnia Daniel Heinsius [...] emendavit, Lugduni Batavorum.

Hill, J., 1774, Theophrastus's History of stones, including the Modern History of the Gems described by that author and of many other native fossils [...], London.

Kentmann von Torgau, J., 1565, Johanni Kentmanni Dresdensis Nomenclatura rerum fossilium quae in Misnia praecipue et in aliis quoque regionibus inveniuntur, *in* Gessner, K., De omni rerum fossilium genere, gemmis, lapidibus, metallis et huiusmodi libri quinque plerique nunc primum editi opera Conradi Gesneri, Tiguri, Jacobus Gesnerus.

Marabini, S., Donati, L., and Vai, G.B., 2003, Ulisse Aldrovandi's printing contract 1594, *in* Vai, G.B., and Cavazza, W., eds., Four Centuries of the Word Geology: Ulisse Aldrovandi 1603 in Bologna, Minerva Edizioni, Bologna, p. 113–125.

Ruellius, J., 1547, Pedanii Dioscoridis Anazarbei, De materia medica libri sex, Johanne Ruellio Suessionensi interprete. His accessit preater farmacorum simplicium catalogum, copiosus ferme medelarum sive curationum Index: Lugduni, apud Theobaldum Paganum.

Vai, G.B., 2003, Aldrovandi's Will: introducing the term 'Geology' in 1603, *in* Vai, G.B., and Cavazza, W., eds., Four Centuries of the Word Geology: Ulisse Aldrovandi 1603 in Bologna, Minerva Edizioni, Bologna, p. 64–111.

MANUSCRIPT ACCEPTED BY THE SOCIETY 17 JANUARY 2006

Geological Society of America
Special Paper 411
2006

Geology and the artists of the fifteenth and sixteenth centuries, mainly Florentine

David Branagan*
School of Geosciences, University of Sydney, FO 9, New South Wales, 2006, Australia

ABSTRACT

The role of Leonardo da Vinci as the originator of landscape painting and his significance as a pioneer of geological thought and practice have been discussed by numerous authors. Leonardo was not alone, and the Florence region was a center, in the late fifteenth century, for artists interested in landscape and geology. Some art historians have emphasized the influence of Jan van Eyck on Florentine painters, placing special emphasis on his *Stigmata of Saint Francis*. The rock exposure in this painting is said to have been copied by many Florentine artists. However, there are many rock exposures around Florence that provided sites for observant artists.

Francesco Botticini's *Assumption and Crowning of the Virgin* shows the Arno valley landscape, with the city of Florence in the far distance, but readily recognizable.

Illuminated manuscripts have been relatively unstudied by geological historians. A large illustration by Gherado and Monte di Giovanni (ca. 1490) depicts a portion of Florence and its walls, partly obscured by an extraordinary small hill of carefully depicted graded or alternating bedded rocks, surmounted by a waterfall, this hill forming the centerpiece of the painting. It is possibly unique from an artistic point of view.

The interest in geological features shown by so many Florentine artists of the period foreshadowed the important geological principles set out so clearly by Steno a century or more later; based on his observations in the region, Steno laid the foundations for the theoretical development of modern stratigraphy. Indeed, the writings of Leonardo seem to have clearly anticipated Steno's thoughts.

Keywords: Leonardo, van Eyck, Florentine artists, rock outcrops, landscape, Botticini, illuminated manuscript, di Giovanni, graded and alternating bedding, Steno, geological principles.

GEOLOGY IN DETAIL AND AT LARGE (JAN VAN EYCK VS. LEONARDO)

To simplify the discussion in this paper, the representation of geology in paintings is taken to occur either as detail or on a large scale. Of course, in practice, this dichotomy is by no means absolute. Geology in paintings has usually been discussed only in relation to paintings from the seventeenth century on, with the emphasis mainly on the large scale—landscapes (geomorphology) (see for instance Stafford, 1984; Langdon, 1996; Bedell, 2001).

This paper will examine these two aspects (detail and large scale) of the depiction of geology in paintings of the late fifteenth and early sixteenth centuries, with particular attention to the

*E-mail: dbranaga@mail.usyd.edu.au.

Florence region, the birthplace of Leonardo da Vinci. Leonardo has rightly been given credit for his contributions to geology, many of which extend well beyond the influence of his landscapes (see for instance: Oberhummer, 1909; Sarjeant 1980; Zammattio et al., 1981; Vai, 1995; Lorenza, 2000).

JAN VAN EYCK, DETAIL, AND HIS POSSIBLE INFLUENCE IN FLORENCE

Montgomery (1996) drew the attention of historians of geology to a particularly striking example of detailed geology in a work attributed to Jan van Eyck, *The Stigmata of St. Francis* (ca. 1428) (Fig. 1). Showing in the foreground (right, as viewed) is a rock exposure some 2 m thick, which is very "real" indeed and could readily be the basis for a stratigraphic section. Montgomery's study was based on the version in Philadelphia. This is smaller (12.4 × 14.6 cm) than the version in Turin, which is more than twice as large (29.2 × 33.4 cm). Now regarded as the original (Reynolds, 2000, quoting Spantigati [1997], Butler [1997], and van Asperen de Boer [1997]), the Turin version was probably matched by a similar sized copy, now missing.

Branagan (2003) pointed out some of the problems of deciding the authenticity of the St. Francis paintings, as to whether they were done by Jan van Eyck, or at least partly by Hubert van Eyck. Geirnaert (2000) took up this problem, suggesting that at least one of the paintings probably came from van Eyck's workshop, while Reynolds (2000) questioned the attribution to Jan van Eyck more strongly.

Branagan (2003), again in the context of the history of geology, discussed this painting (in its several versions) and some related works. He, as well as Vai (2003), specifically considered some aspects of geology, such as faulting and weathering, which are depicted in the background of the painting. Unfortunately, the location of the detailed section has yet to be ascertained, and whereas it is most probably in Belgium, a Spanish location cannot be ruled out in view of Jan van Eyck's travel there.

Nuttall (2000) discussed Jan van Eyck's influence on Italian painters in the years shortly after his death in 1441, when van Eyck's paintings were being eagerly collected by rich Italians (such as King Alfonso of Naples) that had learned of van Eyck's work from other Italians living in the Netherlands, for whom van Eyck had worked. In Florence, the Medici were certainly among owners of van Eyck's work.

More significant to the present study, Nuttall also pointed out the influence of van Eyck's style on Florentine painters. This influence possibly came about by the showing of one of the copies of *The Stigmata of St. Francis*, which had been taken to the Holy Land from Bruges, in 1470, on the pilgrimage of Anselm Adornes (a descendent of the Ardorno family of Genoa). It was exhibited in Florence on his return, early in 1471 (see Geirnaert, 2000). Geirnaert pointed out that Adornes belonged to an informal cosmopolitan group that read Italian humanist literature, a group which previously included van Eyck. This group also became involved in some of the bodies devoted to the rapidly spreading "observant movement" of the Franciscans, in which friars and their followers lived an austere life apart from the main conventual groups of Franciscans. Despite Adornes' copy, it seems likely that a version of the painting was already known in Florence a year or so prior to his visit (Nuttall, 2000).

Art historians, such as Panhans (1974) and Rohlmann (1993), have suggested that the painting had influence beyond its small size, and they claim to identify at least ten "quotations" from the *Stigmata* landscape in works by "virtually every major Florentine painter of this generation" (Nuttall, 2000, p. 176). The detailed geological rock face shown graphically in the *Stigmata*, discussed *in extenso* by Montgomery (1996), appears to have been quite closely imitated, notably by Botticelli and Filippino Lippi in their *Adoration of the Magi*, dated 1472, when Filippino was in Botticelli's workshop. Nuttall attributes this work to Filippino rather than to Botticelli, as she sees further examples of van Eyck's influences, possibly at times occurring unconsciously, in other Lippi works, such as *The Vision of Saint Bernard* (dated ca. 1486) (Florence, Badia Panel). Perhaps she sees too much influence of the van Eyck outcrop in works by Ghirlandaio (*Meeting of the young Christ and John the Baptist* (ca. 1470, Staatliche Museen, Berlin) and the *Baptism of Christ* by Andrea del Verrochio (possibly with the young Leonardo).

The influence, if such it was, possibly extended much farther than Florence, as can be seen in works by Bellini, done in Venice, and Petrus Christus, whose *St. Anthony and Donor* (with background rocks, left side, as viewed) is in the Copenhagen Museum. Indeed, if the *St. Anthony* is as early as 1426, as suggested by some writers, it raises questions as to who followed whom.

While van Eyck's outcrop might indeed have influenced Florentine artists, the art historians interestingly make little or

Figure 1. Jan van Eyck's *Stigmata of St. Francis* (ca. 1428), Turin, Galleria Sabauda.

no comment about the reality of this feature. For historians, the outcrop might well be, in artistic terms, purely an interesting artifact that has proved worthy of copying. In view of its suggested influence, it is extraordinary that the site for van Eyck's cliff exposure has not been authenticated, and that there are few comments about its likely occurrence. On the other hand, there are numerous exposures of rock in the Florence region itself that could have provided sites for observant artists. Thus, I suggest, that despite some similarities, Gombrich (1976, p. 33–34) and other art historians influenced by him have drawn too long a bow in ascribing Leonardo's "Arno sketch" (described by some as the first authentic landscape) to an enlarged reworking of a section of van Eyck's small work. Nevertheless, the comment by Nuttall (2000), that she "would go so far as to suggest, tentatively, that what we think of as quintessentially Leonardesque rock forms might perhaps owe something to the *Saint Francis*," needs to be remembered when considering the development of "geological art." But, compared with the abundance of outcrops in Tuscany, it would be somewhat ironic if this major influence derived from the art of a region (Flanders) where outcrops are generally sparse.

Although the emphasis has been on *The Stigmata of St Francis*, van Eyck's interest in landscape and rock, if not in geology, did not stop there. This is illustrated by the fine landscape and detailed rock pile he shows on the right side (as viewed) of his relatively large *Saint Barbara* panel, whether a finished work or not (Billinge et al, 2000), and the gentle landscape (in the middle distance), backed by alpine scenery, in *The Virgin and Chancellor Rollin* (ca. 1433) in the Louvre.

Vai (1995) has specifically discussed one aspect of Leonardo's detailed attention to geology in the painting of the *Virgin and Child with St. Anne* (1506–10, or possibly as late as 1513), displayed in the Louvre, in which he recognized the depicted rock types as typically occurring in the Florence to Romagna Apennine region. Branagan (2003) suggested that perhaps a small fault is shown disrupting the thin successions. Leonardo's observations for this painting date back to at least 1501, for we see much of the same detail in the sketches for this painting in the unfinished *St. Anne* (National Gallery, London), which Zammattio et al. (1981) suggested was done several years earlier (ca. 1498).

THE LARGER VIEW, MAINLY THE ARNO VALLEY

In contrast with the discussions of detail above, we move on to examine landscape geology. One of the best-known, early, "true" geological landscapes is Konrad Witz's *Miraculous Draft of Fishes* (1444), which uses an accurately depicted view of the Lake Geneva region, as illustrated and discussed by Branagan (2003), based on some comments in a letter from T.G. Vallance (1977). While Witz does not seem to have continued this striking development in his later paintings, it was only a few years later that Florentine artists took up the same theme, using their local landscape as a "real" background for well-known religious topics. Which other painters of this era and region were attracted to the local landscape to the extent that they depicted it realistically, even when they were ostensibly painting religious subjects? We will consider just two examples.

Saint Sebastian

Marshall (2002) discusses the cult of honoring St. Sebastian, "the chief celestial defender against the ravages of bubonic plague," one of the most popular saints of the fifteenth and sixteenth centuries whose cult was particularly strong in Florence. Sebastian, an officer of the imperial cohort, was condemned and pierced with arrows "like a hedgehog" in the persecution by Diocletian, in Rome (A.D. 284). Left for dead, Sebastian was found miraculously alive by the Christians who came to bury him. Nursed back to health, he refused to leave Rome, but sought out the emperors (Diocletian and Maximillian) to "reproach them for their persecution of the servants of Christ." His reproaches had no effect; he was flogged to death and later buried by Christians on the Via Appia outside the walls of Rome. His feast is commemorated on 20 January. The reasons for Sebastian becoming a defender against the plague are not clear and need not concern us here, but the cult was in place as early as the seventh century.

Marshall (2002) points out that there are varying opinions as to the stance adopted by various artists to the martyrdom, some perhaps drawing inspiration from classical antiquity to glorify the human figure, whereas others attempted to depict either an historical event or a devotional image.

While more than eighty martyrdoms depicting St. Sebastian were painted in the fifteenth and sixteenth centuries (Marshall, 2002), only a few show significant landscape backgrounds. These include works by Pietro Perugino (Louvre) (with no archers); Vicenzo Foppa (Milan, Museo di Castello Sforzesco) (which has a large cliff on the left, in the near background); an anonymous Bologna artist (1497) (Bologna Capella Vaselli, S. Petronio) (many figures obscure the landscape); and Luca Signorelli (1498) (Città di Castello, Pinoteca). The famous painting by Botticelli (Berlin, Staatliche Museen, Gemäldegalerie) has no archers and is a much more personal depiction. It was painted ca. 1473, about the same time as the Pollaiuolo (see below), and originally was hung on a pillar in Santa Maria Maggiore, Florence. However, although the backgrounds of the above paintings are real enough, there are few features that enable one to state the locations with any degree of confidence.

The *Martyrdom of St. Sebastian* (now in the National Gallery, London) by Antonio and Piero Pollaiuolo (1475) is another matter. It is special in many ways, being, as it is, several paintings within one—the martyrdom itself and the detailed background, which could virtually stand alone (see, for instance, Hartt, 1977). It is a very large painting commissioned by the Florentine merchant Antonio di Puccio Pucci (1418–1484), a Medici henchman, for a family oratory (a chapel in the Servite church of SS Annunziata, Florence) built in the early 1450s. The commission was delayed as the chapel was remodeled, and Pucci had financial difficulties for some years. The church was the site of international pilgrimages thanks to a famous and "much-venerated miraculous image" (Marshall, 2002) of the Annunciation. From the 1440s, the Medicis took on the patronage of the shrine, ensuring that visiting dignitaries would visit, and possibly, in many cases, also see the Pollaiuolo painting, placed there a few years later.

Marshall (2002) points out that the architectural details (triumphal arch) on the left of the painting pay allegiance to the Roman site of Sebastian's death. However, she notes that the heads portrayed on the arch are Moors, as are two of the soldiers on horseback, injecting a historical and Florentine note, as the coat of arms of the Pucci family used the Saracen as the family emblem, suggesting a lineage from the important Saraceni family. The suggested link, however, was spurious, as the lineage did not extend from the Saracenis to the Puccis.

Of the river scene forming the "second picture," or rather the upper right portion of the whole work, there seems little doubt that it is based on the Arno. The painting probably depicts an area downstream from Florence, looking upstream, but from what vantage point? Can this site be identified today, and how truthful is the depiction?

The inclined cliffline on the right, although reminiscent of many presentations of similar slight overhangs (e.g., Filippino Lippi's *Madonna and Pietà*, Alte Pinacotek, Munich), seems perhaps more real than many such depictions. Although it lacks the detail of Leonardo's layers at the feet of the Virgin and St. Anne, the view is suggestive of real beds dipping at an angle into the cliff. Interestingly, a very similar overhang on the same side of a river exists in Antonio Pollaiuolo's slightly earlier (ca. 1470) nonreligious, allegorical *Hercules and Deiamara*. Is it the same cliff? Possibly. Such cliffs occur close to the Arno where it enters the Montelupo Gorge, a little west of Signa.

Francesco Botticini's *Coronation of the Virgin*

There are some interesting similarities, but many differences, between the Pollaiuolos' *St. Sebastian* and the extraordinary *Assumption and Coronation of the Virgin* by Francesco Botticini (di Giovanni), also in the National Gallery, London (Fig. 2). Presently hung high above an arch in the section joining the Sainsbury wing and the old gallery, its details are not easily seen. Although recognized in his own period, Botticini's significance was not really appreciated until the late nineteenth century, much of his work previously being incorrectly attributed for many years (Venturini, 1994).

This painting, like the Pollaiuolos', was commissioned by a Florentine merchant, apparently as an altarpiece in the family chapel in the Church of San Pier Maggiore (until 1783). However, there is some doubt about this intended placement, as the

Figure 2. Francesco Botticini's *Assumption and Crowning of the Virgin* (ca. 1470), London, National Gallery. Arno Valley and Florence in distance.

painting is much too large to have fitted into the chapel. Likewise, it does not seem to fit in the Palmieri chapel of the Badia of Fiesole, the home of the Palmieri family. Perhaps it was just hung in one of the Fiesole churches, rather than in a chapel. For many years, the work was wrongly attributed to Botticelli. In the manner of the Pollaiuolos' *St. Sebastian*, it is two paintings in one—a religious one underlain by a real landscape. The scene depicts the donors, Matteo Palmieri and Niccolosa Serragli, on either side of the apostles. This earthly group is overshadowed by several extraordinary coronas of angels, saints, prophets, patriarchs, and the Virgin. There is possibly a hint of heresy in the mingling of human saints with the angels, the latter said theologically always to be on a higher plane. The living group is on a bare rocky rounded dome, apparently between two river valleys, although it could possibly be a single valley with a large meander obscured by the hilltop.

The work has been dated by many as ca. 1470–76, but Nuttall (1992) confidently dated it as 1466. After earlier, less-than-successful cleanings, the work was restored between 1954 and 1959.

There seems little doubt that the city in the distance behind the male donor is Florence. How real is the landscape portrayed? If we visit the vicinity of Artimino, (or probably on a hillside a little to the south), perhaps we can see this view, little changed, except for the course of the Arno. The bridge shown on the river below the donor and perhaps several kilometers distant is at Signa, ~20 km downstream from Florence. For many years this was the only bridge on the Arno between Florence and Pisa, and it was an important crossing from Siena on the way to Prato and Pistoia. The bridge has a similar design to that depicted by Leonardo in the *Mona Lisa*, identified by Starnazzi (2005) as at Buriano, well upstream on the Arno (near Arezzo).

We have to keep in mind that the course of the Arno above (and also below) the Montelupo Gorge has been considerably modified since the sixteenth century by engineering works. Although not precise, Leonardo's map of 1503–1505 (Fig. 3) gives a good idea of the meandering nature of the river between Florence and Montelupo at that time. This sixteenth-century character can only be realized today during exceptional flooding. Some of the artificial changes to the river's course, which were begun in the sixteenth century, are discussed by Coli et al. (2004) and Rubellini et al. (2004). The apparent separate river valley to the right (as viewed) in the Botticini painting can be reasonably accepted as the Vignone, the tributary of the Arno which flows northwesterly from San Martino alla Palma to join the Arno near Signa, and which sits close to the relatively straight edge of the high range flanking the Arno Valley in this area.

While Botticini shows other signs in his paintings of an interest in landscape, as in his *Saint Sebastian* (New York, Metropolitan Museum of Art), *Madonna adoring the Child* (Florence, Cassa di Risparmio di Firenze), *St. Andrew the Apostle* (Florence, Galleria Academia), and *Madonna and Child in a Landscape* (Cincinnati, Museum of Art), the landscapes in these paintings are of minor significance compared with the *Assumption* painting.

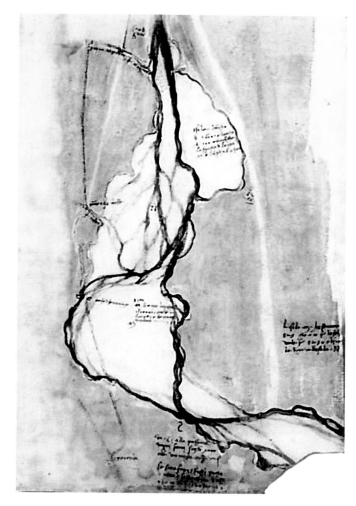

Figure 3. Leonardo's *Arno River map* (ca. 1505), Windsor Castle.

A VIEW FOR ITS OWN SAKE

Whether we regard it as a work of art, or just a practical "view," the Carta della Catena (Fig. 4), an anonymous work, dated around 1470 (post 1461, when the lantern of the Duomo was completed) (Florence, Museo di Firenze Com'era), is an impressive landscape, showing the entire walled city of Florence centered on Brunelleschi's wonderful dome, completed nearly thirty years earlier (Hamerman and Rossi, 1989; King, 2000). The painting takes us beyond the city, past nearby settlements to the heights of Fiesole and the ranges beyond. All in all, it is a celebration of the joint wonders of human creativity and the majesty of nature. The position of the artist when painting this work can be confidently placed on the Monte Oliveto, not far outside the original walls of the city.

We can see another relatively close view of the city and the Arno in Biagio d'Antonio's *Angels in a Tuscan Landscape* (Florence, Bartolini–Salimberi Collection), painted some forty years earlier (ca. 1430) when Brunelleschi's dome was still being built. But in this work, the background hills are not treated

Figure 4. Anonymous's *Carta della Catena*, Florence and surrounds (1470), Florence.

to any degree, and the view is but a detail, even though the landscape is given its due in the title. This would have been painted from about the same spot (Monte Oliveto).

Taking us well into the sixteenth century is another panoramic view, that of the imperial exercise of 1529–1530 (possibly by Giovanni Stradano) and illustrated some years later by Giorgio Vasari (Fig. 5). Drawn from a high perspective, from beyond San Miniato al Monte, it shows fairly truthfully the range stretching from east of Fiesole west to beyond Prato, (consisting mainly of Tertiary flysch deposits), with the Arno meandering through the plain west of Florence toward Signa. In the southern part of the panorama there are indications, in places, of likely outcrops.

Leonardo's map of the Arno (discussed above) can be usefully compared to the present, greatly modified course shown on the recent geological map (Fig. 6). While other artists of the time might have taken the river course for granted, Leonardo no doubt wondered about how the river flowed through a plain covered by alluvium, then took a winding course through a rocky gorge at Montelupo, before emerging on to another plain. At the time, there was an awareness of earth movements associated with earthquakes, which were not infrequent in the region. Leonardo's writings tell us that he thought about erosion, and, as Vai (1995) has pointed out, he clearly linked phases of erosion with tectonism. Such a link led him to an understanding of both disconformities and unconformities, features which appear to be depicted in some art of the period (Branagan, 2003), although it is unlikely that the artists had the insight of Leonardo concerning these natural phenomena. These observations, clearly recorded in Leonardo's writings (Vai, 1995, p. 18–19), predate Steno and Hutton.

It is also interesting that Leonardo chose a route for his proposed canalization of the Arno from Florence to Livorno via the Serravalle gap, just south of Pistoia, which, although long, avoided the Montelupo Gorge and would have encountered only a short distance in material older than alluvium, a point also noted by Kemp (2004, p. 148) (Fig. 6).

Although the Arno is said to have been a major influence on the theories of Nicholas Steno a century later, strictly speaking the observations which were the basis of his theory were probably made in tributaries, such as the fault-bounded Val d'Era and Elsa valleys flowing northerly from Volterra and San Gimignano into the lower reaches of the Arno, below Empoli. Here, the influences of tectonism are more obvious (with movements possibly on a lesser scale than in the Florence-Pistoia basin, where thick alluvium hides much of the Plio-Pliocene succession in a half-graben—a down-faulted block with a well-defined fault only on one side). In the "lower Arno" region, below Empoli, the relations between three units are more clearly discerned. These consist of (1) rocks older than late Miocene, generally folded, (2) Neogene and Pleistocene sediments, and (3) Holocene alluvial deposits. Principe (2004) shows sections (Fig. 7) that fit the theoretical concepts set out by Steno and are embodied in the recent reconstruction of the Apennines (Vai and Martini, 2001).

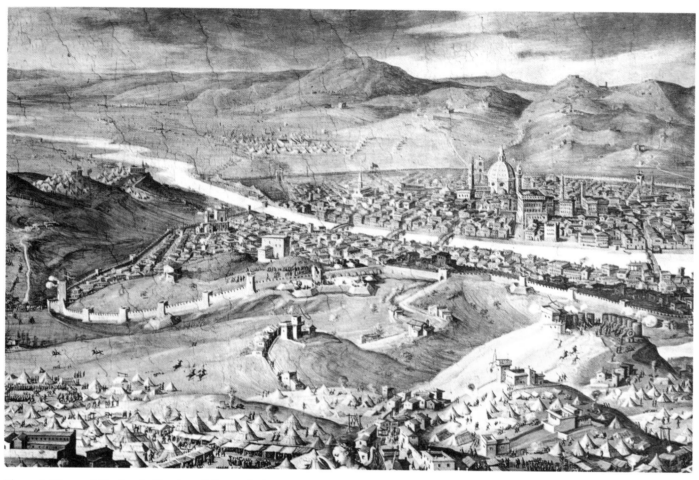

Figure 5. Giovanni Stradano's *The Arno Valley and Florence* (1529), Florence.

DETAIL ONCE AGAIN—GRADED BEDS AND "CYCLES"

Some of the limy rocks and serpentinite used in the Duomo and other important buildings in Florence came from quarries near Prato and Pistoia (Malesani et al., 2003). However, these rocks can hardly concern us in this paper, except to note that many artists made use of such colored "ornamental" stone (which could be polished) in paintings of interior scenes, on walls, altar pieces and floors—a small, but noticeable use of "geology" in art.

Two different sandstones can be recognized in many buildings in Florence. They are generally referred to as *Pietraforte* and *Pietra Serena* (this unit was described by Vasari) (Geology Museum, Florence, 2004; and Carmignani and Lazzarotto, 2004). The former rock type was more commonly used as a load-bearing material, the latter for decorative purposes. The *Pietraforte* material came from quarries south of the Arno River, the *Serena* from the Fiesole area, where good outcrops can still be observed in road cuttings.

The *Pietraforte* is of Cretaceous age, consisting of alternating sand and argillite, with lenses of coarser microconglomerate. In places it is rather cherty and shows sharp jointing, and there are distinct breaks between the coarse and fine units. The *Serena* is geologically slightly younger (late Oligocene–early Miocene). It is a quartz-felspar sandstone in a turbidite succession, in which the coarse material often grades almost imperceptibly into the finer (graded bedding), before a break in sedimentation.

Although the original environments of deposition of these two rocks were different, in the broad sense, each shows an alternation of coarse (mainly sandstone) and finer (shales) grain size. Artists would have noticed the alternations but may not have wondered much about the differences between the two rock successions, which is what gives clues to the geologist as to their different origins. Whereas to an artist the two successions might not have appeared greatly different, the different rates of weathering could have been evident when used as a building stone, and the *Pietraforte* tends to have thicker sandstone beds.

Although the graded-bedded sequence, so typical of the rock sequences of the Florence-Fiesole region (Fig. 8), and the alternating sand–shale cyclic succession were almost certainly not understood as geological phenomena, except by Leonardo

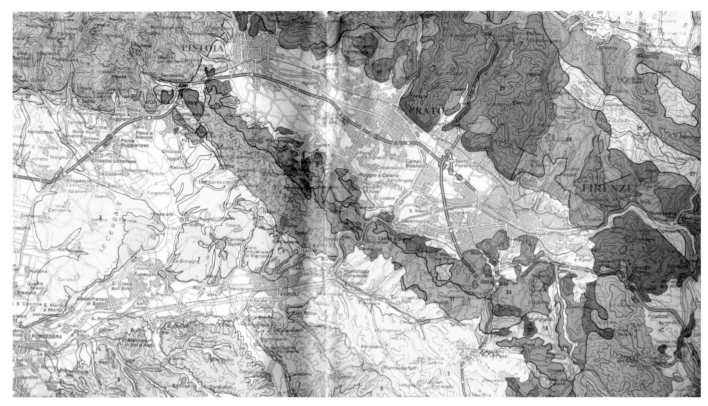

Figure 6. Geological map of Arno Valley and surrounding areas (Carmignani and Lazzarotto, 2004).

(according to Vai [1986, 1995]), the interest artistically must have been strong, for this rock variation was taken up by numerous painters of the era. This variation is not unlike the van Eyck outcrop, of which the significance has possibly been overemphasized, as discussed above. Perhaps one of the best known depictions of the alternating beds is in Botticelli's *Pallas and the Centaur* (Florence, Uffizi Gallery), painted ca. 1482, which, besides the grading, shows the control of the rock surfaces by jointing (sharp edges), suggesting that he might have been depicting a *Pietraforte* outcrop (Fig. 9).

Interestingly, the graded-alternating bedding was used a number of times by the Venetian painter Giovanni Bellini in his representations of *St. Jerome in the Desert*. The variations suggest that Bellini might have been influenced more by the *Pietra Serena* model. Tempestini (1992) discussed three versions of St. Jerome by Bellini (Florence, Uffizi; London, National Gallery; and Washington, National Gallery), attributing them to a range of dates from the 1470s to 1505. Branagan (2003) also discussed an earlier version that authorities date as possibly 1452. He pointed out the apparent anomaly that Bellini had resided in Venice almost his entire life and had had only a limited opportunity to view outcrops exposing successions somewhat similar to those of the *Pietra Serena*, which occur in the Valdagno region near Vicenza. However, Bellini seems to have made the most of his limited opportunities, as shown in his *Ecstasy of St. Francis* (1485) (New York, The Frick Collection), which has connections with real outcrops in the Verona region. Vai (1986; 2004, personal commun.) pointed out that Andrea Mantegna and other non-Florentine artists also depicted similar geological features, but they might have had more opportunities to observe rock outcrops, than, I believe, did Bellini.

Whether the observations and depictions of geology by Florentine artists became well-known to Venetian artists by exchange or word of mouth is uncertain, but there was doubtless more than a little communication, particularly in regard to the sources of pigments for color, and experimentation with techniques, which were rapidly tried and adopted, if successful. Some aspects of this story are told in the fascinating novel by James Runcie, *The Colour of Heaven* (2003, p. 285), which discusses the successful search for the source (lapis lazuli) of ultramarine blue, which Runcie claims, "encouraged the development of landscape, perspective and depth in Italian painting" (and was linked also to the development of spectacles).

ILLUMINATED MANUSCRIPTS

As mentioned in Branagan (2003), to date there has been little study by geologists of the numerous landscapes that appear in illuminated manuscripts, many of which originated in Florence. Good examples of the genre are found in (a) the Urbino Bible (Vatican Library— MSS Vat Lat. 1 and 2) (notably folio 27), produced for Frederico da Montefeltro, the Duke of Urbino.

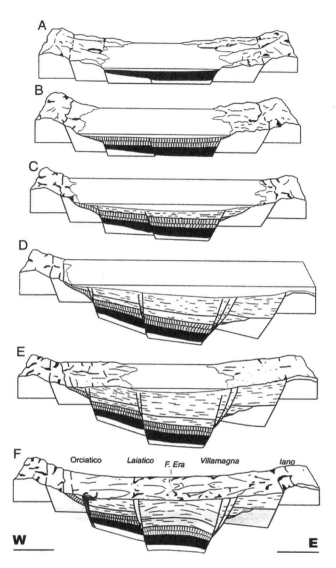

Figure 7. Claudia Principe's cross sections of the Val D'Era region, illustrating the history recognized by Steno (Principe, 2004). Stages of paleogeographic evolution of the Val d'Era area. (A) upper Miocene (gray—substratum; black—lacustrine deposits); (B) upper Miocene evaporate episode; (C) lower Pliocene sands and clays; (D) lower-middle Pliocene; (E) middle Pliocene; (F) present day situation (from the Geological Map of Italy 1:25,000, Foglio Volterra).

Figure 8. G.B. Vai's photo of graded-bedded turbidite succession, Tuscany, Apennines (2004)

These are large volumes (596 × 442 mm). Folio 27 from volume 1, containing part of the Old Testament in St. Jerome's translation, shows the entrance of the Israelites into Egypt. The picture is in landscape form (i.e., elongated horizontally), itself an unusual presentation in manuscripts. It was largely the work of Francesco d'Antonio del Chierico. Various personages (such as Jacob and his wife) can be identified, and there are camels, but the large cast of people has a peaceful, fertile, rural background of rolling, partly timbered hills, areas of cultivation, and towns in the distance, all of which is reminiscent of the Italian landscape and certainly a far cry from the scenery one might expect to see in Egypt. However, Fantoni (1997) suggested that the artwork in these volumes is not typical of the Italian style of the period (1477).

The Vatican Library houses another treasure prepared for the Duke of Urbino between 1478 and 1480. This manuscript of the Gospels (Urb. Lat. 10, Folio 175), was begun by Guglielmo Giraldi and completed by Franco de' Russi (School of Ferrara). It shows St. John writing, backed by a rugged, cliffed hillside with the sea behind him on the right. The cliff on the right side consists of scalloped joint blocks, in the style used by Mantegna, amongst others. Immediately on the left of St. John, as viewed, the blocks are a light brown, and they tilt (dip into the cliff) realistically. The cliff on the far left shows at the top what could be regarded as an unconformity with flat-lying, alternating light and dark beds resting on vertical layers, also showing alternating variation.

Figure 9. Sandro Botticelli's *Pallas and the Centaur* (1482), Florence, Galleria degli Uffizi

Figure 10. Gherado and Monte di Giovanni's *Hebron and cliff* (ca. 1490), Florence, Biblioteca Medicea Laurenziana.

A third variation of the manuscript can be seen in the Nautical Atlas of Andrea Bianco in the Biblioteca Marciana, Venice (It. Z. 76 [4783]). The illuminated manuscript dates from 1436. Folio 10 is a circular planisphere, centered on Jerusalem, which covers from the Atlantic Ocean to the Indian Ocean. As is to be expected, the known parts of Europe are recognizable, but the factual geographic data become scarcer away from this area and from the Middle East, to the Far East, where the Earthly Paradise is set, and from where the four important rivers—Tigris, Euphrates, Nile, and Ganges—were said to emanate. Many localities are marked not just by name, but by drawings and captions taken from medieval legends. Some aspects of such planispheres are discussed by Stokes (1967).

THE CORVINUS MANUSCRIPT

From the Florentine view, the Psalter and New Testament manuscript in the Biblioteca Medicea Laurenziana, Florence (Plut. 15. 17, Folio 3), is of particular interest to the present discussion.

This manuscript was the third part of a large, three-volume Bible being prepared for Matthias Corvinus, King of Hungary. It was left incomplete after Corvinus died in 1490 (Fantoni, 1997). This volume contains the Psalter and the New Testament. It was the only volume to be completed, the first was only partly decorated, and the second has no illumination at all (Fantoni, 1997, p. 228–229). When Corvinus suddenly died, the manuscript was still in Florence. It was acquired by the Medici Library together with other unfinished manuscripts that had been commissioned by the king.

Fantoni (1997) wrote: "the text of this exceptionally large manuscript, measuring 535 × 370 mm, was copied in two columns by Sigismondo de' Sigismondo, one of the most active scribes in Florence in the late fifteenth century, probably between 1489 and 1490. The two sumptuous leaves at the beginning, illuminated in Florence by Gherado and Monte di Giovanni, are considered to be their masterpieces." [Fol. 3, detail].

Folio 3, which contains the beginning of the Psalter, written in gold on a purple ground, is richly illuminated. There is a wide border around the page decorated with putti, lions and devices, and a number of medallions showing biblical figures.

The biblical story depicted represents a battle between the Israelites, led by King David, and the Philistines, in front of the town of Hebron, identified by its name on the city wall. Some recent scholars believe that the battle scene probably paid tribute to Corvinus and alludes to the struggle against the Turks, in Corvinus's time, whereby Corvinus is represented by King David. Whether or not David is a depiction of Corvinus, one other important depiction is clear: the city of Florence has been used to represent Hebron, as it is possible to make out the Palazzo Vecchio and the Loggia dei Lanzi.

However, for the geologist, much of the city is obscured by an extraordinary small mount of carefully depicted, alternating, bedded rocks, similar to those shown by Botticelli's *Pallas and the Centaur*. Surprisingly, Fantoni (1997) makes no comment about such a feature forming almost the centerpiece of the work, with its waterfall flowing from the top of the rock mass (Fig. 10).

Note the reality of the rock mound and its virtual independence from the city beyond. It is one of the most striking geological illustrations of the late fifteenth century. The two other illustrated leaves show a number of similar rock faces but in more conventional situations. What influenced the artists to make these pictures, in particular the isolated rock mound? There is clearly more to be discovered.

CONCLUSIONS

This paper can give only a brief assessment of the complex relations between Florentine artists and the landscape of Florence and the Arno Valley. However, it introduces a number of themes that might be fruitfully explored by historians of geology, in conjunction with art historians, as it seems that cooperation has much to offer researchers in both fields.

The interest in geological features shown by so many Florentine artists of the late fifteenth and early sixteenth centuries foreshadowed the important geological principles set out so clearly by Steno a century or more later, based on his observations in the Arno Valley (but mainly in its tributaries and largely downstream from Empoli), thus laying the foundations for the theoretical development of modern stratigraphy.

ACKNOWLEDGMENTS

I am indebted to Gian Battista Vai for his encouragement of my research and for valuable discussion. Thanks are also due to the conveners of the INHIGEO (International Commission for the History of the Geological Sciences) 2004 field trip in Italy, the late Nicoletta Morello and Ezio Vaccari, for the opportunity to visit so many outstanding geological sites and institutions, where much wonderful art and early science can be seen and appreciated in original conditions. The Medici Library, Florence, at short notice, kindly granted me access to material related to the Corvinus manuscript. The School of Geosciences, University of Sydney, and INHIGEO also gave financial assistance. I am grateful for the positive criticisms offered by referees Kennard B. Bork and Massimo Coli.

REFERENCES CITED

Bedell, R., 2001, The anatomy of nature: Geology and American landscape painting, 1825–1875: Princeton and Oxford, Princeton University Press, xiii plus 185 p.

Billinge, R., Verougstraete, H., and Van Schoute, R., 2000, *The* Saint Barbara, *in* Foister, S., Jones, S., and Cool, D., eds., Investigating Jan van Eyck: Turnhout, Belgium, Brepols, 251 p.

Branagan, D.F., 2003, Rock and stone on canvas—real or imagined or geology and painting: Fifteenth century beginnings? *in* Serrano Pinto M., ed., Proceedings of INHIGEO Meeting, Portugal 2001. Geological Resources and History: Aveiro, Tipografia Minerva, p. 159–180.

Butler, M., 1997, An investigation of the Philadelphia "St Francis receiving the stigmata," *in* Van Asperen de Boer, J.R.J, Spantigati, C., and Butler, M., eds., Jan van Eyck: Two paintings of "St Francis receiving the Stigmata": Philadelphia, Philadelphia Museum of Art, p. 28–46.

Carmignani, L., and Lazzarotto, A., eds., 2004. Carta geologica della Toscana, 1: 250,000. Direzione Generale delle Politiche Territoriali e Ambientali: Servizio Geologico, Firenze.

Coli, M., Agili, F., Pini, G., and Coli, N., 2004, Firenze: Its impact on the landscape: Florence, Department of Geoscience, University of Florence, brochure created for the 32nd International Geological Congress.

Fantoni, A., 1997, *in* Crinelli, L., ed., Treasures from Italy's great libraries: New York, The Vendome Press, 288 p.

Geirnaert, N., 2000, Anselm Adornes and his daughters. Owners of two paintings of Saint Francis by Jan van Eyck?, *in* Foister, S., Jones, S., and Cool, D., eds., Investigating Jan van Eyck: Turnhout, Belgium, Brepols, p. 163–168.

Geology Museum, Florence, 2004, Notes provided in the museum display cabinets, exhibition of building stones of the Florence region.

Gombrich, E.H., 1976, Light, form and texture in fifteenth century painting, *in* Gombrich, E.H., ed., The Heritage of Apelles, Studies in the Art of the Renaissance: Oxford, p. 19–35.

Hamerman, N., and Rossi, C., 1989, The Apollo project of the golden Renaissance: Brunelleschi's Dome: 21st Century Science and Technology, v. 2, no. 4, p. 24–39.

Hartt, F., 1977, A history of painting, sculpture, architecture, v. 2—Renaissance, baroque, modern world: London, Thames & Hudson, 527 p.

Kemp, M., 2004, Leonardo: Oxford, Oxford University Press, xviii plus 286 p.

King, R., 2000, Brunelleschi's dome: The story of the great cathedral in Florence: London, Chatto & Windus, 184 p.

Langdon, H., 1996, Landscape painting. *in* Turner, J., ed., The dictionary of art: New York, Grove Dictionaries, v. 18, p. 700–720.

Lorenza, D., 2000, Léonard de Vinci, Artiste et Scientifique: Pour la Science, Les Génies de la Science, Trimestriel Mai–Août, no. 3.

Malesani, P., Pecchioni, E., Cantisani, E., and Fratini, F., 2003, Geolithology and provenance of materials of some historical buildings and monuments in the centre of Florence (Italy): Episodes, v. 26, p. 250–255.

Marshall, L., 2002, Reading the body of a plague saint: Narrative altarpieces and devotional images of St. Sebastian in Renaissance art, *in* Bernard, J., Muir, B.N.J., ed., Reading texts and images: Essays on medieval and Renaissance art and patronage in honour of Margaret M. Manion: Exeter, University of Exeter Press, p. 237–272.

Montgomery, S., 1996, The eye and the rock: art, observation and the naturalistic drawing of earth strata: Earth Sciences History, v. 15, no. 1, p. 3–24.

Nuttall, P., 1992, The patrons of chapels at the Badia of Fiesole: Studi di Storia dell'Arte, 3, p. 97–112.

Nuttall, P., 2000, Jan van Eyck's paintings in Italy, *in* Foister, S., Jones, S., and Cool, D., eds., Investigating Jan van Eyck: Turnhout, Belgium, Brepols, p. 169–182.

Oberhummer, E., 1909, Leonardo da Vinci and the art of the Renaissance in its relations to geography: The Geographical Journal, v. 33, no. 5, p. 540–569.

Panhans, G., 1974, Florentiner Maler verarbeiten ein eyckisches Bild: Wiener Jahrbuch für Kunstgeschichte, v. 27, p. 193.

Principe, C., 2004, From Volterra to Vinci, *in* Morello, N., and Vaccari, E., organizers, Italian Institutions and Geological Sites in the History of Geosciences: Florence, INHIGEO Post–IGC Congress Fieldtrip, 29 August–3 September 2004, Guidebook.

Reynolds, C., 2000, The King of painters, *in* Foister, S., Jones, S., and Cool, D., eds., Investigating Jan van Eyck: Turnhout, Belgium, Brepols, p. 1–16.

Rohlmann, M., 1993, Zitata flämischer Landschaftmotive in Florentiner

Quattrocentomalerei, *in* Poeschke, J., ed., Italienisch Frürenaissance und nordeuropisches Mittelalter: Munich, Hirmer, p 242–243.

Rubellini, P., Corti, F., and Paris, E., 2004, Historical analysis of flood events in the city of Florence: Florence, Geological Survey of the Municipality of Florence, Display at 32nd International Geological Congress.

Runcie, J., 2003, The Colour of Heaven: London, Harper Collins, 289 p.

Sarjeant, W.A.S., 1980, Geologists and the history of geology; An international bibliography from the origins to 1978: New York, Arno Press, v. 3, p. 2334–2339.

Spantigati, C., 1997, The Turin van Eyck "St. Francis receiving the Stigmata," *in* Van Asperen de Boer, J.R.J, Spantigati, C., and Butler, M., eds., Jan van Eyck: Two paintings of "St Francis receiving the Stigmata": Philadelphia, Philadelphia Museum of Art, p. 13–27.

Stafford, B., 1984, Voyage into substance; art, science, nature and the illustrated travel account, 1760–1840: Cambridge, Massachusetts, MIT Press, xxiii plus 645 p.

Starnazzi, C., 2005, Leonardo e la terra di Arezzo: Firenze, Calosci, 390 p.

Stokes, E., 1967, Fifteenth century earth science: Earth Science Journal, v. 1, no 2, 1, 19 p.

Tempestini, A., 1992, Giovanni Bellini: Catalogo completi dei Depinti: Firenze, Cantini, 319 p.

Vai, G.B., 1986, Leonardo, La Romagna e la geologia, *in* Marabini, C., and Della Monica, W., eds., Romagna: vicende e protagonisti: Bologna, Ed. Edison, 1, p. 30–52.

Vai, G.B., 1995, Geological priorities in Leonardo Da Vinci's notebooks and paintings, *in* Giglia, G., Maccagni, C., and Morello, N., eds., Rocks, fossils & history. Proceedings of the 13th INHIGEO Symposium, Pisa-Padua (Italy), September–October 1987: Firenze, Edizioni Festina Lente, p. 3–26.

Vai, G.B., 2003, I viaggi di Leonardo lungo le valli romagnole: riflessi di geologia nei quadri, disegni e codici, *in* Leonardo, Machiavelli, Cesare Borgia: Arte, Storia e Scienza in Romagna (1500–1503): Roma, De Luca Editori d'Arte, p. 37–47.

Vai, G.B., and Martini, I.P., eds., 2001, Anatomy of an orogen: The Apennines and adjacent Mediterranean basins: Dordrecht, Kluwer Academic Publishers, 633 p.

Vallance, T.G. 1977, Letter (from Geneva) to David Branagan, 7 December 1977. Original in possession of the author.

Van Asperen de Boer, J.R.J., 1997, Some technical observations on the Turin and Philadelphia versions of "St Francis receiving the Stigmata," *in* Van Asperen de Boer, J.R.J, Spantigati, C., and Butler, M., eds., Jan van Eyck: Two paintings of "St Francis receiving the Stigmata": Philadelphia, Philadelphia Museum of Art, p. 51–63.

Venturini, L., 1994, Francesco Botticini: Firenze, Edifir, 246 p.

Zammattio, C., Marinoni, A., and Brizio, A.M., 1981, Leonardo the scientist: London, Hutchinson, 192 p.

MANUSCRIPT ACCEPTED BY THE SOCIETY 17 JANUARY 2006

Ulisse Aldrovandi and the origin of geology and science

Gian Battista Vai*
William Cavazza*
Dipartimento di Scienze della Terra e Geologico-Ambientali, Università di Bologna, via Zamboni 67, I-40127 Bologna, Italy

ABSTRACT

The Italian naturalist Ulisse Aldrovandi (1522–1605)—often reductively considered as a mere encyclopedist and avid collector of natural history curiosities—lived an adventurous youth and a long maturity rich of manuscripts, books, and outstanding achievements. He assembled the largest collections of animals, plants, minerals, and fossil remains of his time, which in 1547 became the basis of the first natural history museum open to the public. Shortly after that, he established the first public scientific library. He also proposed a complete single classification scheme for minerals and for living and fossil organisms, and he defined the modern meaning of the word "geology" in 1603.

Aldrovandi tried to bridge the gap between simple collection and modern scientific taxonomy by theorizing a "new science" based on observation, collection, description, careful reproduction, and ordered classification of all natural objects. In an effort to gain an integrated knowledge of all processes occurring on Earth and to derive tangible benefits for humankind, he was a strenuous supporter of team effort, collaboration, and international networking. He anticipated and influenced Galileo Galilei's experimental method and Francis Bacon's utilitarianism, providing also the first attempt to establish the binomial nomenclature for both living and fossil species and introducing the concept of a standard reference or type for each species.

His books and manuscripts are outstanding contributions to the classification of geological objects, and to the understanding of natural processes such as lithification and fossilization, thereby also influencing Steno's stratigraphic principles. The importance given to careful observation induced Aldrovandi to implement a uniformitarian approach in geology for both the classification of objects and the interpretation of processes.

Aldrovandi influenced a school in natural history that reached its climax with the *Istituto delle Scienze* of Bologna in the seventeenth and eighteenth centuries with scientists such as Cospi, Marsili, Scheuchzer, Vallisneri, Beccari, and Monti in geology, and Malpighi, Cassini, Guglielmini, Montanari, Algarotti in other fields.

Keywords: collection, natural history, new science, public museum, public library, scientific library, classification, scientific taxonomy, teamwork, utilitarianism, binomial nomenclature, standard reference, type, lithification, fossilization, geology, observation, experiment, uniformitarianism.

---

*vai@geomin.unibo.it; cavazza@geomin.unibo.it.

Vai, G.B., and Cavazza, W., 2006, Ulisse Aldrovandi and the origin of geology and science, *in* Vai, G.B., and Caldwell, W.G.E., The Origins of Geology in Italy: Geological Society of America Special Paper 411, p. 43–63, doi: 10.1130/2006.2411(04). For permission to copy, contact editing@geosociety.org. ©2006 Geological Society of America. All rights reserved.

INTRODUCTION

Known worldwide as a polymath, encyclopedist, collector, and natural philosopher, Ulisse Aldrovandi (1522–1605) wrote the word "giologia" [geology] for the first time in its proper sense in his will published in 1603 (Vai, 2003a). This fact, however, was long neglected both in Italy and elsewhere. It was only between the end of the seventeenth century and the beginning of the eighteenth century that the word geology started emerging in common use (Gortani, 1931, 1963; Adams, 1938; Dean, 1979; Carapezza, 1984; Ellenberger, 1988, p. 191; Oldroyd, 1996, p. 323; Vai, 2003a, p. 74).

Several sources attribute the first use of the term "geologia" to Richard de Bury (1473, according to Adams, 1938/1954). The Scottish bishop, however, introduced the term with a meaning completely different from today's to indicate earthly philosophy as opposed to theology, the study of spiritual facts. De Bury's suggestion never took off and left Aldrovandi an open field to define the term geology in the sense that it has been used ever since. It is mystifying that some researchers and dictionaries still indicate the Swiss naturalists Jean-André Deluc and Horace-Bénédict de Saussure as the definers of the modern meaning of the term geology in the years 1778–1779 (e.g., Encyclopedia Britannica, fifteenth edition).

Naming a discipline is no mean achievement. However, one could ask whether this was the only contribution Aldrovandi gave to the development of geology. And more generally, what was Aldrovandi's contribution to the origin of modern science? Answering these two relevant questions is the aim of this article.

TRANSMISSION OF KNOWLEDGE VS. ORIGINAL RESEARCH

The experimental approach is considered as marking the beginning of modern science and the starting point for its exponential growth. Such an approach is commonly referred to as "Galilean science," after the late Renaissance genius who gave elegant operational and theoretical bases to the scientific method. Nevertheless, experimentation and speculation have always gone hand in hand as complementary facets of the curiosity of humankind and its need to understand nature. In fact, the development of culture is punctuated both by technological discoveries, such as fire control, stony tools, writing, implementing the wheel, building the arch, etc., and by conceptual assessments in religion, painting, poetry, legislation, and philosophy. Even the development of logic occurred by the concurrent use of both deductive and inductive approaches.

During the Middle Ages, the body of natural science knowledge developed earlier by the ancient Greek and Latin civilizations was for the most part simply transmitted or adapted to the expanding new monotheistic religions. Even the Muslim culture—although very active in mathematics, astronomy, and education—brought about hardly any development in the experimental sciences, with the notable exception of medicine.

During this period, technological evolution continued, as witnessed by the Christian cathedrals and Muslim mosques, technological wonders of their ages. Culture at large evolved as well, for example, with the birth of the cities and universities in Italy and Europe during the twelfth and thirteenth centuries. However, this was not accompanied by scientific conceptual elaboration, which at the time was focusing mainly on theology, philosophy, and literature, leaving arts and mechanics to the craftsman.

The turning point occurred in the fourteenth and fifteenth centuries in the affluent Italian society of the humanists when craftsmen and savants started to interact. Universities founded by the Church and think-tanks sponsored by princes or *Signori* promoted the development of a network of productive *botteghe* (workshops) and planning *studioli* (little court studios) where artisan experience backed up the new theoretical paradigms of savants. Among the protagonists of this productive period we would like to remember Leon Battista Alberti (1404–1472), Filippo Brunelleschi (1377–1466), Nicolò Cusano (1401–1464), Paolo Uccello (1397–1475), Piero della Francesca (1416–1492), Luca Pacioli (1445–1517), Aristotele Fioravanti (1420–1486), Vanoccio Biringuccio (1480–1537), Marsilio Ficino (1433–1499), Giovanni Pico della Mirandola (ca. 1469–1533), Paolo dal Pozzo Toscanelli (1397–1482), and Nicolaus Copernicus (1473–1543), who was a student in Bologna and Padua. Many of these were savants and craftsmen at the same time and gave special attention to observation and experimentalism.

Such attention to observations and experiments, coupled with the discovery of the New World, implied *de facto* to overcome the limits of the past authorities and Aristotelianism, without refusing Aristotle himself. The late Renaissance scientific blooming, first in Italy and then in Europe, was not sudden. Instead, it had been prepared more than a century before Galilei. Even an isolated and unique genius such as Leonardo benefited from the special innovative cultural environment of the second half of the fifteenth century (Table 1).

The role of a scientist like Ulisse Aldrovandi (1522–1605) can be correctly understood only in a similar perspective, beyond the dangers of either hagiography or underevaluation as a simple collector, encyclopedist, and erudite polymath. Aldrovandi is the paradigm of the scientists, naturalists, and natural philosophers of the second half of the sixteenth century (Fantuzzi, 1774, 1781; Costa, 1907; Frati, 1907b; Mattirolo, 1897; Olmi, 1976, 1985; Findlen, 1989, 1994; Simili, 2001; Tugnoli Pàttaro, 1977, 1981, 2000, 2001; Tosi, 1989; Vai, 2003a). Aldrovandi clearly transcended simple collecting or encyclopedism. He marks the passage from the Medieval monastery library, devoted to collecting, keeping, and copying ancient codices, to the modern library, which aims to integrate its holdings with new

---

Table 1. Chronologic distribution of geoscientists arranged by cultural areas from the beginning of the thirteenth century to the beginning of the nineteenth century.

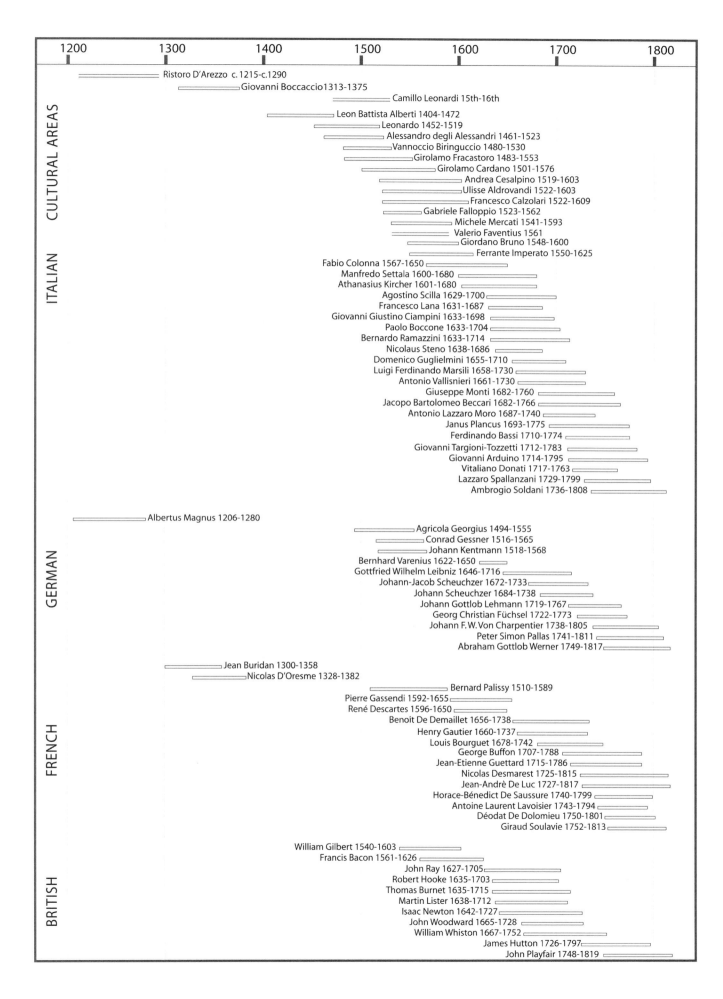

contributions derived from the first experimental researches in the field (Vai, 2003a). Developing new experimental research *and* critically transmitting past knowledge was the centerpiece of Aldrovandi's manifesto.

SUMMARY OF ALDROVANDI'S LIFE

Ulisse Aldrovandi was born in Bologna on 11 September 1522, in one of the most prominent families in the city oligarchy (Fig. 1). He was cousin to Ugo Boncompagni, who became Pope Gregory XIII, champion of the Catholic Reformation and promoter of the reform of the antiquated Julian calendar. The new system was named after him (Gregorian calendar). Aldrovandi's youth was marked by travels and adventures, punctuated by private home education and early employment. When he was only twelve years old, he went alone incognito to Rome, fascinated by the appeal of the great ancient city. Sixteen years old, he moved again to Rome in 1538, came back to Bologna, and continued without stopping at home, traveling for months across France and Spain, ultimately ending up in Compostela. Upon his return to Bologna, he took regular courses in the humanities and law at the university, became a notary in 1542, and continued studying with the most distinguished teachers of the time. In 1547, he started displaying in his home museum the natural objects he had collected during his first travels.

On the verge of receiving his law degree, he rededicated himself to philosophy and logic, and in 1548 he moved for a couple of years to the University of Padua, where he studied medicine and mathematics. In 1549, he was imprisoned in Rome on a groundless accusation of heresy. Although released soon due to a timely amnesty, he demanded that the trial be held anyway in order to demonstrate his innocence. While waiting

Figure 1. (A) Rare engraving of the young Ulisse Aldrovandi derived from a portrait by Tiziano (private collection). (B) Detail of a bronze statue of Pope Gregory XIII, the Bolognese Ugo Boncompagni, cousin of Aldrovandi, Bologna, Palazzo d'Accursio (photo Vai).

for the trial, Aldrovandi drafted his first publication, *On the ancient Roman statues which are seen throughout Rome in various places* (1550), printed in 1556 in *The antiques of the city of Rome*, by Lucio Mauro. He also met and befriended Guilleaume Rondelet (1507–1566)—a specialist in the study of fishes—in a fish market in Rome. This encounter was quite influential and marks a turning point in Aldrovandi's personal history; he finally settled on the study of natural history, thus putting to an end the multifarious—and mostly unproductive—activities that characterized his early life.

Back to Bologna, Aldrovandi obligingly graduated in philosophy and medicine in 1553, and—in a rapid sequence of academic achievements—obtained the chair of logic in 1554, the chair of extraordinary philosophy in 1555, the teaching of medical botany (*de simplicibus*) in 1556, and the chair of ordinary philosophy in 1559. The academic environment suited Aldrovandi perfectly. When not occupied by his teaching duties, he devoted his time to studying all aspects of natural history, often by leading fieldtrips with colleagues and students to the Apennines, the Alps, and the Po Plain. He soon established a full network of correspondence and exchanges with the leading natural scientists in Europe, and started planning research travel to the Americas. His museum had already displayed for the first time natural specimens from the New World. Aldrovandi's classes in the natural sciences, a broader field than traditionally was taught by a single professor, were so popular that in 1561 his students successfully petitioned the University of Bologna to transform the chair into a full professorship of natural sciences (*lectura philosophiae naturalis ordinaria de fossilibus, plantis et animalibus*), held by Aldrovandi until 1600.

As a collaborator of Cardinal Gabriele Paleotti—one of the masterminds of the Counter Reformation—he joined the opening session of the Council of Trent in 1562. In 1568, he established the Bologna Botanical Gardens—the origins of which he had already assembled in his own property—and was nominated its first director. Correspondence and exchanges of samples with the Grand Duke of Tuscany, Francesco I, began in 1577. Besides performing research in the natural sciences, he experimented in embryology, anatomy, and medicine, and he wrote a score of manuscripts most of which were published posthumously or remained unpublished.

His major methodological work, *Discorso naturale* (*Natural discourse*) (Biblioteca Universitaria di Bologna [BUB], Fondo Aldrovandi [FA], ms. 91, c. 503r–559r) was written in 1569–1570. In the 1560s, he had already begun to write the *Syntaxis plantarum*, in 1578 the *Historia naturalis*, and in 1580 the *Bibliologia*. In 1587, the *Ornithologia* was planned, in 1588 *De lucentibus* was completed, as were the first volumes of the *Syntaxis plantarum* and the *Pandechion Epistemonicon*, or *Selva universale delle scienze*. A large plan of printing his many naturalistic works was established by contract with a Venetian publisher in 1594. The first volume of his major zoological work, *Ornithologia*, appeared in 1599, the second in 1600, and the third in 1603.

On 10 November 1603, Aldrovandi (Fig. 2) published his will bequeathing his museum and library to the Bolognese Senate, which had generously supported his long academic career. The will is not just a legal document but summarizes also the core of Aldrovandi's thoughts. It is also a landmark document for the Earth sciences as it introduces the word *geology* and contains its first modern definition.

Aldrovandi died on 4 May 1605.

Aldrovandi's personal library clearly testifies to both his depth of knowledge and his originality (Fig. 3). It contained more than 4,000 volumes, of which 3,800 were printed, 28 were codices and 332 were copies (Frati, 1907a; Ghigi, 1907). In addition, it included a score of Aldrovandi's original books printed during his life and posthumously (13 quarto volumes totaling almost 9,900 pages), and an impressive series of 870 manuscripts, representing the digested revision of the whole of human knowledge, with nomenclature, description (often with illustrations), and the classification of more than 10,000 plants, animals, minerals, rocks, and fossils never before completely classified in a scientific way. His collections also included 18,000 prints and

Figure 2. Detail of a marble bust of Ulisse Aldrovandi sculpted for the third centennial of his death, upon request by Giovanni Capellini, president of the Second International Geological Congress, held in Bologna in 1881 (photo Vai).

Figure 3. The old Ulisse Aldrovandi teaching a class of students at his home. Detail of a fresco by Vittorio Maria Bigari in the main hall of the Aldrovandi (Montanari) Palace (1748) in Bologna (photo by Vai).

drawings (made before 1595). His museum—divided into rooms and fully ordered and classified—contained at least 20,000 objects (plants, animals, and 4,000 fossils), some of which were the holotypes described and drawn in his printed works or in his manuscripts. His botanical garden of medicinal plants was one of the first (1551–1553), and at the time was the richest and the best classified in the world. His herbarium contained more than 9,000 dried plants, and was the largest ever assembled. This was the huge legacy, assembled mostly at Aldrovandi's expense, of a lifetime of dedicated work, and left as public property to the Bologna Senate for the scientific benefit of scholars worldwide and for the teachers of the University of Bologna.

OBSERVATION, EXPERIMENTAL RESEARCH, AND THE VALUE OF NATURAL OBJECTS

Throughout his long life Aldrovandi never lost intellectual curiosity toward all kinds of natural items and artifacts. He was well aware and proud "never to have described any things unless having touched them with his own hands and personally investigated the specimen's anatomy" (BUB, FA, ms. 21, v. IV, c. 36–72, 79–82).

Direct observation of objects in the field and their experimental anatomical dissection opened the way to modern science. Aldrovandi's experimental approach was clearly theorized after more than twenty years of field and museum practice in *Discorso naturale* (BUB, FA, ms. 91, c. 503r–559r), dated 1572–1573 according to Tugnoli Pàttaro (1977), but substantially written in 1569–1570, as autographically attested in c. 538r. A few direct quotations from this basic Aldrovandi work, never before translated into English, follow to show the level of theoretical foundations of the new experimental science he had achieved decades before Galilei and Bacon.

[…] because I knew the true sublunar philosophy to be nothing but the true knowledge of perfect and imperfect objects, both animated and inanimated, which are continuously under our senses (c. 507v).

I started considering the nature and differences, both outer and inner, of each natural thing, realizing that the true philosophy was to know openly generation, temperature, nature and faculty of each thing by means of experience. I was aware by Aristotle's witness in the second book of *Posteriora*, that experience originates memories, and memories give rise to universals, which are principles of the Arts and Sciences. True universals are based on the first matters devoid of which the universals are illusion of the bare intellect. Everyone knows that no thing reaches our intellect unless through our outer senses in the process of induction (as witnessed by our Philosopher [Aristotle]) (BUB, FA, ms. 91, c. 507v).

Clearly this natural philosophy of inanimate objects such as earths, stones, gems, marbles, metals, pebbles, and other mixed things, which are found in the bowels of the Earth and at its surface, cannot be known unless by diverse wandering (c. 507v).

[...] with great personal expenses and bodily exertion I had to see with my own eyes not only all of Italy, but also all of France and Spain and part of the German land with the only aim to have knowledge of these major mysteries of nature which shine in plants, animals and other subterraneous things, [...] (c. 508r).

My Natural History was truthfully written, not having ever described any thing not previously seen with my own eyes and touched with my own hands, examining the anatomy of both external and internal parts: and preserving each, one by one, in my little natural world, so that everyone at any time can observe and contemplate, all [things] being stored as paintings and samples in our Museum, collected not only while I was student but principally after my Doctorate in various trips for the benefit of scholars (BUB, FA, ms. 91, c. 508r–508v) (Fig. 4).

Aldrovandi was aware that he had observed, collected, and described many more animals and plants than Aristotle (e.g., five thousand versus two thousand plants):

[...] it seems to me I have observed many more animals and plants than the Philosopher [Aristotle]; I could observe and

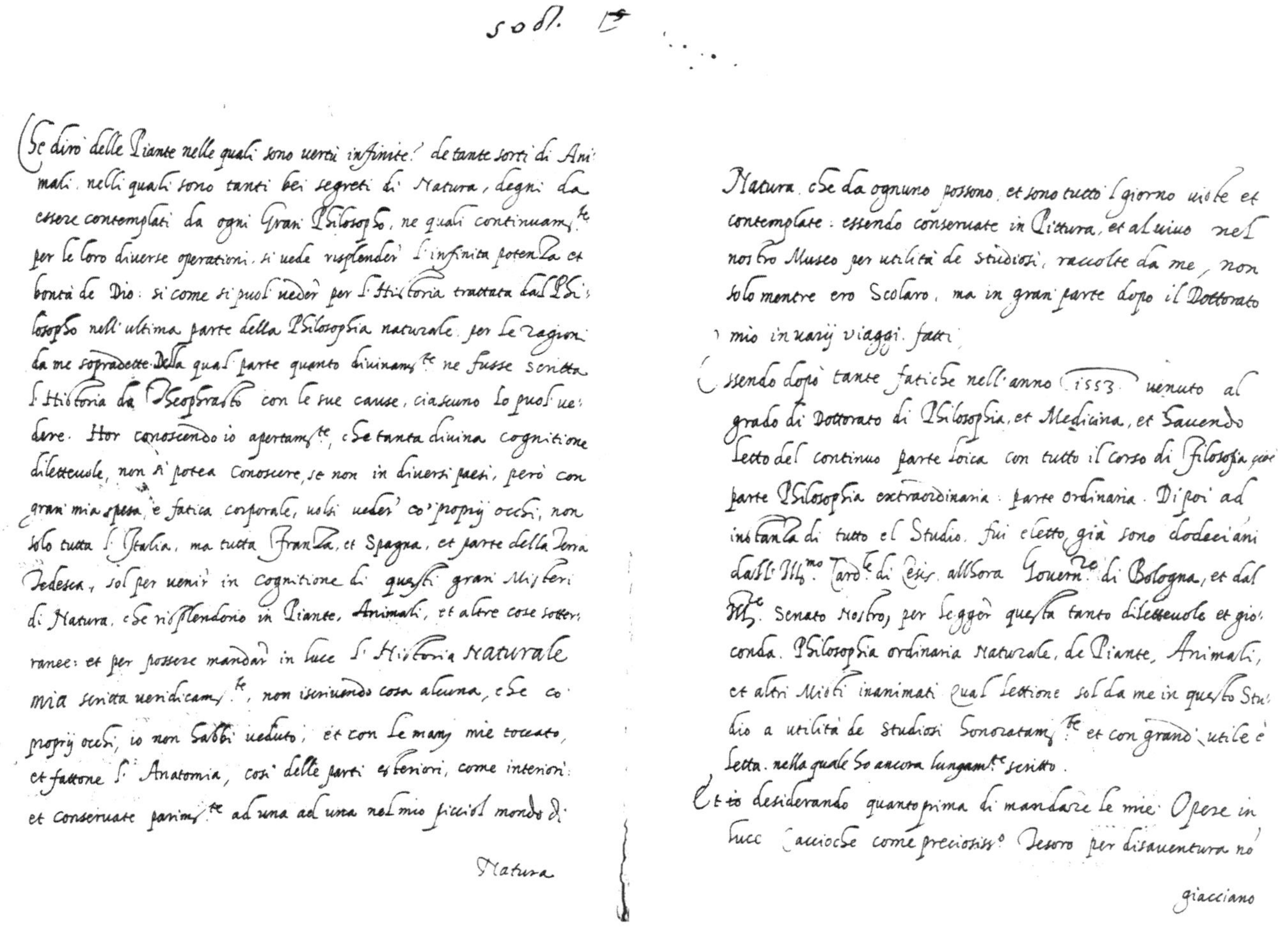

508.

Che dirò delle Piante nelle quali sono uertù infinite? de tante sorti di Animali, nelli quali sono tanti bei segreti di Natura, degni da essere contemplati da ogni Gran Philosopho, ne quali continuam.te per le loro diuerse operationi, si uede risplender l'infinita potenza et bontà de Dio: si come si puol ueder per l'Historia trattata dal Philosopho nell'ultima parte della Philosophia naturale, per le ragion da me sopradette. Della qual parte quanto diuinam.te ne fusse scritta l'Historia da Theophrasto con le sue cause, ciascuno lo puol ueder. Hor conoscendo io apertam.te, che tanta diuina cognitione dilettevole, non si potea conoscere, se non in diversi paesi, però con gran mia spesa, e fatica corporale, uolsi ueder co' propij occhi, non solo tutta l'Italia, ma tutta Francia, et Spagna, et parte della Terra Tedesca, sol per uenir in cognitione di questi gran Misteri di Natura, che risplendono in Piante, Animali, et altre cose sotterranee: et per possere mandar in luce l'Historia naturale mia scritta uendicam.te, non iscriuendo cosa alcuna, che co propij occhi, io non habbi ueduto; et con le mani mie toccato, et fattone l'Anatomia, così delle parti esteriori, come interiori: et conseruate parim.te ad una ad una nel mio picciol mondo di Natura

Natura, che da ognuno possono, et sono tutto 'l giorno viste et contemplate: essendo conseruate in Pittura, et al uiuo nel nostro Museo per utilità de studiosi, raccolte da me, non solo mentre ero Scolaro, ma in gran parte dopo il Dottorato in varij viaggi fatti.

Essendo dopò tante fatiche nell'anno 1553 uenuto al grado di Dottorato di Philosophia, et Medicina, et hauendo letto del continuo parte Loica con tutto il corso di Filosofia, cioè parte Philosophia extraordinaria, parte ordinaria. Di poi ad instanza di tutto el Studio, fui eletto, già sono dodeci anni dall'Ill.mo Card.le di Cesis. allhora Gouern.re di Bologna, et dal Ill.mo Senato nostro, per leggere questa tanto dilettevole et gioconda Philosophia ordinaria naturale, de Piante, Animali, et altri Misti inanimati, qual lettione sol da me in questo Studio a utilità de studiosi honoratam.te et con grande utile è letta. nella quale ho ancora lungam.te scritto.

Et io desiderando quantoprima di mandare le mie Opere in luce (acciochè come preciosiss.o Thesoro per disauentura no giaciano

Figure 4. Original quotation from Aldrovandi's manuscript 91, *Discorso Naturale*, c. 508r–508v) (photo Vai).

collect together ideas and samples of the species produced by the great God for our usefulness out of twelve thousand different things, in spite of not having obtained any assistance as he had from such a great king [Alexander the Great] (c. 510r).

Speaking about minerals Aldrovandi remarks:

[…] it takes a lot of labor, expenses and danger for life to acquire knowledge in this science, because to learn it well and perform the true anatomy one needs to be present in the mines and furnaces, otherwise one can not learn it. I have gone deliberately inside them for checking so many beautiful speculations I had recorded in MY NATURAL HISTORY (c. 524r).

Aldrovandi was able to identify thousand of mistakes accumulated in many centuries of natural history, and he welcomed the introduction of the public teaching of this discipline in the universities as a means of improvement:

We consider in fact that such a public lecture, introduced in the Studios (universities) in our times, will bring innumerable benefits. Students coming from various parts of Europe to listen to this and other disciplines will maintain and expand it continuously because each of them having learned this knowledge will find new things once back home, and so the discipline will augment. I can witness it because many of my own students have sent to me new things I never saw before. The species of natural objects as to our knowledge are infinite for the man who cannot go everywhere. Additionally, the individual man cannot be perfect. However, many can find out different things, which the sciences together will then make perfect. This is the usefulness deriving from this public lecture. (c. 530r–c. 530v)

This same point is made once more with similar words in Aldrovandi's ms. 21, vol. IV. In a time of extreme individualism of the savants, Aldrovandi underscored instead the advantages of international cooperation in scientific research and also planned scientific expeditions to the New World with the sponsorship of the pope:

The best, easiest and surest ways to attain a perfect knowledge in this natural philosophy […] were to send excellent people in various parts of the world […] to ascertain any least dubious thing found therein […] such as roots, rubber, woods, seeds, fruits and other similar things which, being parts of selected plants, are difficult to recognize without knowing their mother. […] By doing so, many dubious descriptions of the ancient [authors] would be checked, and innumerable others not yet discovered would be known, because the diversity of sites and climates produce innumerable various things (c. 534v–535r, c. 535v).

All of the above quotations are from a single manuscript, *Discorso naturale*, and indicate a lucid and coherent project to reopen the process of knowledge of natural objects after many centuries of stasis. The new natural science has to be active in the field in any part of the Old and especially the New World, collecting any objects, dissecting, describing, representing by painting, and keeping reference samples within dedicated collections ordered in adequate public scientific museums. Direct observations, comparison, classification, and experiments were the tools to check the validity of the ancient Greek, Roman, and Arabic works (unfortunately devoid of illustrations), to show thousands of mistakes made by the Medieval transmission, and to expand the limited number of objects known in the past.

In Aldrovandi's concept, Aristotle and the other famous ancient authors were reference sources to be checked and questioned, not absolute authorities. The Aristotelian peripatetic controversy was not yet sharp or was not yet felt by Aldrovandi because he knew that he had already surpassed the knowledge Aristotle had reached in the study of natural objects.

Additionally, Aldrovandi underlined that, in the frame of God's creation, the value of all natural objects, both animate and inanimate, is to serve the needs of humankind. A close scientific knowledge of the useful properties of natural objects prompted him to qualify them as divine—especially the plants—in his *Discorso naturale*.

FROM PRIVATE COLLECTING TO THE PUBLIC MUSEUM OF NATURAL HISTORY

Encyclopedic collections have been popular from ancient to modern times, yet different types of such collections need to be distinguished. Lapidaries, herbariums, and bestiaries, mainly in the form of inventories, were commonly used in Medieval times for practical purposes. Monks, abbots, and, later, princes kept promiscuous book (codex) collections sometimes assembled inside true philosophical and theological libraries, for internal or private use. This was also instrumental to preserve and transmit the past written cultural heritage.

In the fifteenth century, enlightened princes founded the first permanent general libraries aimed at practical use for select middle-class people, such as the Biblioteca Malatestiana in Cesena—established by Novello Malatesta in 1452 with a truly public admission—and the Biblioteca Laurenziana in Florence.

At the same time, princes throughout different parts of Europe began to build their own private *Studiolo* or *Cabinet* which were beautifully adorned by rare natural objects and books. The studiolos were status symbols aimed at impressing the peers of the prince during their visits.

At the beginning of the sixteenth century, the time was ripe to originate a new type of collection for general or specific scientific purposes. This new approach led to the founding of the first natural history museum by Aldrovandi in 1547, in Bologna (Aldrovandi, *in* Fantuzzi, 1774; Vai, 2003a, p. 68). Soon after, in

1569, the conceptual assessment of the role of museums for the development of science was envisaged (see *Discorso naturale*).

The new museums of natural history were composed of collections of humble natural objects and not works of art or artifacts. This was in contrast with the Medieval lists and inventories of animals, plants, and stones (bestiaries, herbariums, lapidaries), which were sometimes speculative and often confused. Such evolution implied the systematic study and classification of those objects, and the birth of disciplines for describing and interpreting the essence and the role of the items collected.

In Italy in the early fifteenth century, the scientific and cultural momentum related to Humanism and the concurrent economic growth led to a wide debate on the meaning of a special category of objects—fossils (*fossilia*)—in natural history and on the relevance of mining for economy. During the early Renaissance, foreign students attending the Italian universities spread the debate across Europe. Universities and religious institutions from monasteries to parishes were involved. Anatomists, abbots, alchemists, herbalists, artists, and priests wrote descriptions, treatises, cosmogonies, and theories of Earth and its history, mostly focused on the origin of fossils. This new theoretical interest made fossils a treasured target for permanent collections designed for knowledge purposes—far beyond the scope of traditional lapidaries, herbariums, and bestiaries.

Soon after the Aldrovandi Museum was established in Bologna (1547), several other naturalists arranged large natural history collections, such as Francesco Calzolari (1522–1609) in Verona (from 1554), Michele Mercati (1541–1593) in Rome (from 1571–74 or 1573–76), and Ferrante Imperato (1550–1625) in Naples (from 1573–75) (Lancisi, 1717; Mattirolo, 1897; Gortani, 1931; Accordi, 1980, 1981; Vai, 2003a). All of these were Aldrovandi's correspondents and friends, as were Johannes Kentmann (BUB, FA, ms. 136, II, c.140v-144r), founder of a mineralogical collection in Torgau, illustrated in 1565 (Kázmér, 2002), and Konrad Gesner, who founded a collection in Zurich in 1564. Aldrovandi was the most curious and influential in this circle. He was open-minded and used disciplinary, encyclopedic, and even all-encompassing classifications in his collections, always at the highest level of quality and innovation. Aldrovandi played the role of theorist and leader of this group, and was the most respected among the founders of the first naturalistic museums (BUB, FA, ms. 38[2]; Mattirolo, 1897, p. 68; De Toni, 1907; Olmi, 1985, 2001; Findlen, 1989, 1994; De Bellis, 1998; Vai, 2003a).

The intimate relationship between the objects of his collections and the books of his vast library is of paramount importance to understanding Aldrovandi's innovative approach (Fig. 5). For Aldrovandi, the museum and library were a single entity. Besides collecting books from other authors concerned with natural history, Aldrovandi's library included all the books and manuscripts Aldrovandi himself had written to describe and illustrate his collections. From this viewpoint, Aldrovandi's library can be considered the first scientific library of the modern world.

Unlike the court and *studiolo* collections of the time, the Aldrovandi Museum had been designed as a *public* institution. It was based in a *public* palace, at the public expenses of the Bologna Senate (after 1603), and included a library designated "public" directly by him (Vai, 2003a). Even when Aldrovandi's collections were kept in his private residence (1547–1603), he used them systematically for public lecturing on natural philosophy to the students of the University of Bologna, at least beginning in 1561. Findlen's (1989, p. 71) opinion that the Aldrovandi Museum was not open to the public cannot be shared, even considering the evidence this author provides in the same paper (p. 72–73) and later (Findlen, 1994). Moreover, even Findlen's distinction made between Aldrovandi's and Marsili's museums concerning their public nature is not founded. Marsili simply adapted Aldrovandi's ideas and methods to his new collections, having no intention of subsuming the Aldrovandi Museum into his own in the Bologna *Istituto delle Scienze*. On the other hand, one can only agree with Findlen (1989) that the Ashmolean Museum in Oxford, opened in 1683, was the first democratic public museum open not only to students and learned people, but also to the layman (Vai, 2003a).

Aldrovandi knew in advance how many people were to visit his museum and did everything he could to secure an adequate building to store it:

> I would be pleased if those Gentlemen and literary men who visited and will be visiting the Museum after my death continue to write their names in my two books I have provided. [...]
>
> [I beg] the Senate to persuade the Pope, for the honor of the City and of the Testator, to favor and help in building within the [city] Palace or somewhere else four or five bright and beautiful Rooms for the Museum and Library to accommodate [the materials] in the given order, together with my Portrait I have left; and to beg His Holiness to donate funds for the expansion and decoration, because in the field of Letters the major enterprise is to expand and erect a public Library (Aldrovandi's will, 10 November 1603, ASB, Atti Notarili, 6/1, v. 3063; *Ulisse Aldrovandi's Will* 1603, *in* Fantuzzi, 1774).

PRINCIPLES OF CLASSIFICATION AND SCIENTIFIC METHOD

Similar to Georgius Agricola (1494–1555), Konrad von Gesner (1516–1565), and others of his time, Aldrovandi generated lists of innumerable data on different subjects, including and beyond mineralogy. In spite of these similarities, however, a basic difference appears: Aldrovandi alone systematically attempted to organize, to simplify, and to sum up the data, suggesting genetic and interpretative relations through hierarchical classification charts and dichotomous subdivision tables.

This was witnessed by Buffon (1707–1788) in his *Histoire Naturelle* (1749–1804, t.I, c. 26) (Fantuzzi, 1774, p. 104):

Figure 5. Detail of Aldrovandi's personal library with students of different ages listening to his lesson, as painted by Vittorio Maria Bigari in the main hall of the Aldrovandi (Montanari) Palace (1748) in Bologna (photo Vai).

…Aldrovandi was the most industrious and learned of all Naturalists … his books have to be regarded … as the best about the whole Natural History. The plan of his works is well done, his distributions are full of sense, his subdivisions clear cut, his descriptions correct.

This is even more important because Buffon was not biased toward Aldrovandi having criticized his excess of credulity and verboseness (see also below under the forelast subheading).

Aldrovandi produced classifications in all natural sciences as well as in many fields of the humanities, as, for example, libraries (BUB, FA, ms. 97, c. 440–443, 650–651) (Vai, 2003a). His true obsession was to bring "order in sciences," and he even thought that "how to place the books in a well-ordered library" was worthy of scientific treatment. He was probably the first to formulate a golden rule for a profitable bibliographic investigation based on alphabetic files:

> to be later able to refer to carefully studied books, please make note of various sentences or other related matters in that book always writing them down in alphabetic order, so that at any time in the future, one can find the same matters written by different [authors] according to their opinion, especially those noticed by reading that are not expressed in the illustrations; moreover, there are many books showing no illustrations, which must be annotated in the said order, to be usable when needed. (BUB, FA, ms. 97, c. 650–651)

The major innovations Aldrovandi brought to the classification of natural sciences are (1) making reference to illustrations, (2) the concept of type specimen for species description, and (3) the use of binomial nomenclature.

Reference to illustrations was pivotal for Aldrovandi's principles of classification and scientific method. Aldrovandi was aware that "many books show no illustrations." His books, instead, have plenty of perfectly *natural* marvelous figures. According to Aldrovandi, to be properly classified, all species of studied natural objects need to be drawn (Battistini, 2003; Vai, 2003a), and the drawings have to supplement the description in the books devoted to classification. But this is not enough.

For Aldrovandi, the final standard of reference in the classification of scientific objects (animals, plants, fossils) is a selected sample preserved "forever" in "his public museum." Aldrovandi wrote of "*veri essemplari delle loro specie representate*" (Olmi, 2001, p. 24) or "true specimens of their represented species." This was a truly momentous statement for the history of the natural sciences because it was the first time that someone expressed the concept of having a material specimen serve as a type or objective reference for a species and standard for its classification.

Modern binomial nomenclature was used systematically in Aldrovandi's taxonomic works for naming and defining genera and species of both living and fossilized beings and for clustering different species of the same genus (*Musaeum Metallicum*, 1648; BUB, FA, ms. 91, c. 522r–524v; Sarti, 2003, p. 156; Vai, 2003a, p. 93). An alphabetic inventory of inanimate fossil samples of his museum (ca. 1595) (BUB, FA, ms. 135) lists different species of *Chama*, such as *Chama subrinosa*, *Chama fasciata*, etc., anticipating by more than a century the Linnaean method. An ad hoc commission of the 2nd International Geological Congress was trusted to discuss and submit to the congress in 1881 a drafted resolution for establishing a starting date for implementing the principle of priority in the binomial nomenclature used in paleontological publications. Despite the fact that the congress ultimately preferred to start with Linnaeus' *Systema Naturae* (1776 edition), one of the options examined and approved by the commission was to start with Aldrovandi's publications (Vai, 2004), thus recognizing a modern validity to his nomenclature and classification. Perhaps this occurred because only a minor part of Aldrovandi's paleontological works have been published. Moreover, his collections were not yet restored in 1881 after Napoleon's war spoliation at the beginning of that century.

ALDROVANDI'S SCIENTIFIC HIGHLIGHTS IN GEOLOGY

Aldrovandi was a pioneer in methodology of modern science and he opened many new paths in different natural and human sciences. His contribution to geology goes well beyond the authorship of its very name. Aldrovandi was the first to observe, describe, and represent microscopic fossil beings on a polished surface of a calcareous marble-like block (*Musaeum Metallicum*, p. 887–888). Aldrovandi must have magnified the surface ~7–8 times with the help of a lens, long before the discovery of the microscope (Fig. 6). The fossils represented are different genera of benthic and planktic foraminifera (Vai, 2003a, p. 93). About a century later, and with the use of a microscope, Robert Hooke (1635–1703) unwittingly drew for the second time an extant foraminifer, *Rotalia beccari*, which he described as a "small

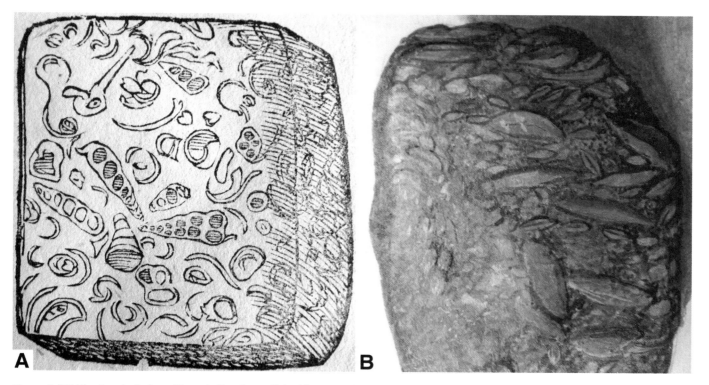

Figure 6. (A) The first depiction of foraminifera in a polished limestone block from the Aldrovandi Museum. Aldrovandi magnified the block 7–8 times with the help of a lens. From *Museum Metallicum* (posthumous print in 1648), p. 888 (Museo Geologico Giovanni Capellini, Bologna). (B) The first depiction of macro-foraminifera (Nummulitic limestone) in Aldrovandi's *Tavole Acquarellate* (hand-made water-colored plates) (Fondo Aldrovandi, Tavole, v. 7, by permission of Biblioteca Universitaria di Bologna) (photo Vai).

shell" of the British seas, meaning that it belonged to the shelly fauna. About a century and a half after that, Jacopo Bartolomeo Beccari formally discovered foraminifera in the sand from the beach at Rimini along the coasts of the Adriatic Sea (Vai, 2003b, p. 240). Thus, the species of *Rotalia*, mentioned above, was named after Beccari. This lens-aided "observation" of studied objects separates Aldrovandi from the "Aristotelians" and makes him the precursor of Galilei in both the technique of observation and scientific method (Battistini, 2000; Vai, 2003a).

Aldrovandi applied the uniformitarian principle over two centuries in advance of James Hutton and Charles Lyell. A good example of this is provided by the modern description and interpretation he gave of concretions encrusting fossil shells (Fig. 7):

> We figure different forms of Conchites rhomboides. The first facies drawn under nr. 1 and 2 is designated Concha rhomboids, not that one simple described in the Issue about Molluscs, but rather a Concha rhomboids bearing epigenetic stoney tubercles [...]

> The rhombic shells have epigenetic stony tubercles [...] So one should believe such warts to have been produced by some marine salt liquid which can easily lithify; and the warts attached to the shell grew to the size painted in the picture by continually adding new liquid. Thus its designation is Concha rhomboides tuberosa (*Musaeum Metallicum*, p. 468–469).

Aldrovandi's collection comprised fossil samples representative of foraminifers (both micro- and mega-forms), coelenterates (with single and colonial corals and tabulate corals), echinoids, brachiopods, bivalves, gastropods, ammonites, nautiloids, belemnitids, scaphopods, fishes, mammals (both marine and continental), plant remains, fossil traces and concretions (Sarti, 2003; Vai, 2003a), as well as possibly archaeocyathids.

Aldrovandi argued quite clearly about fossilization processes and the perfect reproduction of every feature of the living forms that such processes can achieve. Examples from elongate teeth-like shells or dentalia from a slender swampy plant and from fish teeth are quoted below after *Musaeum Metallicum*:

> The other image represents a fennel stem because it contains foramen, knots, veins, and nerves that are all seen in the vegetable stem; the fennel stem is hardened in stone (p. 854–855) (Fig. 8).

> These teeth-like shells are called Dentales. Similar exemplars are found in the mountains, not in bony, but in stony form (p. 847).

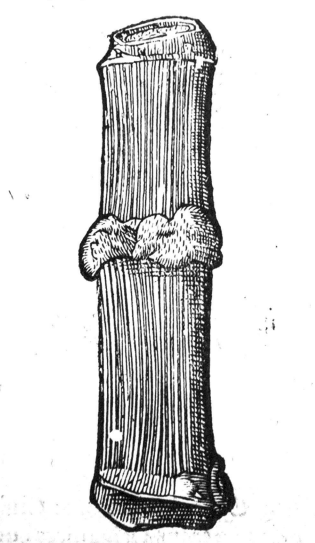

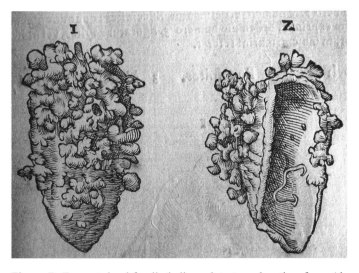

Figure 7. Encrusted subfossil shells and text explanation from Aldrovandi's *Museum Metallicum*, p. 468–469 (Museo Geologico Giovanni Capellini, Bologna) (photo Vai).

Figure 8. Fossilized fennel stem and text explanation from Aldrovandi's *Museum Metallicum*, p. 454–455 (Museo Geologico Giovanni Capellini, Bologna) (photo Vai).

Both quotes cast serious doubt on the opinion referred to by many authors about Aldrovandi's concept of fossils (Figs. 9–13). According to them, Aldrovandi thought the organic-like fossils excavated in the mountains to be a mysterious facsimile and not a lapidified replacement (plus infilling and cementation) of true shells or skeletons once living in the same places of excavation. Bartolomeo Ambrosini—who curated the posthumous publication of several of Aldrovandi's manuscripts—perhaps believed it, and thus modified Aldrovandi's text making the *Musaeum Metallicum* (1648) ambiguous and inconsistent: Aldrovandi, instead, did not believe it, as witnessed in the caption to a mammoth tooth figured in Aldrovandi's *Tavole Acquarellate* (Fig. 14).

Moreover, the notion of a fossil shell clearly distinct from its inner and outer mold (or cast) is clear to Aldrovandi who, describing a figured *Cochlites minor* remarked, "however, it is wholly composed of arenaceous matter"(*Musaeum Metallicum*, p. 471). And about the fossilization of fish teeth:

[…] in the hollow of the animal brain, a soft corruptible matter was contained: thus, to retain rot in the bone until petrifaction, it was necessary to fill the bone with a drying admixture […] (*Musaeum Metallicum*, p. 814)

The last two quotations show how close Aldrovandi was to anticipating Steno's concepts explaining fossilization in *De Solido* (Morello, 2003).

After Leonardo (Vai, 1986; 1988, p. 13, 22; 1995; 2003c), Aldrovandi was one of the first authors to use, in written form, conventional formation names that became commonly used in stratigraphy and regional geology two centuries later. This happens, for example, with the term "macigno," which was later widely employed in Tuscany after Targioni Tozzetti in 1751 (Vai, 1995) and was used by Aldrovandi for a sandstone outcropping in the region neighboring Bologna (*Musaeum Metallicum*, p. 454) (Vai, 2003a). To quote Aldrovandi, "it was similar to our arenaceous stone, which the Bolognese people call macigno."

ALDROVANDI'S INFLUENCE ON THE EUROPEAN SCIENTIFIC DEVELOPMENT

Several historians, especially on the occasion of the third centennial of Aldrovandi's death, have claimed his influence on Galilean experimentalism (Aldrovandi, 1907; Berlingozzi, 1907; Costa, 1907). There is no evidence that the two great scientists were in direct contact, in spite of a quite long lifetime overlap. Different facts suggest that Galilei (Fig. 15) knew about Aldrovandi and may have read his published works. Aldrovandi was particularly familiar with both the Pisa and Padua universities, where his teacher, Luca Ghini, had been a professor. In turn, Galilei was a student in Pisa and aimed at obtaining a teaching position in Bologna (Battistini, 2003, p. 30).

Galilei's best seller *Sidereus nuncius* (1610) was dedicated to Cosimo II, Grand Duke of Tuscany, whose father Ferdinando I was in correspondence with Aldrovandi asking for an exchange of naturalistic material to enlarge his own collections (BUB, FA, ms. 6, vol. I, six letters of Aldrovandi from September 1577 to July 1579, and five answers of the Grand Duke; see Ghigi, 1907; Tosi, 1989). Thus, Aldrovandi's writings and publications were available, as soon as they appeared, in the Grand Duchy library, which was open to Galilei. It is rather reasonable that Galilei's passionate observation of nature and his interest in the study of worthless objects was stimulated by Aldrovandi's natural philosophy and scientific method. The way and the aims followed by Aldrovandi in assembling his museum were not different from those inspiring Galilei in devising his own experiments. The research interests of the two savants were so different and mutually exclusive as to avoid a need in Galilei's works—a generation later—to quote or acknowledge an Aldrovandian influence, which is instead quite clear from a historical perspective.

In other respects, Aldrovandi and Galilei were quite different. Aldrovandi was a liberal and inclusive innovator in science, Galilei was more rigorous and intransigent. Throughout his life, Aldrovandi avoided a clear-cut anti-Aristotelian stance and still used a Scholastic logic, although biased toward direct observation, to discover Aristotelian mistakes and start post-Aristotelian science. Conversely, Galilei was a fierce anti-Aristotelian.

Aldrovandi's renowned works may have had an even larger influence on the education of Francis Bacon (1561–1626) (Fig. 16). In spite of little attention given to this matter by Bacon's biographers (e.g., Rossi, 1974, 2002; Jardine, 1974), Aldrovandi's most significant legacy is found in Bacon (Vai, 2003a). Aldrovandi's manuscripts and published works anticipated the thoughts, methodology, and even the key words and titles later expressed by Bacon. Bacon knew the learned Italians of his time well, and he exchanged correspondence and works with them, as many thinkers did during the European Renaissance and Baroque. As an example, Bacon was fascinated by the cosmologic Platonism of his contemporaries Bernardino Telesio and Tommaso Campanella.

There are many points of convergence between Aldrovandi's and Bacon's thoughts. Both thinkers professed natural philosophy as a whole, without distinction between the scientist and the humanist. Both wrote "histories" and treatises about a diversity of subjects and disciplines. Both were concerned with the expansion of sciences (*De augmentis Scientiarum*) and their classification. Bacon worked on the *Instauratio Magna Scientiarum*, a titanic unfulfilled enterprise, as Aldrovandi did on his encyclopedia. Bacon's last rhetorical dream, the *Sylva Silvarum* (collection of collections) is a perfect equivalent duplication in essence and title of Aldrovandi's *Selva Universale delle Scienze* [*Sylva Universalis Scientiarum*] or *Pandechion Epistemonicon* (BUB, FA, ms. 105). Beyond the title, the similarities extend to method and key ideas. Aldrovandi first, and Bacon just a generation later, claimed the "new science" to be based on the "direct observation of nature" (see quotations above), following an inductive method "based on experience" and cooperation (teamwork). The new science had to be "global in scope," and the Aldrovandi Museum, a public institution for scientific research, was used for teaching the

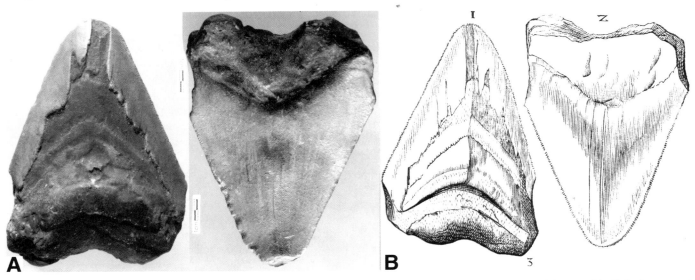

Figure 9. Fossil shark tooth (*Carcharodon megalodon* Ag.) (A) from the Aldrovandi Museum, figured together with ceraunids (lightning stones) in the *Museum Metallicum* (p. 611), but "better referred to as glossopetres" (shark teeth) in Aldrovandi's opinion. Notice the drawings (B) derived from the specimen still preserved (Museo Geologico Giovanni Capellini, Bologna) (photo Ferrieri and Vai).

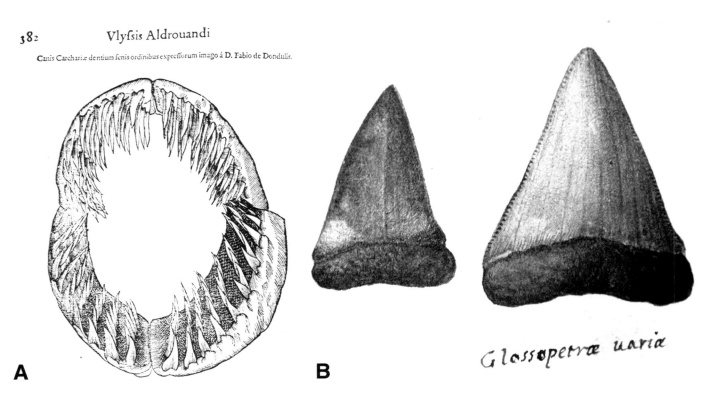

Figure 10. The skeletal head of a modern shark (A) from the Aldrovandi Museum (figured in *De Piscibus*, Aldrovandi 1618, p. 382) and various glossopetres painted in Aldrovandi's *Tavole Acquarellate* (B) (hand-made water-colored plates) (Fondo Aldrovandi, Tavole, v. 7, by permission of Biblioteca Universitaria di Bologna) (photo Vai).

Figure 11. Permian brachiopod *Productus geinitzianus* Kon. (A) from the Aldrovandi Museum, classified as *Hieracites* in *Museum Metallicum* (p. 445, Fig. 1), and a spiriferoid brachiopod (B), now lost, painted in Aldrovandi's *Tavole Acquarellate* (hand-made water-colored plates) (Fondo Aldrovandi, Tavole, v. 7, by permission of Biblioteca Universitaria di Bologna) (photo Vai).

new science to students coming from all parts of Europe. Once back in their own countries, those students would find new natural objects not yet investigated, so that "science would *expand* (the same verb used by both thinkers) for the use of scholars and Christian nations."

Bacon's works, almost all published during his lifetime, had a much wider audience and dissemination than Aldrovandi's publications, letters, and manuscripts. Moreover, Aldrovandi's works circulated mostly among scientists and natural philosophers, whereas Bacon's works also reached the strictly philosophical circles. This might explain why such important derivative relations of Bacon from Aldrovandi were not noticed by Bacon's biographers (Rossi, 1974; Jardine, 1974).

The immediate and quite spontaneous support given by the Bologna scientific academies to Bacon's manifesto of the Royal Society in the second half of the seventeenth century and the beginning of the eighteenth century (Vai, 2003b) demonstrates how Aldrovandi's lesson about the "new experimental science" had influenced both Bacon and the Bologna scientific school. Among the Bologna brain trust were Ferdinando Cospi (1606–1686), Giandomenico Cassini (1625–1712), Geminiano Montanari (1633–1687), Marcello Malpighi (1628–1694), Domenico Guglielmini (1655–1710), Francesco Maria Grimaldi (1618–1663), Luigi Ferdinando Marsili (1658–1730), Anton Maria Valsalva (1666–1723), Iacopo Bartolomeo Beccari (1682–1766), Giuseppe Monti (1682–1760) and many others representing most of the fields of science.

Figure 12. Various *Belemnites* samples painted in Aldrovandi's *Tavole Acquarellate* (hand-made water-colored plates) (Fondo Aldrovandi, Tavole, v. 7, by permission of Biblioteca Universitaria di Bologna) (photo Vai).

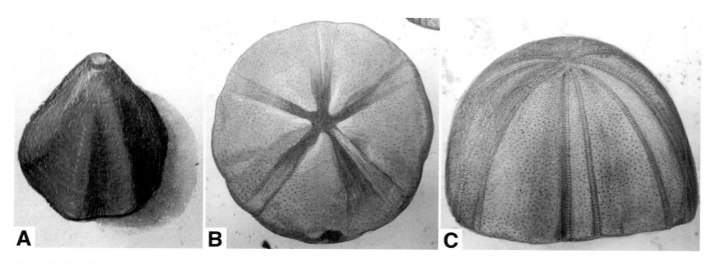

Figure 13. Fossil brachiopod *Terebratula* (A) and echinoids (B, C) painted in Aldrovandi's *Tavole Acquarellate* (hand-made water-colored plates) (Fondo Aldrovandi, Tavole, v. 7, by permission of Biblioteca Universitaria di Bologna) (photo Vai).

Figure 14. Front view of a mammoth tooth painted in Aldrovandi's *Tavole Acquarellate* (hand-made water-colored plates) (Fondo Aldrovandi, Tavole, v. 1, by permission of Biblioteca Universitaria di Bologna). The magnified caption reads, "Tooth of a marine wild beast which is usually excavated from the interior of the Earth in Russia and Prussia after having been converted into lapideous matter," telling how Aldrovandi was convinced of the organic origin of fossils (photo Vai).

Figure 15. Portrait of Galileo Galilei, Archivio storico Università di Bologna (photo Mattei-Zannoni).

Comparisons between the geological vision of Aldrovandi and Nicolaus Steno (1638–1686), who was about three generations younger, are quite uncommon. Steno, who is unanimously considered as one of the founding fathers of modern geology, was called to Florence in 1666 by the Tuscan Grand Duke Ferdinand II (in power 1621–1670) to be his "physician" and assistant for his naturalistic collections. Aldrovandi also had close relations to Tuscany and the Grand Duchy (Tosi, 1989; Olmi, 2001). He was in correspondence with the Grand Dukes Francesco I and his younger brother Ferdinand I exchanging samples, materials, and information to enrich the Florence Cabinet (see above). After Aldrovandi's death, the keepers of the Aldrovandi Museum maintained contacts with the Grand Duchy. Aldrovandi's *Monstrorum Historia* was published posthumously in 1642, with a dedication to the same Ferdinand II who was the mentor of Steno. It is thus reasonable to assume that Aldrovandi's other books and copies of some of his important manuscripts were available in the Duchy library.

Steno shared Aldrovandi's experimental attitude, the priority given to personal observation, the inductive method, and the logical-critical check of any previous scientific assessment either from ancient (Aristotle) or modern (Descartes) savants (Vai, 2003b, p. 225–226).

The two scientists have several geological points of contact and share very close conclusions on several specific issues, for example, the process leading to the lithification and fossilization of fish and wood remains or the genesis of concretions. These are two of the key points developed in Steno's *Prodromus* (1669), which may indicate an Aldrovandian influence not explicitly acknowledged, possibly because of the preliminary scope and the extremely synthetic style of the *Prodromus* (Vai, 2003a).

Once in Florence, Steno was in contact with Bologna University, especially with Marcello Malpighi, a famous anatomist and one of the successors of Aldrovandi. This suggests that Steno had many opportunities to study Aldrovandi's works and be stimulated by them. Steno wrote a letter to Wilhelm Leibniz (1646–1716) in 1677 to explain to him the rationale for his conversion to the Catholic faith ten years earlier. This letter contains, in a few sentences, the legacy of Aldrovandi's scientific thoughts and the

Figure 16. Portrait of Francis Bacon (Archivio Storico Università di Bologna, photo Mattei-Zannoni).

method, which was accepted and best interpreted by Steno (Vai, 2003b), that was extended through the Bolognese school to most of the Italian geological community in the seventeenth century.

In fact, Aldrovandi's influence on the development of science and geology in Italy was enormous and diffuse. A "liberal diluvianism" in geology and a "balanced integration of science, philosophy, and religion" in the wider field of the new science are examples of his impact (Vai, 2003b). The Aldrovandian integration trend resulted in a continued gentlemen's agreement between archbishops, cardinals, and popes of Bolognese origin on one side and the city's leading thinkers on the other. This was established before and was maintained even after Galilei's trial. It was referred to as "Emilian erudition" or "Galilean Catholicism" (Raimondi, 1978) and was derived from an earlier "Aldrovandian Catholicism" (Vai, 2003b, p. 228). This open-minded Catholic scientific attitude developed around Bonaventura Cavalieri (1598–1647) and other pupils of Galilei, and it was propagated by savants such as Marcello Malpighi, Luigi Ferdinando Marsili, Benedetto Bacchini (1651–1721), Jean Mabillon (1632–1707), Ludovico Antonio Muratori (1672–1750), ultimately expanding from Bologna and Emilia into Tuscany and most of Italy.

Aldrovandi's approach also might have influenced the more opportunistic "metaphysical neutrality" adopted by Bacon and the Royal Society of London in the more dogmatic Anglican realm (Cavazza, 1990; Vai, 2003b, p. 226).

The "liberal Diluvianism," so visibly represented in Italy by the scientists of the *Istituto delle Scienze* of Bologna and practiced in many other regions of Italy except Tuscany during the seventeenth to eighteenth century, was also a heritage of Aldrovandi's approach, harmoniously integrating science, philosophy, and religion as typically accomplished by Steno.

Aldrovandi's influence can even be traced to Kircher, who implemented Aldrovandi's lesson about the value of having a wealth of illustrations to enrich and qualify books. Also, Kircher's "geocosm" (1656) (Vai, 2004) is a clear scientific neologism deriving from Aldrovandi's "geology."

OBSERVATION AND MYTH IN ALDROVANDI'S NATURAL HISTORY

From the second half of the eighteenth century until recent times, particularly outside Italy, Ulisse Aldrovandi's role in the history of science was no longer adequately considered. Aldrovandi was dismissed as a laborious and patient encyclopedist and collector, destitute of the critical faculties, who had hardly made any attempt in his monumental work to distinguish between the true and the fabulous, the important and the trivial. It was acknowledged that his work contained information of great interest to the naturalist, but part of it was considered of no scientific value.

Buffon was exasperated by the fact that Aldrovandi, in several of his works, placed detailed descriptions, narrative, trivial information, myths, and legends side by side. For example, Aldrovandi's extraordinary collection of topics in his work on snakes (*Serpentum et draconum historiae*, 1640) included heraldic applications, mythological creatures, dreams, use in human diet, mating habits, and fables as subheadings of a single chapter. Buffon was moved to conclude with an eighteenth century frustration we could share today, "Let it be judged after what proportion of natural history is to be found in such a hotchpotch of writing. There is no description here, only legend." A clarification to that harsh assessment is in order (see also above under the fifth subheading).

Aldrovandi's work must be placed in the backdrop of a system of knowledge quite different from that which governed the work of late eighteenth century European scientists such as Buffon. Aldrovandi, remarks Foucault (1970), "was neither a better nor a worse observer than Buffon; he was neither more credulous than he, nor less attached to the faithfulness of the observing eye or to the rationality of things. His observation was simply not linked to things in accordance with the same system or by the same arrangement of the *episteme* [i.e., the collections of beliefs and assumptions which create worldviews and practices in a particular epoch]." In other words, in the age of Aldrovandi, there was still no clear-cut distinction between detailed observations and narrative or myths, "it was all *legenda*—things to be read" (Foucault, 1970), which Aldrovandi duly reported in his obsessive struggle toward completeness. Yet a number of novel approaches set Aldrovandi far apart from his contemporaries.

The volumes published either during his life (*Ornithologiae*, 1509–1603; *De animalibus insectis*, 1602; *De Piscibus*, 1613) or posthumously (e.g., *Museum metallicum*, 1648) typify Aldrovandi's approach to the natural sciences. In several instances, these are the first complete works that attempt to correct and integrate the preexisting body of literature with as much direct observation as possible, in the spirit of a new, modern, scientific, and experimental attitude. The scientific and explanatory approach particularly emerges in the wealth of detailed illustrations, which the author conceived as a necessary complement to the text rather than as an ornamental addition. And the illustrations are always complemented by the samples preserved in the museum as types. On the other hand, as a scientist of the sixteenth century, he was necessarily dependent, for the parts related to exotic species and objects, on the accounts of earlier or foreign naturalists, thus introducing some degree of inaccuracy.

The presence of deformed humans, dragons, unicorns, and other mythological creatures in a few of Aldrovandi's works has attracted much criticism over the centuries. This led to the definition of Aldrovandi as a scholar of vast erudition but also somewhat more accepting and uncritical of tales and myths about human and animal monstrosities than one might expect. Once again, this evaluation does not take into account the cultural atmosphere of the time, when facts and narrative were not yet clearly separated, savants and patrons alike were fascinated by bizarre natural specimens (*lusus naturae*, jokes of natures), and the emotional neutrality of modern science was not yet operational. This should not detract attention from Aldrovandi's concern with actual observation and with careful descriptions from specimens, which set him apart from his contemporaries even in his taste for oddities. For example, recent research has demonstrated that at least some of the "monsters" included in *Monstrorum Historia* (1642) are meticulous descriptions and illustrations of dermatological pathology (e.g., Ruggieri and Polizzi, 2003).

Aldrovandi's detailed color tables had an immediate influence, not only on naturalists, but on others as well. In the field of sixteenth-century painting, while the taste for Mannerism developed in various parts of Europe, the masters most attentive to natural objects were the painters who were active in Flanders and in northern Italy, areas which share primacy in the birth of still-life painting. In the second half of the sixteenth century, it was thanks to Ulisse Aldrovandi and his relations with the Florentine Court of the Medici that the naturalistic image developed in the figurative arts (e.g., in the paintings of Bartolomeo Passarotti, 1529–1592, and Jacopo Ligozzi, 1547–1627). It was then that painters attempted to create images as faithful as possible to the real thing.

Throughout his life, Aldrovandi demonstrated his bent (or rather obsession) for taxonomy—from the simple list of Roman antiquities written as a young man to the complex classifications contained in the publications and manuscripts of his maturity. Inheritor of the classical tradition, he forged an observational approach to nature that often left him at odds with the persistent authority of Aristotle, Pliny, and other classical authors. Aldrovandi personifies the transition between the mere collection of interesting natural objects and modern, critical science. As such, he has had bad fortune with some historians of science who found him difficult to categorize: last of the Medieval and Renaissance collectors or first of the modern scientists?

CONCLUSIONS

Before the dangerous breakup of the humanistic cultural unity that lead to the modern division of science from the humanities, Aldrovandi practiced and theorized a new science based on observation, experimentalism, and logical-critical review of the whole past knowledge. He was among the first to take into consideration humble natural and artificial objects worthy of scientific study and taxonomic classification. He also had visions of their correct and widened use for the education and practical needs of mankind.

In this sense, Aldrovandi anticipated the bulk of the scientific revolution usually epitomized as Galilean or, even more, Baconian. Their major differences lie in the strategic emphasis given to evolution versus revolution, in the naturalistic-qualitative versus mathematical-quantitative approach, in the nonideological versus philosophical-political aim.

The scientific method of Aldrovandi was the same followed later by Bacon. Both Bacon and Aldrovandi implemented a probabilistic (statistical) approach requiring large collections of data, whereas Galilei's method was based on a deterministic mathematical treatment of less data organized in processes with description followed by a check.

Aldrovandi set priorities in many fields of the natural sciences (such as geology, botany, zoology, anatomy, and embryology) and humanities (museology, bibliology, biblioteconomy, teaching, publishing, art history, religion, epistemology, and science methodology).

Aldrovandi's most outstanding material achievement, intimately connected with his theoretical elaboration, was to establish the first natural history museum associated with the first scientific library of worldwide scope. Both the museum and library were open to the public for direct educational purposes, and they formed a whole *Studio*, which was conceived as a site of scientific research to develop "the new science." Thousands of natural objects were carefully preserved in the museum and became the standard types of the related species after having been studied, compared, described, classified, and ordered inside specific disciplines and knowledge domains by Aldrovandi, who largely anticipated Linnaeus' approach. The Aldrovandi Museum, and the related *Studio,* served as a prototype for an international net of knowledge and the exchange of ideas continuously propagated by the means of Aldrovandi's correspondents and students located throughout all of Europe and the New World. Aldrovandi's entire life and work can be characterized by his amazingly modern global view of planning the foundation and advancement of science.

Aldrovandi's peers in the late Renaissance were well aware of his leadership and contributed actively to expand his thoughts and scientific method. His material achievements, the Museum and *Studio*, remained outstanding scientific institutions for centuries. Aldrovandi's use of types and binomial nomenclature in classifying the remains of living and fossil organisms, his key role in instituting museums of natural history and scientific libraries, and his impact on the epistemology of science and the scientific method are all duly recognized and acknowledged by the large majority of experts now, but they have not always been. Unfortunately, indeed, less than a tenth of Aldrovandi's manuscripts were printed, and those were not even the most important contributions, thereby limiting the acknowledgment of his priorities and of the core of his heritage. Aldrovandi cannot be identified with any revolutionary or spectacular discovery, yet his achievements as a teacher, the author of landmark volumes and manuscripts, and as cultural and research manager and innovator of his times earns him a special place among the fathers of modern science.

ACKNOWLEDGMENTS

We are indebted to Biancastella Antonino, director of the Biblioteca Universitaria di Bologna (BUB), for providing access to Aldrovandi's manuscripts, stored in the Fondo Aldrovandi (FA), and for granting permission to publish original reproductions from them. Reviewers have greatly improved our text.

REFERENCES CITED

Accordi, B., 1980, Michele Mercati (1541–1593) e la Metallotheca: Geologica Romana, v. 19, p. 1–50.

Accordi, B., 1981, Ferrante Imperato (Napoli, 1550–1625) e il suo contributo alla storia della geologia: Geologica Romana, v. 20, p. 43–56.

Adams, F.D., 1938, The birth and development of the geological sciences: Baltimore and London, Williams & Wilkins Company, v + 506 p. (reprinted 1954).

Aldrovandi, L., 1907, Parole pronunciate il 12 Giugno 1907: Ulisse Aldrovandi nel III Centenario dalla sua morte, Bologna, Stabilimento Poligrafico Emiliano, 4 p.

Battistini, A., 2000, Galileo e i gesuiti. Miti letterari e retorica della scienza: Milano, Vita e Pensiero, 419 p.

Battistini, A., 2003, Bologna's four centuries of culture from Aldrovandi to Capellini, *in* Vai, G.B., and Cavazza, W., eds., Four centuries of the word geology: Ulisse Aldrovandi 1603 in Bologna: Bologna, Minerva Edizioni, p 13–63.

Berlingozzi, 1907, Aldrovandiana. Echi delle Feste Bolognesi nel Giugno 1907: Montevarchi, Tipografia Ettore Pulini, 18 p.

Carapezza, M., 1984, La trame concettuali della geologia: Energia e Materie Prime, v. 36, p. 53–57.

Cavazza, M., 1990, Settecento inquieto. Alle origini dell'Istituto delle Scienze di Bologna: Bologna, Il Mulino, 281 p.

Costa, E., 1907, Ulisse Aldrovandi e lo Studio Bolognese nella seconda metà del secolo XVI: Bologna, Stabilimento Poligrafico Emiliano, 95 p.

Dean, D.R., 1979, The word geology: Annals of Science, v. 36, p. 35–43.

De Bellis, N., 1998, Ordine e reperibilità degli oggetti nelle collezioni mineralogiche di Ulisse Aldrovandi: Museologia scientifica, v. 14, p. 279–297.

De Toni, G.B., 1907, Spigolature aldrovandiane VII. Notizie intorno ad un erbario perduto del medico Francesco Petrollini (anteriore al 1553) e contribuzione alla storia dell'erbario di Ulisse Aldrovandi: Nuovo Giornale botanico italiano, v. 14, n. 4, p. 506–518.

Ellenberger, F., 1988, Histoire de la géologie: Paris, Lavoisier Tec Doc, v. 1, viii + 352 p.

Fantuzzi, G., 1774, Memorie della vita di Ulisse Aldrovandi medico e filosofo bolognese, Bologna, per le stampe di Lelio Dalla Volpe, 264 p.

Fantuzzi, G., 1771, 1781, Notizie degli scrittori bolognesi: Bologna, I tomo, p. 165–190.

Findlen, P., 1989, The museum: its classical etymology and Renaissance genealogy: Journal of the History of Collections, v. 1, p. 59–78.

Findlen, P., 1994, Possessing nature: museums, collecting, and scientific culture in early modern Italy: Berkeley, University of California Press, 449 p.

Foucault, M., 1970, The order of things: An archaeology of the human sciences: London, Tavistock, 365 p.

Frati, L., (coll. Ghigi, A. e Corbelli, A.), 1907a, Catalogo dei manoscritti di Ulisse Aldrovandi: Bologna, Nicola Zanichelli, 288 p.

Frati, L., 1907b, La vita di Ulisse Aldrovandi cominciando dalla sua natività sin' a l'età di 64 anni vivendo ancora, in Intorno alla vita e alle opere di Ulisse Aldrovandi, *in* Studi per il III centenario dalla morte (collective vol.): Bologna, Libreria Treves, p. 1–27.

Ghigi, A., 1907, Catalogo abbreviato dei manoscritti di Ulisse Aldrovandi ordinati per materie, *in* Frati, L., (cur.), Catalogo dei manoscritti di Ulisse Aldrovandi: Bologna, Nicola Zanichelli, p. 211–244.

Gortani, M., 1931, Bibliografia geologica italiana. Parte prima. Opere di carattere generale. Cap. I, Per la storia: Giornale di Geologia, v. 6, p. 9–38.

Gortani, M., 1963, Italian pioneers in geology and mineralogy: Journal of World History, v. 7, no. 2, p. 503–519.

Jardine, L., 1974, Francis Bacon: Cambridge, Cambridge University Press, 267 p.

Kázmér, M., 2002, Johannes Kentmann's minerals from Hungary in 1565: Földtani Közlöny, v. 132, no. 3, p. 457–470.

Lancisi, G.M., ed., 1717, Metallotheca Vaticana Michaelis Mercati [1574]: Romae, apud Io. Mariam Salvioni, lxiv + 378 p.

Mattirolo, O., 1897, L'opera botanica di Ulisse Aldrovandi (1549–1605): Bologna, R. Tipografia Frat. Merlani, 137 p.

Mauro, L., 1556, Le antichità della città di Roma: Roma, appresso Giordano Ziletti, 316 p.

Morello, N., 2003, The question on the nature of fossils in the 16th and 17th centuries, *in* Vai, G.B., and Cavazza, W., eds., Four centuries of the word geology: Ulisse Aldrovandi 1603 in Bologna: Bologna, Minerva Edizioni, p. 127–151.

Oldroyd, D.R., 1996, Thinking about the Earth: a history of ideas in geology: London, Athlone, and Cambridge, Massachusetts, Harvard University Press, 410 p.

Olmi, G., 1976, Ulisse Aldrovandi. Scienza e natura nel secondo Cinquecento: Trento, Quaderni di storia e filosofia della scienza, Bologna, Clueb, 129 p.

Olmi, G., 1985, Science, honour, metaphor: Italian cabinets of the sixteenth and seventeenth century, *in* Impey, O., and MacGregor, A., eds., The origins of museums: Oxford, Clarendon Press, p. 1–16.

Olmi, G., 2001, Il collezionismo scientifico, *in* Simili R., (cur.) Il teatro della natura di Ulisse Aldrovandi: Bologna, Editrice Compositori, p. 20–50.

Raimondi, E., 1978, Il barometro dell'erudito, *in* Scienza e letteratura: Torino, Einaudi, p. 57–84.

Rossi, P., 1974, Francesco Bacone: dalla magia alla scienza: Torino, Einaudi, 366 p.

Rossi, P., 2002, Francis Bacon. I grandi della scienza: Le Scienze, v. 30 (dic. 2002), 95 p.

Ruggieri, M., and Polizzi, A., 2003, From Aldrovandi's "*Homuncio*" (1592) to Buffon's girl (1749) and the "Wart Man" of Tilesius (1793): antique illustrations of mosaicism in neurofibromatosis?: Journal of Medical Genetics, v. 40, p. 227–232, doi: 10.1136/jmg.40.3.227.

Sarti, C., 2003, The geology collections in Aldrovandi's Museum, *in* Vai, G.B., and Cavazza, W., eds., Four centuries of the word geology: Ulisse Aldrovandi 1603 in Bologna: Bologna, Minerva Edizioni, p. 153–167.

Simili, R., ed., 2001, Il teatro della natura di Ulisse Aldrovandi: Bologna, Editrice Compositori, 143 p.

Tosi, A., 1989, I volti della scienza: il ritratto dello scienziato tra il XVI e XIX secolo, *in* Simili, R., ed., Il teatro della natura di Ulisse Aldrovandi: Bologna, Editrice Compositori, p. 121–128.

Tugnoli Pàttaro, S., 1977, La formazione scientifica e il "Discorso naturale" di Ulisse Aldrovandi, Trento, Quaderni di storia e filosofia della scienza: Bologna, Clueb, 115 p.

Tugnoli Pàttaro, S., 1981, Metodo e sistema delle scienze nel pensiero di Ulisse Aldrovandi: Bologna, Clueb, 251 p.

Tugnoli Pàttaro, S., 2000, Osservazione di cose straordinarie. Il De observatione foetus in ovis (1564) di Ulisse Aldrovandi: Bologna, Clueb, 305 p.

Tugnoli Pàttaro, S., 2001, Filosofia e storia della natura in Ulisse Aldrovandi, in Simili R., ed., Il teatro della natura di Ulisse Aldrovandi, Bologna, Editrice Compositori, p. 9–19.

Vai, G.B., 1986, Leonardo, la Romagna e la geologia, in Marabini, C., and Della Monica, W. eds., Romagna. Vicende e protagonisti: Bologna, Edison ed., v. 1, p. 30–52.

Vai, G.B., 1988, A field trip guide to the Romagna Apennine geology. The Lamone valley, in De Giuli, C, and Vai, G.B., eds., Fossil vertebrates in the Lamone valley, Romagna Apennines: Faenza, Litografica Faenza, Field Trip Guidebook, International Workshop, p. 7–37.

Vai, G.B., 1995, Geological priorities in Leonardo da Vinci's notebooks and paintings, in Giglia, G., Maccagni, C., and Morello, N., eds., Rocks, fossils and history: Proceedings 13th INHIGEO Symposium, Pisa-Padova, 1987, Firenze, Festina Lente Edizioni, p. 13–26.

Vai, G.B., 2003a, Aldrovandi's will: introducing the term 'geology' in 1603, in Vai, G.B., and Cavazza, W., eds., Four centuries of the word geology: Ulisse Aldrovandi 1603 in Bologna: Bologna, Minerva Edizioni, p. 64–111.

Vai, G.B., 2003b, A liberal Diluvianism, in Vai, G.B., and Cavazza, W., eds., Four centuries of the word geology: Ulisse Aldrovandi 1603 in Bologna: Bologna, Minerva Edizioni, p. 220–249.

Vai, G.B., 2003c, I viaggi di Leonardo lungo le valli romagnole: riflessi di geologia nei quadri, disegni e codici, in Leonardo, Machiavelli, Cesare Borgia: Arte, Storia e Scienza in Romagna (1500–1503): Roma, De Luca Editori d'Arte, p. 37–47.

Vai, G.B., ed., 2004, Athanasii Kircheri Mundus Subterraneus in XII Libros digestus, Editio Tertia, 1678, Bologna, Arnaldo Forni Editore, 22 + xxiv + 366 + vi +x + 507 +xi p. (photostatic edition with forward by the editor, and presentations by Nicoletta Morello and Umberto Eco).

MANUSCRIPT ACCEPTED BY THE SOCIETY 17 JANUARY 2006

Kircher and Steno on the "geocosm," with a reassessment of the role of Gassendi's works

Toshihiro Yamada*

University of Tokyo, Graduate School of Arts and Sciences, Department of History and Philosophy of Science, 3-8-1, Komaba, Meguro-ku, Tokyo, 153-8902 Japan

ABSTRACT

Examining the works of Athanasius Kircher and Nicolaus Steno allows similarities and differences to be drawn between their theories of Earth. This is aided by paying particular attention to the role of the French atomist Pierre Gassendi. With his friend Nicolas-Claude Fabri de Peiresc, Gassendi had a significant impact on Kircher's career and his thinking, and his work was read and noted by Steno in his student years in Copenhagen. Later, in the 1667 treatise *Canis*, Steno also appraised Gassendi's ideas on the origin of stones. Kircher's experiences of volcanism and earthquakes, gained during his expedition into southern Italy in 1637–1638, led him to formulate his theory of Earth in the early 1640s, when his *Magnes* was to be published. Completion of his theorizing about Earth was delayed, however, until publication of *Mundus subterraneus* (1665), in which he developed his concept of the "geocosm." Steno probably met Kircher in 1666, and they are known to have corresponded on theological topics. In his *Prodromus* (1669), Steno criticized Kircher's idea of the "organic" growth of mountains. Steno adopted Descartes' idea of "collapse tectonics" and the formation of strata. Kircher's influence on Steno should not be neglected, however, given Steno's substantial excerpts from Kircher's *Magnes* in his manuscript. In fact, although Steno rejected the idea of a plastic force in his *Prodromus*, he may well have used Kircher's idea on magnetism to explain the growth of mineral crystals. Thus, given the usual wide acceptance of Cartesian influence on Steno, the historiography of geosciences may be appropriately and usefully revised by considering the role of the works of such figures as Gassendi and Kircher.

Keywords: Descartes, Gassendi, classification of terrestrial objects, Kircher, geocosmos, Collegio Romano, Etna, Vesuvius, Chaos manuscript, *Magnes, Mundus subterraneus*, magnetism, Peiresc, Steno, crystal growth, action of fluids, theories of Earth.

INTRODUCTION

René Descartes' (1596–1650) theory of Earth in his *Principles of Philosophy* (*Principia philosophiae*, 1644) has been assumed to have been the principal mechanical scheme for seventeenth- and eighteenth-century theories of Earth. In terms of this scheme, Nicolaus Steno (1638–1686) (Fig. 1), for example, has been seen as a Cartesian, as have Burnet, Leibniz, and Buffon. Indeed, after representing his theory of matter and space, Descartes summarized his ideas about Earth so clearly in part 4 of his *Principia* that historians of science have regarded it as being clearly influential on Steno (Burke, 1966; Metzger, 1969; Schneer, 1971; Roger, 1982; Emerton, 1984; Laudan, 1987, p. 38–43; Oldroyd, 2000, p. 618; Vaccari, 2002, p. 12–16). In

---

*E-mail: tosmak-yamada@muf.biglobe.ne.jp.

Yamada, T., 2006, Kircher and Steno on the "geocosm," with a reassessment of the role of Gassendi's works, *in* Vai, G.B., and Caldwell, W.G.E., The Origins of Geology in Italy: Geological Society of America Special Paper 411, p. 65–80, doi: 10.1130/2006.2411(05). For permission to copy, contact editing@geosociety.org. ©2006 Geological Society of America. All rights reserved.

Figure 1. Portrait of Nicolaus Steno (private collection, reproduced by courtesy of Angelo Livi, Firenze).

Figure 2. Portrait of Athanasius Kircher, from *Mundus* (1665) (reproduced by courtesy of the History of Science Collections, University of Oklahoma Libraries).

addition, Descartes adopted a version of the Copernican system, despite his cautious and ambivalent language on the topic. Obviously, Descartes provided a supposedly deductive account of Earth's formation on the basis of a heliocentric cosmology, in which a sun, covered by sunspot material, is captured by another solar system and becomes one of its planets. Thus, an Earth-like planet, as a member of the solar system, has subterranean fire.

If we adopt this kind of successful Cartesian theory as belonging to the mainstream of theories of Earth in the scientific revolution, the views of Athanasius Kircher (1602–1680) (Fig. 2) tend to be regarded as "primitive" and "organic" or simply ignored. Moreover, Kircher's cosmology was based on scholastic Aristotelian doctrines and the eclectic system of Tycho Brahe. Thus, we can easily understand why the Cartesian Steno criticized Kircher's theory of Earth.

On the other hand, Steno's geological contribution can be aligned with Pierre Gassendi's (1592–1655) (Fig. 3) work and the southern French intellectual tradition, as the geohistorian François Ellenberger has suggested (Ellenberger, 1988, p. 224–232); and on closely examining Gassendi's philosophical contributions, various matters pertaining to studies of Earth are found therein (Halleux, 1982, p. 117, 130; Yamada, 2003, p. 81–82; Yamada, 2004). It seems reasonable, therefore, to revisit the historiography of the emergence and development of the early modern geological thought, with particular attention to the Kircher-Steno relationship.

In such a context, certain relationships seem perfectly obvious. For example, on the one hand, Steno certainly took up Descartes' ideas of strata formation and collapse tectonics. On the other hand, Steno's interest in meteorological and terrestrial er's magnetism book; fourth, by surveying Kircher's work on the subterranean world; and last, by offering a comparison of the work of Kircher and Steno with suggestions of the possible influence by the former on the latter.

KIRCHER IN SOUTHERN EUROPE: THE BEGINNING OF HIS STUDY OF EARTH

After publishing his maiden work on a physical inquiry into the "magnetic art" in 1631, Kircher fled from the vortex of war in Germany to the shelter of southern France (Kangro, 1973; Reilly, 1974; Godwin, 1979, p. 9–15; Bach, 1985, p. 1–55; Fletcher, ed., 1988, p. 1–15). As a quasi-political refugee, Kircher arrived at Avignon and was received there by the politician, humanist, and friend of Galileo, Nicolas-Claude Fabri de Peiresc (1580–1637) (Fig. 4). Peiresc, who was an antiquarian as well as a patron of the new science and a student of the problem of longitude, inspired Kircher to study ancient Egyptian documents (Miller, 2004). Later, in his works on Egyptian studies, especially *Egyptian Language Restituted* (*Lingua Aegyptiaca restitute*, 1644) and *Pamphilian Obelisk* (*Obeliscus Pamphilius*, 1650), Kircher referred to Peiresc's role with respect (Gassendi,

Figure 3. Portrait of Pierre Gassendi (reproduced by courtesy of the Library of the University of Tokyo).

phenomena is clearly indicated by his so-called "Chaos manuscript" (Steno, 1659; Ziggelaar, ed., 1997), written during his student years in Copenhagen. In parts of this manuscript, we can find extensive transcriptions from Kircher's *Magnes* (second edition, 1643)—a work in which many terrestrial phenomena are treated.

To be more specific, what seems to be most urgently called for is a re-evaluation of Kircher's position in the history of geosciences, especially from the perspective of the relevance of his work to that of Steno. This is done here, first, by reviewing the background of Kircher's geoscientific works, including the Kircher-Gassendi relationship; second, by tracing the "imprints" of Gassendi in Steno's manuscript and published work; third, by examining Steno's transcription from the Kirch-

Figure 4. Portrait of Nicolas-Claude Fabri de Peiresc (reproduced by courtesy of the Center for Historical Social Science Literature, Hitotsubashi University, Tokyo).

1657, v. 2, p. 283–291). Kircher was also introduced to Gassendi through Peiresc, and Gassendi called Kircher "a man of truly great learning (*vir eruditionis oppido magnae*)" (Gassendi, 1658, v. 5, p. 313a).

During his Mediterranean voyage of 1633, which was part of Kircher's journey to Vienna to take up the job of imperial mathematician, a shipwreck forced him to visit Rome, where he was graciously welcomed by the Collegio Romano and succeeded to the professorial post of Christoph Scheiner (1575–1650). The Collegio Romano was the principal Jesuit center for education in the tradition of the famous mathematician and astronomer Christoph Clavius (1538–1612). The center influenced missionary work and secular education and was the hub of the Jesuits' global network (Sasaki, 2003, ch. 1). Undoubtedly, the position made Kircher's name as the "polyhistor" of the age, helped by the Collegio's numerous visiting scholars, the connections with all parts of Europe, and the concentration in Rome of information from all parts of the known world. The information about the data of terrestrial magnetism given in his work (1641, 1643) is even now referred to in scientific papers (Jonkers et al., 2003).

From his position in Rome, Kircher had the opportunity to visit southern Italy, where he made important excursions from 1637 to the end of 1638. Kircher and his companions crossed the Messina Strait to Sicily and then went to the islands of Sicily and Malta. In Sicily, he visited the historically important site of Syracuse, investigating the probable position of the mirrors of Archimedes and examining the possibility of burning enemy ships with them (Middleton, 1961, p. 535–536). On the return journey to Rome, he saw eruptions of Etna and Stromboli and ventured to the top of Vesuvius, descending into the crater in peril of his life. He also experienced a large earthquake in Calabria, about which he later wrote that his mind and those of his companions had been shaken by an event too great to be described in words (Kircher, 1665, v. 1, p. 221–222).

Evidently, these experiences triggered Kircher's plan ca. 1640 to write about Earth, as will be discussed later. The first of such writings appeared as the second part of his *Ecstatic Journey* (*Iter exstaticum*, 1657) subtitled as a prolegomenon to the "Subterranean World." The great twelve-volume book on the *Subterranean World* (*Mundus subterraneus*) was published in 1665 and 1678 (Fig. 5), in which Kircher developed his geocosmos concept about the globe. Additionally, the early encyclopedic work *Loadstone or on the Magnetic Art* (*Magnes sive de arte magnetica*, 1641, 1643, 1654), in which he discusses meteorological and terrestrial phenomena, deserves attention.

Besides these works, other articles in Kircher's published writings are interesting from the standpoint of the theory of Earth: for example, the natural history of China in *China Illustrated* (*China illustrata*, 1667), which reproduced the stone monument unearthed at Xi'an in 1625 supposedly indicating a former Nestorian trend in China; *Latium* (1671), the description of the ancient city, reporting the discovery of the Alba as a lake formed in a volcanic crater; and *Noah's Arc* (*Arca Noë*, 1675), which expounded the narrative of Genesis and contained a world map

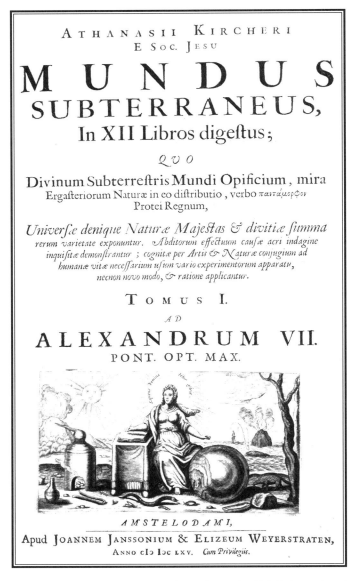

Figure 5. Detail from the frontispiece of *Mundus* (1678) (reproduced by courtesy of the Museo Geologico Giovanni Capellini, Bologna).

representing the lands supposedly covered by the flood (Godwin, 1979; Fletcher, 1988, p. 179–195). All of these texts had numerous illustrations.

Part 4 of the six parts of *China Illustrated* contained information on regional geography and natural history, based on overseas information especially from the Jesuit missions (Kircher, 1667; Van Tuyl, 1987; Szczesniak, 1952). The last chapter of this part discussed Chinese stones and minerals and began with the statement, "in the economy of stones and minerals ... sports of nature [*Naturae ludibria*] are found, as the Atlas of China and other writings inform us" (Kircher, 1667, p. 205). In the same work, Kircher introduced the word "geologist [*geologus*]," in reference to one who investigates stones—a very early use of the term. Indeed, this may well be the first use of the word geologist, following up on the first use of the term "geology (*giologia*)," which has been attributed to a manuscript by Ulisse Aldrovandi in 1603 (Vai, 2003). Kircher also discussed a kind of selenite and stated that Chinese scholars held that the stone was generated according to the waxing and waning of the Moon (Kircher, 1667, p. 205–206)—hence the name.

Although post-Enlightenment readers of Kircher's works have tended to regard them as fantastic or even absurd, his empirical contributions based on personal experience should not be neglected. The sources of, and influence on, Kircher's work also need to be carefully examined (see Collier, 1934; Thorndike, 1958; Strasser, 1982; Leinkauf, 1993; Morello, 1998; Cohen, 1998; Magruder, 2000, p. 526–540; Morello, 2001; Gould, 2004).

GASSENDI'S ROLE IN SEVENTEENTH-CENTURY EARTH STUDIES

When Steno was a student in Copenhagen, he was evidently influenced by Cartesian thought. At the same time, however, he studied Gassendi's work (Fig. 6), under the guidance of the alchemical polymath, Ole Borch (Borrichius) (1626–1690). In Steno's Chaos manuscript (1659), we can identify various passages copied from Gassendi's *Animadversiones* (Gassendi, 1649) (Chaos manuscript, col. 161–184; Sunday 26–Tuesday 28 June 1659). The manuscript was a student notebook that records Steno's reading and his thoughts about what he read as well as private miscellaneous matters (Ziggelaar, 1997, p. 393–447). Steno's mentor, Borch, called Steno's attention to Gassendi's work and lent him his manuscript of the *Animadversiones*, but there is no way of knowing whether Steno's transcription was an exact copy of Borch's manuscript or the result of Steno's selection. Anyway, the excerpt from Gassendi's work was the longest one in the Chaos manuscript, and it revealed great interest and diligence, for Steno accomplished the transcription of more than twenty columns in just three days, without making any other private comments or notes.

Steno and Borch also revealed their interest in geoscientific phenomena in Gassendi's voluminous book *Animadversiones*. In fact, their manuscript contains topics in physics, such as the atmosphere, loadstones, the beginning and end of the world, the origin of humans, the shape of Earth, meteorological phenomena, earthquakes, thermal springs, and so forth. Mentioning Augustine's famous passage on time in *Confessiones*, book 11, Steno or Borch excerpted and summarized Gassendi under the heading "the origin of the world (*origo mundi*)" (On the appearance of the world [*De exortu mundi*] in Gassendi, 1649, v. 1, p. 630 f.). Also it was noted that "Epicurus rightly holds that the world once had a beginning but less correctly [he says] that it had infinitely many, and that these have come about by the fortuitous coalitions of atoms" (Ziggelaar, 1997, p. 435; Gassendi, 1649, v. 1, p. 630). Then, "it is true," Steno wrote, "in Holy Scripture it is said that the world has come forth from 'unseen' matter as from

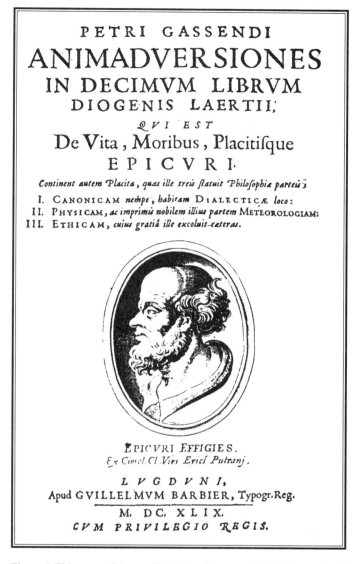

Figure 6. Title page of Gassendi's *Animadversiones* (1649) (reproduced by courtesy of the Library of Shimane University, Matsue, Japan).

chaos" (Ziggelaar, 1997, p. 435; Gassendi, 1649, v. 1, p. 630). In the section on "meteors," Steno or Borch made notes from the articles on clouds, rain, floods, winds, earthquakes, the source of the Nile, sea salt, ice, and snow (Ziggelaar, 1997, p. 438–446; Gassendi, 1649, v. 1, p. 981–1114). It is evident and noteworthy that Steno or Borch, or both, were particularly interested in cosmogonic and meteorological phenomena.

Later in Paris, after rigorous medical training at Leiden University, Steno inclined toward an anti-Cartesian position, in part at least because of his anatomical knowledge, which was at odds with that of Descartes. Decisive in this regard was Steno's lecture on the anatomy of the brain, at Thévenot's house in 1665 (Scherz, 1965, p. 61–103), in which he overtly criticized Cartesian physiology. In his *Canis Carchariae dissectum caput* (*Shark's Head Dissected*), Steno (1667) also referred to Gassendi's ideas on the formation of stones. This was one of the first papers that Steno dedicated to Ferdinand II, the Grand Duke of Tuscany, in the Medici court. Ferdinand was a disciple and a patron of Galileo. Steno said in acknowledgment of Gassendi:

> What varieties of diet accomplish in the humors of the microcosm, so alteration in the sun and moon and various other changes could produce in the humors of the earth. Gassendi, the "glory of France [*Galliae lumen*]," supports this assertion with the clearest of examples in his learned work, in which he explains the origin of stones (Steno, 1667, p. 103; Scherz, 1969, p. 107; cf. Gassendi, 1658, v. 2, p. 115a).

What, then, were Steno's opinions of the origin of fossils? In his *Canis*, he discussed fossil objects that resembled the parts of animals in eleven descriptions (*historia*) and six conjectures (*conjectura*). First, let us direct attention to the ninth description:

> Whether they are dug out of harder or softer soil, bodies resembling different parts of aquatic animals are not only very like each other but are also very like the animal parts to which they correspond; there is no difference of any kind in the course of the ridges, in the texture of the lamellae, in the curvature and windings of the cavities, and in the joints and hinges of bivalves (Steno, 1667, p. 92; Scherz,, 1969, p. 97).

This view that fossil objects resemble parts of living animals in their anatomical particulars and functions is also seen in the observations of Peiresc and Gassendi (Ellenberger, 1988, p. 228). They pointed to the resemblances between the detailed lines or serrations of fossil leaves and those of living ones. From these descriptions, Steno asserted that the objects resembling animal parts could not generate themselves in Earth today (conjectures 1 and 2); that Earth was once covered with waters (conjecture 3); that Earth was formerly mixed or mingled with water (conjecture 4), from which sedimentation occurred (conjecture 5); and finally that fossil objects dug from Earth that resemble the parts of animals are indeed the parts of former living animals (conjecture 6) (Steno, 1667, p. 92; Scherz, 1969, p. 108).

Next, we shall say something about Gassendi's theory of Earth, since his philosophical system encompassed "Earth studies," as Descartes' *Principia* had much to do with Earth studies as well as cosmology. Although the meteorological content of the *Animadversiones* was broadly in the Aristotelian tradition, we find in the appendix a short geographical description headed "Extract from a Treatise on the Terrestrial Globe (*Exscriptum ex tractatu de globo telluris*)" (p. xxvii–xxxv). Moreover, in another appendix (p. 47 ff.), there is a small-scale *Syntagma* titled *Philosophiae Epicuri syntagma*, which also contained logics, physics, and ethics, as did the *Animadversiones*' main text. But this appendix sketch was really a miniature of *Syntagma*, including just a single two-page article, "On Inanimate Terrestrial Things," in which we encounter the terms juice (*succus*), metals (*metalla*), rocks and stones (*saxa lapidesque*), and plants (*plantae*).

In fact, in Gassendi's posthumous works, such as *Syntagma philosophicum* (1658), one finds a useful "summa" of Gassendi's philosophical encyclopedia (Gassendi, 1658, v. 1–2; Bernier, 1684; Thomas, 1967; Mandon, 1969). In the second part of this Epicurean "philosophical triad," there are three further sections pertaining to physics: natural things in general, celestial things, and terrestrial things. Here our attention is drawn to the third section, which is divided into two parts: "On Inanimate Terrestrial Objects (*De rebus terrenis inanimis*)" and "On Animate Terrestrial Objects or on Animals (*De rebus terrenis viventibus, seu de animalibus*)." The primary division filled four books: first, the globe itself dealing with the shape and measurement of Earth, the source of springs and rivers, the flux and reflux of the tides, "juices" in Earth, subterranean heat and earthquakes, and sea salt (book 1); second, things usually called meteors, dealing with wind, clouds and rains, snow, lightning and thunder, rainbow, and aurora (book 2); third, stones and metals, dealing with the generation of stones, gems, petrified objects and shells, loadstones, and metals and their transmutation (book 3); and fourth, plants (book 4).

If we leave aside the volume on plants, these subjects as a whole corresponded closely to the objects in part 4 of Descartes' *Principia*. Thus, Gassendi reorganized the sublunary things referred to in the appendix to the *Animadversiones* of 1649 into two parts: celestial objects (*res caelestes*) and terrestrial objects (*res terrenes*). The former included comets and new stars, whereas the latter included atmospheric phenomena. For fossil or mineral descriptions, Gassendi dealt with the juices contained within Earth in chapter 5 of book 1 and again in book 3. Fossils, forming the mineral kingdom, were divided into stones, metals, and minerals. Minerals, moreover, are subdivided into earth, "concretional juice" (*succus concretus*), and "mixed minerals" (Gassendi, 1658, v. 2, p. 33b–34a). The so-called "fatty juice" (*succus pinguis*) could form sulphurs and bitumens (Gassendi, 1658, v. 2, p. 40a) (see also Morello, this volume, chapter 2).

The matter theory of Gassendi was, however, rather different from that of Descartes (Bloch, 1971; Hirai, 2003a). Probably influenced by the old Aristotelian doctrine of "minima," though in a modified version, Gassendi assumed the existence of minimal units ("seeds [*semina*]" or "molecules"), which supposedly possessed the properties of the particular material in the macro state whereas Descartes thought of variously shaped particles, imagined nonetheless to have features that would give rise to the appropriate macro properties (e.g., smooth, spherical particles for mercury). According to Gassendi, atoms were concreted into seeds or molecules, and these could be resolved into atoms. Even if we can't sense atoms directly, we can feel the properties of the materials formed by assemblages of molecules or seeds. Thus, for instance, the seeds or molecules of salt are small cubes and those of alum small octahedra, so small, indeed, that they could only be seen with a microscope (Gassendi, 1658, v. 2, p. 36b).

In book 3, dealing with stones and metals, Gassendi asserted that, in forming stone as a mixture of various substances, some "lapidifying force" (*vis lapidifica*) or "seminal force" (*vis semi-*

nalis) is in action, which, besides heat or cold, produces unification and solidification. Gassendi compared the formation of stones to the formation of seeds in plants: like grains of barley and wheat in ear, so the formation of rock crystal and amethyst occurs in a "matrix" rock (Gassendi, 1658, v. 2, p. 114a). Clearly, Gassendi considered the force that regulates the formation of various mineral crystals to be similar to the force or action of a plant growing from its seed. Also, a type of crystal found in Rians-en-Provence was rhombohedral and could be broken into many small rhombohedra when crushed. Gassendi suggested that this was like the bulb of a bulbous plant, which, when split into separate parts, produces several new bulbs through the agency of the seminal force in the plant (Gassendi, 1658, v. 2, p. 114b).

Gassendi also seemed to assume that large structures like mountains are constructed by the action of seminal forces. He thought of solutions charged with lapidifying juice or "semen," which might be injected into gravels and consolidate them into rocky masses or hills. Also, he thought of similar lapidifying semen or juice, causing stones to be concreted in the thermal springs in the Auvergne region and remarkable objects from water drops found on the walls of the grotto of Villa-crosana in Provence (Gassendi, 1658, v. 2, p. 116b–117a). How, then, did Gassendi try to explain the formation of fossils that resembled animals and plants?

As for objects that we now call fossils, Gassendi discussed in his chapter 3 the idea that they are not generated with their particular properties where they are found but are petrified and formed in two ways: one in which there is only a stony crust or cover, so that only impressed sides remain as a "mold"; the other in which the original substance both inside and outside is petrified and is thus preserved. For the first, he cited fossil plants (*Dendrolitha*), such as ivy, poplar, oak, and others, for which Peiresc had recognized that the exact lines or serrations of the living leaves were sometimes preserved in stones. For the second, he mentioned fossil bones (*Osteolitha*) dug from the ground near Aix. According to Ellenberger (1988, p. 228), this was the first reference to the Tertiary animal fossils from the basin of Aix-en-Provence. And third, there were objects such as ammonite and other shells, some of which were dissolved away and only remained as cavities in stone.

Gassendi explained the appearance of fossil objects in the mountains as the result of some huge cataclysm, yet he also adopted the theory of Seneca who reported in his *Natural questions* that there were large subterranean lakes or reservoirs in which many fishes lived (Seneca, 1971–1972, v. 1, p. 244–246; v. 2, p. 100). So, Gassendi pointed to the likelihood that such reservoirs contained salt water and carried creatures like those in the sea. These creatures became desiccated after frequent earthquakes, the water having penetrated into fissures, so that the dead bodies of creatures resembling shellfish would be buried along with the petrifying juice (Gassendi, 1658, v. 2, p. 120b). Thus, one can easily observe the objects in mountains, after digging by man, flooding, earthquakes, or other contingencies.

In Gassendi's biography of Peiresc (1641) (Gassendi, 1657), there had been previous geoscientific reports of fossil objects (Gassendi, 1657; Gassendi, 1658, v. 5, p. 237–362). In 1613, for example, reference was made to the bones of supposed giants (book 3), and in 1630 petrified fish and shells found in the mountains were ascribed to the agency of Noah's flood (book 4). Ultimately, in 1631, there appeared an assertion that a large bone was not that of a giant but the remains of an elephant (book 4) (Gassendi, 1657, v. 1, p. 160–164, v.2, p. 51, 60; Gassendi, 1658, v. 5, p. 279b–280b, 306a, 308b). It seems certain that Peiresc and Gassendi shared similar ideas about the organic origin of fossil objects.

To summarize our comparison with Descartes, Gassendi presented his theory of Earth about ten years after Descartes: while the former had the idea in his *Discourse* of 1637 and developed it in the *Principia* of 1644, the latter made a sketch of his ideas in the *Animadversiones* (1649) and published these ideas posthumously in *Syntagma philosophicum* (1658). They shared the common goal of the reorganization of traditional meteorology, mineralogy, magnetic theory, etc., as part of their general philosophical systems. Descartes clearly enunciated his theory of the origin and structure of Earth, with illustrations representing the supposed mode of formation of the large-scale structures of the planet. It was, however, all highly speculative, despite claims to its being based on "clear and distinct" principles. Gassendi, on the other hand, as well as advancing his well-known Epicurean atomism, also tried to provide a systematic encyclopedic description of Earth, including geography and mineralogy, as a "fruit" of the Renaissance humanist approach. In addition, Gassendi was a historian, writing biographies of Epicurus, Copernicus, and Tycho Brahe (Collier, 1957, p. 135–155; Joy, 1987). Gassendi gave attention to the interpretation of fossils (which Descartes did not mention), realizing their implications early in the 1630s on the basis of his field experience with his friend Peiresc.

Now, despite Steno's usual emphasis on Cartesian influences, we may point to the strong possibility that Gassendi's works provided an important source for Steno's geological accomplishments. Indeed, in the main contribution of Steno—his *Prodromus* of 1669 (Fig. 7), completed about twenty months after the dissection of the shark's head—one can easily find examples of mineral crystals and fossils that had been treated in Gassendi's works but not in Descartes'. As for the giant's bone, which so interested Peiresc, Steno opined that the large cranium and femur belonged to an elephant (Steno, 1669, p. 64–65; Scherz, 1969, p. 198), though he did not deny the possible past existence of large and tall human beings (Steno, 1669, p. 62; Scherz, 1969, p. 196). Additionally, anyone who encounters Steno's admiring expression "great Galileo" (Steno, 1669, p. 50; Scherz, 1969, p. 184) will have to acknowledge that Steno's interests and sympathies were with the empirical methodology of Galileo, Peiresc, and Gassendi, not with the deductive method of Descartes. Thus, it may be fairly concluded that the Cartesian influence on Steno was limited, especially after his sojourn in France.

Figure 7. Title page of Steno's *Prodromus* (reproduced by courtesy of the Biblioteca Universitaria di Bologna)

Yet, the relation between Steno and Gassendi or "Gassendism" is by no means simple, so that we should remain cautious about calling him a "Gassendist." Steno did not adopt Gassendian atomism. He simply expressed the view that, regardless of one's preferred theory of matter, one need not reject his (Steno's) own ideas about terrestrial objects.

> For what I have stated about matter holds everywhere, [regardless of] whether matter is considered to consist of atoms, or of particles which may change in a thousand ways, or of the four elements, or of as many different chemical principles as are needed to meet the variety of opinions among chemists (Steno, 1669, p. 12; Scherz, 1969, p. 147).

Retaining his skeptical attitude, Steno declared that he was never committed to any particular version of the various natural philosophies that were on offer at that time. He had no wish to become a dogmatic natural philosopher, and even empirical enquiry had its own problems:

> Indeed, the advocates of experiments have rarely had the restraint either to avoid rejecting entirely even the most certain principles of nature or to avoid considering their own self contrived principles as proved (Steno, 1669, p. 9; Scherz, 1969, p. 144).

STENO READS KIRCHER'S *MAGNES* (1643)

Let us now move to consider the relationship between Steno and Kircher. Kircher was acquainted with Peiresc and Gassendi. Given this French, more specifically Provence connection, how can we interpret the Kircher-Steno relationship? Is it possible to make some reassessment of the relation?

A year after the publication of Kircher's geoscientific encyclopedia, the Danish anatomist, Nicolaus Steno, traveled to Italy from France and began his geological studies at the court of Ferdinand II in Tuscany. Steno probably met Kircher at the Collegio Romano in 1666. Not long before, Steno's mentor, Borch, had met Kircher during his stay in Rome from October 1665 to March 1666 (Schepelern, 1983, v. 1, p. xviii). Steno and Kircher are known to have corresponded, but only five of Steno's letters are known to have survived, and for the most part they are on theological, not physical, topics (Scherz, 1952, p. 208–209, 301–302, 314–315, 318–319, 319; respectively dated, 12 May 1669, 28 May 1675, 14 April 1676, 15 September 1676, and 29 September 1676).

Although Steno criticized Kircher's idea of the "organic" origin of mountains, it is certain that he read Kircher's *Magnes* ten years before publication of the *Prodromus*. There is much transcription of *Magnes* in Steno's Chaos manuscript (1659), which treats many topics relating to terrestrial phenomena (cols. 33–48, 50–63, 95–102; Ziggelaar, 1997, p. 113–141, 147–169, 247–249, 254–260) and so it is most unlikely that there was no prior influence of Kircher on Steno. Steno's manuscript of *Magnes* (in the second edition published in Cologne in 1643) should thus be investigated (Kircher, 1643).

Kircher's *Magnes* was first published in 1641 in Rome. It was his first encyclopedic book, apparently intended to succeed William Gilbert's *De magnete* (1600), and it expanded upon the concept of magnetism applied to the whole universe (Baldwin, 1987). But *Magnes* elicited a critical response from Descartes, and this caused Kircher to publish an enlarged and corrected second version in Cologne. Many inhabitants of the republic of letters, such as Mersenne, Gassendi, Boyle, and Huygens, read the volume. A third edition was published in 1654, again in Rome. It is no surprise, therefore, that there was a reader of this book in Copenhagen.

Steno made his transcriptions from this 797-page volume in three stages. The first stage, from 21 to 25 March, included 16 columns (col. 33–48); the second, from 29 March to 3 April, 13 columns (col. 50–62); and the third stage on 13, 15, and 19–22 April, 6 columns (col. 95–96, 99–102). The total was about 35 columns, occupying one fifth of Steno's Chaos manuscript and representing the second largest part of the notes after the material from Gassendi.

Starting his excerpts from the preface to book 1, Steno took note of the passages including the names of Gilbert, Tomaso Garzoni, and Niccolo Cabeo. But, on the next day he jumped to book 3 and made transcriptions from the preface, first prelude, second prelude, part 1, chapter 1, and so forth. Book 3 of *Magnes* treated not only the loadstone or magnetism itself, but also all kinds of things in the universe, using analogies of magnetism under the head of what Kircher called the "Magnetic World [*Mundus magneticus*]." The description was loaded with information from many kinds of experiments, observations, books, and letters, which might readily have stimulated a reader's attention. Steno also seemed to be stimulated by these examples, making lengthy transcriptions, chapter-by-chapter, reaching the first chapter of part 5 on 25 March 1659.

Part I dealt with the movements of Earth, the planets, and the other heavenly bodies. Steno traced the writings of Kircher, who as a defender of geocentric doctrine criticized Gilbert and Kepler, whose theories were not geocentric. The next three parts represented Kircher's studies of Earth, in which one encountered the themes of the magnetism of elements (part 2), heterogeneous parts of Earth including fossils, amber, and metals (part 3), and the supposed influence on the sea of the magnetism of the Sun and Moon (part 4). Steno also transcribed the discussion about the marine currents of the Gulf of Mexico and the East China Sea in part 4.

In part 2, attention is drawn to Steno's note on chapter 3 about the "Magnetic-Meteorological Art (*Ars magnetico-meteorologica*)," especially the section on rain and winds, which was "Worth examining according to Verulum [Francis Bacon] or Descartes" (col. 38; Ziggelaar, 1997, p. 124). This reveals Steno's attitude toward the method of Kircher and the kinds of meteorological phenomena found in various places in the manuscript. In chapter 4, discussing the magnetism of mixed bodies, including mineral crystals, Steno transcribed the passages in which Kircher referred to the "plastic force (*vis plastica*)" that supposedly resided at the center of a body, and the ray(s) (*radius*) that extended in every direction, but not always equally, from the center. In this way, Kircher tried to explain the geometric shapes of crystals. Because of the unequal radiation, particles were thought to be easily added to the body in some directions, less easily in others, so that they produced angular bodies such as the tetrahedron (col. 39; Ziggelaar, 1997, p. 125–126). This plastic principle was also applicable to the seeds of plants and animals. One should keep this point in mind when considering the source of Steno's ideas in the *Prodromus* on crystal formation: inequality of crystal growth and constancy of interfacial angles (see below).

From 26 March, Steno inserted notes from his examination of Hermann Conring's (1606–1681) works: *On Alchemical Medicine* (1648) and *On Fermentation* (1643). A professor at Helmstedt University and court physician to the Swedish Queen, Christina, during the years 1650–1660, Conring attacked Paracelsianism, though he was interested in anatomical practices and alchemical studies. Borch later criticized Conring in his published book *De Hermetica medicina* (1674; Ziggelaar, 1997, p. 486). Steno's conversation with Borch could have prompted his examination of Conring's criticism of Kircher's *Magnes*, book 3, part 2, chapter 2.

After the digression of Conring, Steno resumed his transcription of Kircher on 29 March. In this second stage, he began with chapter 2 of part 5 about plant magnetism; then proceeded to animal magnetism (part 6), and lastly chapter 4 of part 7 on medical magnetism. He made detailed transcriptions, chapter by chapter. In part 5, we find an excerpt about divining rods thought to be useful in searching for metallic veins (col. 52; Ziggelaar, 1997, p. 150). Kircher supposed that different metals responded to different plants under the magnetic influence. Three kinds of animal magnetism were explained by various combinations of "sympathy" and "antipathy." The section on medical magnetism chose the magnetic force of iron as the first topic, and possible medical relationships were considered in notes made up to 2 April.

After another temporal hiatus, spent excerpting the work of the physician Matthias Untzer, the third stage began on 13 April when Steno continued his transcription of part 7, which is chiefly about the human body and health. In chapter 7, part 7, the role of the plastic force (*vis plastica*) was again mentioned, this time with the annotation *Nota bene*, now in the context of animal generation (col. 101; Ziggelaar, 1997, p. 257; Kircher, 1643, p. 730). Lastly, returning to part 2 of book 1, the themes of proper magnetism, such as terrestrial magnetism, disturbances of the magnetic compass, declination, and inclination, were considered. However, the quantity of notes was not great, being mixed with excerpts from book 3—for example, the imagined magnetic attraction of love (*magnetismus amoris*) in part 9. In addition, we see an excerpt from Gassendi's biography of Peiresc. According to Peiresc, there were veins of both water and fire within Earth, the latter being the cause of volcanic action (col. 102; Ziggelaar, 1997, p. 258). Steno's long transcription from Kircher ended on 22 April.

The influence of Steno's mentor, Borch, can be seen behind these transcriptions. For example, Steno noted the critical comment about the signatures (*signatura*) of things (col. 96–97; Ziggelaar, 1997, p. 249; N. 90: the excerpt from book 3, part 7, chapter 5), which led people to have far-fetched ideas about the superficial similarities of things. When one notices the root of a plant being similar to the shape of some organ of the human body, one could (according to the doctrine of signatures or correspondences) infer that the root could cure the disease of the organ. Steno, however, rejected such ideas. He wrote, "These comparisons do more harm than good in the natural sciences" (col. 97; Ziggelaar, 1997, p. 249, N. 90). As the editor of the complete edition of the Chaos manuscript has pointed out, Steno probably discussed these cases with Borch and received comments from him. The same attitude was observed in the transcripts of Gassendi's work (for example, see col. 171; Ziggelaar, 1997, p. 415).

It is certain that Steno (and Borch) had critical ideas about some of Kircher's suggestions, but at the same time Steno was willing to spend a considerable amount of time reading and copying from Kircher's book. Moreover, the selected topics and their detail in the manuscript reveal Steno's interest in meteorological and terrestrial phenomena. This, in turn, indicates the importance of Kircher's *Magnes* to contemporary readers, supplying information of all kinds and ideas about Earth in the period after Gilbert.

FROM *MAGNES* TO *MUNDUS*

As mentioned, Kircher's southern Italian expedition of 1637–1638 caused him to write about Earth. By the early 1640s, he had established the general content of his intended book on this subject, but his studies of Egyptian antiquity delayed his writing. Kircher published *Egyptian Oedipus* (*Oedipus aegyptiacus*) in 1652–1654—a voluminous work of over two thousand pages. In addition, his books on optics and music were written in 1646 and 1650, respectively. Furthermore, according to Gorman, Kircher also had formulated a "geographical plan" (*consilium geographicum*) in the late 1630s to reform and restore terrestrial knowledge (Gorman, 2004, p. 241 and 257, note 7). Apparently all of these programs were to be pursued simultaneously with his writing about Earth.

Turning again to the first edition of *Magnes* (Kircher, 1641), we find, even before the title page, a pictorial representation of relationship among scientific disciplines. Fourteen disciplines, such as theology, astronomy, geography, and medicine, are connected to one another in a peripheral ring, within which is set a triad of worlds that, in turn, encircles the archetypical world (*mundus arche typus*) (Fig. 8). The components of the triad are the world of stars (*mundus sydereus*), the microcosm (*mycrocosmus*), and the sublunary realm (*mundus sublunaris*). These could be related to the traditional macro-micro correspondences and provide a preliminary announcement of Kircher's geocosmology. Already at the beginning of the 1640s, therefore, Kircher seems to have had a plan for a system of geosciences as an independent domain.

In 1657, the prolegomenon to the *Subterranean World* was published as a book titled *Ecstatic Journey, Part Two*. Kircher's manuscript was completed in 1662, and the main outline of the geoscientific plan saw light toward the end of 1664, as can be seen from the correspondence of Henry Oldenburg, secretary of the Royal Society, for example, dated 25 August 1664 to Boyle, September (no date) and 12 October 1665 to Spinoza, respectively (Oldenburg, 1966, p. 206–211, 497–501, and 565–568). The *Subterranean World* had three Latin editions, being published in Amsterdam in 1664–1665, 1665, and 1678 (Fletcher, 1988, p. 185). In addition, there was a Dutch translation in 1682. The 1665 edition was used in writing this paper as it is probably the version used by Steno.

After a dedication to Pope Alexander VII, *Mundus* opens with a rather long preface in two parts. The first one consists of the report on the volcanic activities and earthquakes that Kircher had experienced in southern Italy. These experiences were so remarkable to him that he described the severe Calabria earthquake of 1638 as "resulting in the author's learning about the great secret of nature during fourteen days, jeopardizing his own life" (Kircher, 1665, v. 1, p.**2 v.). He described the results of his excursion to the crater of Mount Vesuvius. He also provided an impressive tableau of the erupting Vesuvius (*Typus montis Vesuvii, prout ab authore anno 1638, visus fuit*), created from the perspective of the summit crater, from which three plumes of smoke were being emitted and inside which lava was shown accumulating.

In the first two books of the *Mundus*, Kircher gave his fundamental concepts for his studies of Earth. Book 1, titled *Centrographicus*, consists of elementary geometry and "physicomathematics." Kircher's physics was Aristotelian, which, while acknowledging the gravitational attraction toward the supposed center of the universe, criticized Copernican theory. Nevertheless, he was familiar with the research of Galileo, Gassendi, Mersenne, and Torricelli. At the same time, Kircher explained the relations between the growth or action of plants and minerals and the direction of gravitation, i.e., toward the center of the universe (Kircher, 1665, v. 1, book 1, section 1, chapter 2, propositions 7 and 8).

Book 2, *Technicus geocosmus*, treats the various actions of the terrestrial globe: the definition of the geocosm, Earth's "idea"; the nature and composition of the world; the Sun and the Moon; the size of Earth; Earth's chains of mountains, volcanoes, terrestrial changes, such as the increase or decrease of the size of mountains, subterranean channels or passages that connected the seas, the heights of mountains and depth of the seas, and terrestrial magnetism.

What was the geocosm for Kircher? As he put it:

> The terrene globe, which we call the Geocosm or the Terrestrial World, is the end and the center of the whole Creation and is also disposed by that Divine Wisdom, worker, skill and industry of things, so that whatever forces lie latent in the universe and whatever hidden properties are latent in the singular globes of stars, all appear crammed into this globe, as it were an "epitome" [of the whole universe] (Kircher, 1665, v. 1, p. 55).

In this way, Kircher envisaged the geocosm as a kind of organic whole, permeated by the Divine Will, with hidden forces everywhere. Under the operation of such forces, he discussed the formation of Earth, including mountains, minerals, plants, animals, human beings, and even demons.

In chapters 8–12, on the mountains of the geocosm, Kircher explained the condition of mountains before the Deluge, submarine mountains, volcanoes, and the changes of mountains. Then in chapters 13–16, he dealt with waters and the atmosphere surrounding the geocosm, also giving an account of the subterranean passages that allowed the underground circulation of waters. Chapter 18 presented a program for the study of subterranean objects, recognizing that the substance of the geocosm was never

Figure 8. Frontispiece of the first edition of *Magnes* (1641) (reproduced by courtesy of the Library of the Tokyo Institute of Technology).

homogeneous. The parallelism between geocosm and microcosm in chapter 19 should also be noted.

In a narrow sense, the next six books (3–8) offered a specific description and explanation of the Kircherian theory of Earth. We find an expanded version of the meteorological parts of *Magnes* (book 3, part 2, chapter 3), which dealt with the role of water and fire in Earth and their circulation, the origin of springs and rivers, and the classification and generation of fossils. Book 3, on "Hydrography," had the theme of water circulation within and on the surface of Earth; whereas book 4, "Pyrography," discussed the supposed network of subterranean fires and winds. To simplify the matter, it was suggested that Earth had two kinds of network, one for the water system, the other for fire. Water and fire were thought to circulate within Earth like the fluid circulation in the microcosm (i.e., the human body). The interaction of water and fire produced such phenomena as hot springs and volcanic eruptions. Kircher even mentioned the mechanical principle of the siphon, which, he thought, was relevant to the movement of water up to mountain reservoirs (Kircher, 1665, v. 1, p. 232).

In book 6, Kircher regards earth (one of the four Aristotelian elements) not as a simple substance but as a "mixed" or "compound" material, whereas salt was supposedly a fundamental principle. Consequently, he regarded minerals and fossils as mixed materials. As to fossil objects, Kircher gave details in book 8. First, he suggested how to group the stones in a table (*Schematismus lapidum resolutorius*) using five proposed criteria for his classification, and he explained the transformation of stones by petrifying juice, the origin of stones and rocks, the effects of stones, and subterranean animals. As would be expected from the fact that Kircher adopted many illustrations from Ulisse Aldrovandi's (1522–1605) *Musaeum metallicum* (1648 posthumous edition), his study on stones was much beholden to the Renaissance tradition of encyclopedic natural history.

The Kircherian character of grouping stones is visible in book 8, section 1, chapter 8, in which he aimed to give an account, according to the same principle, of the formation of both the geometrically regular polyhedral mineral crystals and also the imagined patterns of geometrical figures on the surfaces of stones. Kircher's idea was based on the concept previously stated in the *Magnes*, book 3, part 2, chapter 4 (discussed below in relation to Steno's idea about the formation of mineral crystals). In chapter 9, Kircher summarized five ways of making physical inquiries into stones, providing various images or pictures (Kircher, 1665, v. 2, p. 37–45). Although the author referred to six ways of discussing stones or fossils (*Modus sextus*), only five are listed, and he showed no heading for the fifth one. First, there were things incidentally formed by human imagination; second, things petrified by a seminal or plastic force; third, things that were originally plants or animals but are now petrified; fourth, things attracting similar figures by magnetism; and fifth, things arranged by God and angels. Thus, although historians of geology have represented Kircher as a figure who had asserted the generation of fossil objects by a plastic force, he did, in fact, admit the organic origin of some kinds of fossils. (For an evaluation of Kircher's palaeontological thought, see Gould, 2004.)

The last two parts of the *Mundus*, books 11 and 12, discuss Kircher's chemical philosophy and seminal philosophy, which formed the background to his theory of Earth. He claimed that terrestrial things, such as animals, plants, and minerals, were influenced by the geocosmic power and were generated by the agency of a "universal semen" (*panspermia*). In this context, one can understand his ideas about the spontaneous generation of creatures and the formation of various kinds of subterranean objects by the supposed plastic force (Hirai, 2003b; Rowland, 2004). His manner of thinking was quite traditional (cf. Oldroyd, 1974; Yoshimoto, 1992).

Following its publication, the *Mundus* soon attracted considerable attention. Indeed, even before its publication the book was noticed in the republic of letters. As early as November 1665, the *Philosophical Transactions* carried a review (Anon. [Oldenburg], 1665). Moreover, Kircher's *Mundus* contained numerous ideas, observations, and experimental information providing an encyclopedic overview of Earth. It is copiously illustrated by several, impressive, large plates and numerous smaller drawings. On the other hand, overt criticisms soon began to appear about its problematic contents, as evident in the comments of Moray and Hooke in England and Redi and Steno in Florence (Reilly, 1974, p. 106–107).

Let us now look at Steno's reception of Kircher's massive theory of Earth.

THE KIRCHER–STENO RELATION IN THE HISTORY OF GEOSCIENCES

As mentioned at the outset, historians have tended to interpret Steno's geoscientific contribution as being in the Cartesian tradition. However, having studied the relations between Steno and the works of Gassendi and Kircher, other alternatives now warrant consideration, especially in the relation of Kircher and Steno.

In the 1667 treatise *Myologia,* which included the *Canis* and its geological "digression," Steno mentioned three kinds of "cosms," defending the application of geometrical principles to the study of muscles. He asserted that it would be as valid to introduce mathematical methods into anatomy as into astronomy or geography. Steno wrote:

> And why should not we adopt the way to investigate muscles, that astronomers adopted in the study of heavenly bodies, geographers adopted in the study of the Earth, and if we take an example from microcosm, the writers on optics adopted in the study of eyes? (Steno's dedication to Ferdinand II, *in* Steno, 1667, p. 3r–3v; Scherz, 1969, p. 68–70)

The analogy between Earth and the microcosm was also seen in the passage of *Canis*, referring to the work of Gassendi (Steno, 1667, p. 102–103; Scherz, 1969, p. 106). It is clear, therefore, that Steno intended to elucidate both the phenomena of Earth (the geocosm in Kircher's language) as well as the phenomena of a microcosmic body in a quasi-mathematical manner. This would exhibit the common triadic cosmic scheme shared by both Kircher and Steno.

On the other hand, Steno had strong reservations about Kircher's geocosmic doctrine. Steno's respected friend and colleague, Francesco Redi, criticized Kircher's experiments on spontaneous generation, and Steno and Redi seemed to have had similar ideas on this question. In particular, Steno questioned the organic production of mountains in the Kircherian geocosmic theory. In the passage of *Prodromus* about the origin of mountains, Steno put his criticism in the form of "negative propositions":

> *None* of the mountains grow like vegetables. The rocks of mountains have *nothing* in common with the bones of animals, apart from a certain similarity in hardness, since they agree *neither* in material *nor* in manner of production *nor* in construction *nor* function ... The arrangement of mountain crests or chains, as some prefer to call them, along certain zones of the earth, is in accordance *neither* with reason *nor* experience (Steno, 1669, p. 34; Scherz, 1969, p. 169, italics added).

The Kircherian terms, such as crests (*corona*) and chains (*catena*) of mountains, herein clearly criticized by Steno appeared in book 2 of the *Mundus*.

At the same time, this evidence shows that Steno was well aware of Kircher's geocosmic theory. Judging from the fact that Kircher conscientiously sent his works to the *Accademia del Cimento* in Florence (Cochrane, 1973, p. 253), Kircher did not wish to have trouble with the Academy, and Steno's *Prodromus* was in fact sent to Rome. Incidentally, the naming of Steno's book, the "Prodromus," may have been intended as a parallel to the subtitle of Kircher's book, *Ecstatic Journey* (*Iter exstaticum II qui et Mundi subterranei Prodromus dicitur*), though the term was not unusual for book titles in that period. Steno naturally intended to complete his theory of Earth with a comprehensive account, analogous to Kircher's *Mundus*. The investigations into grottos in 1671 also support Steno's intention, a grotto or cave being a kind of entrance to the subterranean world.

As already acknowledged, Steno took detailed notes from Kircher's *Magnes* in which he excerpted the "sublunary" generation of things, including mineral crystals. The transcriptions contained almost verbatim passages from book 3, part 2, chapter 4 of the *Magnes*, explaining the growth of mineral crystals in particular directions with specific angles by apposition (*appositio*) of particles and the supposed action of magnetism (col. 39; Ziggelaar, 1997, p. 125–126). This idea of geomagnetic action in the formation of crystals also appeared in book 8 of the *Mundus*, which mentioned the hexagonal forms of rock crystal (Fig. 9). Kircher intended to explain the crystal formation by similar particles acting under the influence of a certain "marvelous" magnetism (*miro quodam magnetismo*) as well as by the plastic force

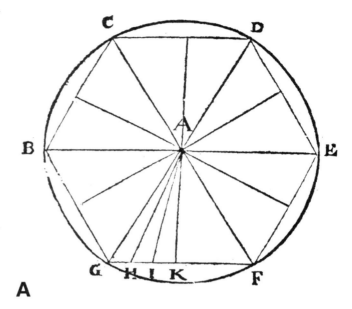

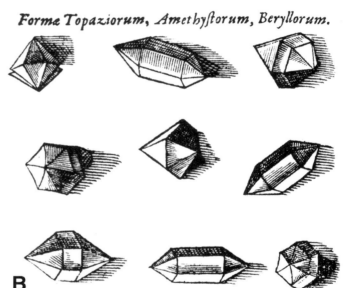

Figure 9. Figures illustrating the manner of growth of mineral crystals (A) and types of crystalline minerals (B) in *Mundus* (1678) (reproduced by courtesy of the Museo Geologico Giovanni Capellini, Bologna).

and the supposed hidden central forces of salt and gems (Kircher, 1665, v. 2, p. 25–26).

All contemporary thinkers who adopted corpuscular theories thought it important to understand how mineral crystals form their regular geometrical figures. It was impossible to explain their formation by the random assemblage of particles. Kepler supposed the controlling power of the figure of snow crystals to be a "formative faculty (*facultas formatrix*)" radiating from the center of the crystal; the faculty supposedly derived from a "terrestrial spirit" (Kepler, 1611; 1966), and Gassendi mentioned this idea of Kepler (Gassendi, 1658, v. 2, p. 81). In England, Robert Hooke, in his *Micrographia,* introduced the terminology of "congruity/incongruity" to explain the phenomenon, as well as the model of packing small spheres (Hooke, 1665, observation 13, esp. p. 85; see Henry, 1989). What about Steno?

Here we recall Steno's use of the metaphor of magnetism to explain the action of fluids in guiding the apposition of particles and the spreading out of the crystalline materials onto a surface. He wrote:

> We must take a double movement into account in the growth of a crystal: one, the movement whereby it comes about that crystalline material is added to certain parts of the crystal and not to others, a movement that I fancy must be attributed to the tenuous permeating fluid, and that is illustrated by the magnetic example I have given; the other, the movement whereby new crystalline material, added to the crystal, is spread out over a plane; this movement must be derived from the surrounding fluid just as, when the iron threads have risen up above the magnet, whatever is stuck off one owing to movement of the air is added to another. To this movement of surrounding fluid I should attribute the fact that not only in a crystal but also in many other angular bodies any given opposite planes are parallel (Steno, 1669, p. 42–44; Scherz, 1969, p. 179) (Fig. 10).

Evidently, the process of crystallization was not random but took place under the "guidance" of the moving "permeating fluid" and the "surrounding fluid" particles with some kind of magnetic force controlling the action of the fluids. This way of thinking—that magnetic action is concerned in the formation of mineral crystals—suggests that Steno adopted Kircher's idea of magnetism being relevant to crystallization, although Steno invoked no central or seminal or plastic force. So far as I am aware, this kind of thinking about magnetic action during the formation of mineral crystals is not known to be present in the works of Gilbert, Kepler, Descartes, or Gassendi.

CONCLUSIONS

Examination of the works and documents of Gassendi, Kircher, and Steno allows the following conclusions to be reached:

First, the intellectual milieu of southern France, represented by such figures as Peiresc and Gassendi, encouraged, or was conducive to, Kircher's historical and terrestrial studies. Second, whereas the tendency to acknowledge the influence of Descartes' work on Steno's theories is common—and in this respect Descartes' ideas on the formation of strata and on "collapse tectonics" are not to be diminished or neglected—a new importance needs to be attached to the influence on Steno of Gassendi's work, especially as it pertains to the formation of

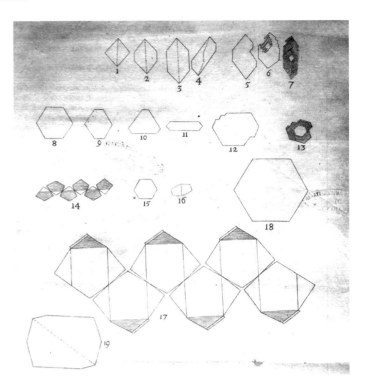

Figure 10. Figures illustrating the manner of mineral crystal growth in *Prodromus* (reproduced by courtesy of the Biblioteca Universitaria di Bologna).

fossil objects. Third, a considerable proportion of the excerpts in Steno's Chaos manuscript (1659) were drawn from Kircher's *Magnes* (1643), suggesting Steno's strong interest in Kircher's magnetic theory, both for meteorological and terrestrial phenomena. Fourth, Kircher's use of magnetism to explain the formation of fossil things under the action of central or seminal plastic force appeared in the *Magnes* (1641) and was retained in *Mundus* (1665). Fifth, in some respects, the Stenonian theory of Earth was a successor of Kircher's theory, despite the fact that Steno's theory is generally being regarded by commentators as modern and progressive while Kircher's is supposedly antique or archaic. Steno and Kircher had common goals in planning a wholesale geoscientific system or "economy"—the Kircherian geocosmic system on the one hand and Steno's general considerations about solids within solids on the other. Moreover, they tried to resolve the problem of the formation or generation of earthy things in this setting, i.e., by developing a theory of Earth. In particular, Steno used the Kircherian idea of magnetic action in the formation of mineral crystals.

Thus, a more complex story about seventeenth century Western thinking about Earth is revealed. Besides the view of a Cartesian Steno, we should now consider a "Gassendist Steno" and even a "Kircherian Steno." These conclusions require a revised interpretation of the other relationships among such figures as Steno, Hooke, Varenius, Erasmus Bartholin, Huygens, Becher, and Leibniz in the history of the geosciences.

ACKNOWLEDGMENTS

I should like to thank Chikara Sasaki for his initiating and supporting my study of Kircher, and Hiro Hirai for his assistance and suggestions in reading the Gassendi and Kircher texts. I am most grateful to David Oldroyd for helping to improve my English and for numerous other useful comments. I sincerely appreciate the facility afforded by Kerry Magruder for providing copies of materials in the library of the University of Oklahoma. I also appreciate the advice and information received from Gian Battista Vai, Hiroo Mizuno, and an anonymous referee.

REFERENCES CITED

Aldrovandi, U., 1648, Musaeum metallicum in libros IIII distributum [Ambrosini, B., ed.]: Bologna, Giovanni Battista Bellagamba, 595 p.
Anonymous [Henry Oldenburg], 1665, Of the *Mundus Subterraneus* of Athanasius Kircher: Philosophical Transactions of the Royal Society of London, no. 6, (Monday, 6 November), p. 109–117.
Bach, J.A., 1985, Athanasius Kircher and his method: A study in the relations of the arts and sciences in the seventeenth century [Ph.D. thesis]: Norman, Oklahoma, University of Oklahoma, 390 p.
Baldwin, M., 1987, Athanasius Kircher and the magnetic philosophy [Ph.D. thesis]: Chicago, University of Chicago, 501 p.
Bernier, F., 1684, Abrégé de la philosophie de Gassendi: Paris, Fayard, 1992, 7 v. (Corpus des oeuvres de philosophie en langue française), originally published as second edition, Lyons.
Bloch, O.R., 1971, La philosophie de Gassendi: nominalisme, matérialisme et métaphysique: La Haye, Martinus Nijhoff, 525 p.
Burke, J.G., 1966, Origins of the science of crystals: Berkeley and Los Angeles, University of California Press, 198 p.
Cochrane, E., 1973, Florence in the forgotten centuries 1527–1800: A history of Florence and the Florentines in the age of the grand dukes: Chicago and London, The University of Chicago Press, 593 p.
Cohen, C., 1998, Un manuscrit inédit de Leibniz (1646–1716) sur la nature des "objets fossils": Bulletin de la Société Géologique de France, v. 169, p. 137–142.
Collier, K.B., 1934, Cosmogonies of our fathers: Some theories of the seventeenth and the eighteenth centuries: New York, Columbia University Press, 500 p.
Collier, R., 1957, Gassendi historien dignois, *in* Collier, R., ed., Actes du congrès du tricentenaire de Pierre Gassendi 1655–1955 (4–7 Août 1955): Presses Universitaires de France, p. 135–155.
Ellenberger, F., 1988, Histoire de la géologie, tome 1: Paris, Lavoisier, 352 p.
Emerton, N.E., 1984, The scientific reinterpretation of form: Ithaca and London, Cornell University Press, 318 p.
Fletcher, J., ed., 1988, Athanasius Kircher und seine Beziehungen zum glehrten Europa seiner Zeit (Wolfenbütteler Arbeiten zur Barockforschung, Bd. 17): Wiesbaden, Harrassowitz, 214 p.
Gassendi, P., 1649, Animadversiones in decimum librum Diogenis Laertii; Qui est de vita, moribus, placitisque Epicuri, Continent autem placita, quas ille treis statuit philosophiae partesj;I. canonicam nempe, habitam dialecticae loco: II. Physicam, ac imprimis nobilem illius partem meteorologicam: III. Ethicam, cuius gratia ille excoluit—caeteras: Lyons, Guillelmum Barbier/reprinted: New York and London, Garland, 1987, 3 v. (Abbreviated *Animadversiones*).
Gassendi, P., 1657, The mirrour of true nobility and gentility, being the life of the renowned Nicolaus Claudius Fabricius, Lord of Peiresk, Senator of the Parliament at Aix, William Rand, trans.: London, Humphrey Moseley, 2 v.
Gassendi, P., 1658, Petri Gassendi Diniensis Opera omnia in sex tomos divisa: Lyons, Sumptibus Lavrentii Anisson & Joannis Baptistae Devenet; reprinted: Stuttgart, Friedrich Frommann, 6 v., 1964. v. 1–2: Syntagma philosophicum (Abbreviated *Syntagma*), v. 5: Viri illustris Nicolai Claudii Fabricii de Peiresc senatoris aquisextiensis vita.
Godwin, J., 1979, Athanasius Kircher: A renaissance man and the quest for lost knowledge: London, Thames and Hudson, 96 p.
Gorman, M.J., 2004, The angel and the compass: Athanasius Kircher's magnetic geography, *in* Findlen, P. ed., Athanasius Kircher: The last man who knew everything: New York and London, Routledge, p. 239–259.
Gould, S.J., 2004, Father Athanasius on the isthmus of a middle state: Understanding Kircher's paleontology, *in* Findlen, P., ed., Athanasius Kircher: The last man who knew everything: New York and London, Routledge, p. 207–237.
Halleux, R., 1982, La litterature géologique française de 1500 à 1650 dans son contexte européen: Revue d'Histoire des Sciences, v. 35, p. 111–130.
Henry, J., 1989, Robert Hooke, the incongruous mechanist, *in* Hunter, M., and Schaffer, S., eds., Robert Hook: New studies: Woodbridge, The Boydell Press, p. 149–180.
Hirai, H., 2003a, Le concept de semence de Pierre Gassendi entre les théories de la matière et les sciences de la vie au XVIIe siècle, Medicina nei secoli arte e scienza: Journal of History of Medicine, new series, v. 15, p. 205–226.
Hirai, H., 2003b, "De novo" or "ex semine": Kircher and the problem of spontaneous generation: paper read at the international conference "Scientific culture in early modern Rome" (Warburg Institute, 10 October, 2003) (manuscript).
Hooke, R., 1665, Micrographia: or some physiological descriptions of minute bodies made by magnifying glasses with observations and inquiries thereupon: London and Illinois, Science Heritage, 1987, 246 p.
Jonkers, A.R.T., Jackson, A., and Murray, A., 2003, Four centuries of geomagnetic data from historical records: Reviews of Geophysics, v. 41, 1006, p. 1–36, doi: 10.1029/2002RG000115.
Joy, L.S., 1987, Gassendi the atomist: Advocate of history in an age of science: Cambridge, Cambridge University Press, 311 p.
Kangro, H., 1973, Kircher, Athanasius, *in* Dictionary of Scientific Biography, v. 7, p. 374–378.
Kepler, J., 1611, Strena seu de nive sexangula, Francofurti ad moenum, *in* Caspar, M., and Hammer, F., eds., Johannes Kepler Gesammelte Werke, v. 4: Munich, C.H. Beck'sche Verlagsbuchhandlung, 1941, p. 261–280, 504–505.
Kepler, J., 1966. The six-cornered snowflake, Hardie, C., ed. and trans.: Oxford, Clarendon Press, 75 p.
Kircher, A., 1641, Magnes sive de arte magnetica opus tripartitum: Rome, Ludovici Grignani (Abbreviated *Magnes*, 1641).
Kircher, A., 1643, Magnes sive de arte magnetica, 2nd edition: Cologne, Jodocum Kalcoven (Abbreviated *Magnes*, 1643).
Kircher, A., 1654, Magnes sive de arte magnetica, 3d edn.: Rome, Vitalis Mascardi (Abbreviated *Magnes*, 1654).
Kircher, A., 1665, Mundus subterraneus, in XII libros digestus: Amsterdam, Joannem Jannssonium à Waesberge & Elizeum Weyerstraten, 2 v (Abbreviated *Mundus*).
Kircher, A., 1667, China monumentis, qua sacris qua profanis, nec non variis naturae et artis spectaculis, aliarumque rerum memorabilium argumentis illustrata: Amsterdam, Joannem Jannssonium à Waesberge & Elizeum Weyerstraten.
Laudan, R., 1987, From mineralogy to geology: Foundations of science 1650–1830: Chicago and London, University of Chicago Press, 278 p.
Leinkauf, T., 1993, Mundus combinatus: Studien zur Struktur der barocken Universalwissenschaft am Beispiel Athanasius Kircher SJ (1602–1680): Berlin, Acadademie Verlag, 434 p.
Magruder, K.V., 2000, Theories of the Earth from Descartes to Cuvier: Natural order and historical contingency in a contested textual tradition [Ph.D. thesis]: Norman, Oklahoma, University of Oklahoma, 871 p.
Mandon, L., 1969, Etude sur le Syntagma philosophicum de Gassendi: New York, Burt Franklin (originally published, 1858), 178 p.
Metzger, H., 1969, La genèse de la science des cristaux: Paris, Albert Blanchard (first edition, 1918), 248 p.
Middleton, W.E.K., 1961, Archimedes, Kircher, Buffon, and the burning-mirrors: Isis, v. 52, p. 533–543.
Miller, P.N., 2004, Copts and Scholars: Athanasius Kircher in Peiresc's Republic of Letters, *in* Findlen, ed., Athanasius Kircher: The last man who knew everything: New York and London, Routledge, p. 133–148.
Morello, N., 1998, Giovanni Alfonso Borelli and the eruption of Etna in 1669, *in* Morello, N., ed., Volcanoes and history (Proceedings of the 20th International Commission on the History of Geological Sciences Symposium, Napoli–Eolie–Catania (Italy) 19–25 September 1995): Genova, Brigati, p. 395–413.
Morello, N., 2001, Nel corpo della Terra: Il geocosmo di Athanasius Kircher, *in* Lo Sardo E., ed., Athanasius Kircher, Il museo del mondo: Roma, Edizioni de Luca, p. 179–196.

Morello, N., 2006, this volume, Agricola and the birth of the mineralogical sciences in Italy in the sixteenth century, *in* Vai, G.B., and Caldwell, W.G.E., The Origins of Geology in Italy: Geological Society of America Special Paper 411, doi: 10.1130/2006.2411(02).

Oldenburg, H., 1966, Hall, A.R., and Hall, M.B., eds. and trans., The correspondence of Henry Oldenburg, v. 2: Madison, Milwaukee, and London, The University of Wisconsin Press, 678 p.

Oldroyd, D., 1974, Some neo-platonic and stoic influences on mineralogy in the sixteenth and seventeenth centuries: Ambix, v. 21, p. 128–156.

Oldroyd, D., 2000, Steno, Nicolaus (Niels Stensen), *in* Wilbur Applebaum, ed., Encyclopedia of the scientific revolution from Copernicus to Newton: New York and London, Garland, p. 618–619.

Reilly, C., 1974, Athanasius Kircher: Master of a hundred arts 1602–1680: Wiesbaden- Rom, Edizioni del Mondo, 207 p.

Roger, J., 1982, The Cartesian model and its role in eighteenth-century "Theory of the Earth," *in* Thomas, M.L., et al., eds., Problems of Cartesianism: Kingston and Montreal, McGill-Queen's University Press, p. 95–112.

Rowland, I.D., 2004, Athanasius Kircher, Giordano Bruno, and the panspermia of the infinite universe, *in* Findlen, P., ed., Athanasius Kircher: The last man who knew everything: New York and London, Routledge, p. 191–205.

Sasaki, C., 2003, Descartes's mathematical thought: Dordrecht, Boston, and London, Kluwer Academic Publishers, 469 p.

Schepelern, H.D., ed., 1983, Olai Borrichii Itinerarium 1660–1665: The Journal of the Danish Polyhistor Ole Borck: Copenhagen, C.A. Reitzels and London, E.J. Brill, 4 v.

Scherz, G., 1965, Nicolaus Steno's lecture on the anatomy of the brain: Copenhagen, Nyt Nordisk, 206 p.

Scherz, G., ed., 1952, Nicolai Stenonis Epistrae et epistolae ad eum datae: Copenhagen, Nyt Nordisk, 2 v.

Scherz, G., ed., 1969, Alex J. Pollock trans., Steno geological papers: Copenhagen, Acta historica scientiarum naturalium et medicinalium, v. 20, 370 p.

Schneer, C.J., 1971, Steno: On crystals and the corpuscular hypothesis, *in* Scherz, G., ed., Dissertations on Steno as geologist (Acta historica scientiarum naturalium et medicinalium, v. 23), p. 293–307.

Seneca, 1971–1972, Naturales quaestiones, Corcoran, T.C., trans. (The Loeb Classical Library): London, Heinemann, and Cambridge, Massachusetts, Harvard University Press, 2 v.

Steno, N., 1659, "Chaos manuscript," Posteriori di Galileo 32, Accademia del Cimento III, Carteggio, v. 17, Scritti di Niccolò Stenone (Gal. 291), fol. 28r–75v. (Biblioteca Nazionale Centrale, Florence).

Steno, N., 1667, Canis Carchariae dissectum caput, *in* Steno, N., Elementorum myologiae specimen: Florence, Stella (Abbreviated *Canis*).

Steno, N., 1669, De solido intra solidum naturaliter contento dissertationis prodromus: Florence, Stella (Abbreviated *Prodromus*).

Strasser, G.F., 1982, Spectaculum Vesuvii: Zu zwei neuentdeckten Handschriften von Athanasius Kircher mit seinen Illustrationsvorlagen, *in* Brinkmann, R., et al., eds., Theatrum Europaeum: Festschrift für Elida Maria Szarota: Munich, Wilhelm Fink, p. 363–384.

Szczesniak, B., 1952, Athanius Kircher's China illustrata: Osiris, v. 10, p. 385–411, doi: 10.1086/368561.

Thomas, P.-F., 1967, La philosophie de Gassendi: New York, Burt Franklin (originally published: 1889), 320 p.

Thorndike, L., 1958, The underground world of Kircher and Becher, *in* Thorndike, L., ed., A history of magic and experimental science, v. 7: New York, Colombia University Press, p. 567–589.

Vaccari, E., 2002, Earth science and geology, *in* Baigrie, B.S., ed., History of modern science and mathematics: New York, Charles Scribner's Sons, v. 3, p. 1–43.

Vai, G.B., 2003, Aldrovandi's will: introducing the term 'geology' in 1603, *in* Vai, G.B., and Cavazza, W., eds, Four centuries of the word geology: Ulisse Aldrovandi 1603 in Bologna: Bologna, Minerva Soluzioni Editoriali, p. 64–110.

Van Tuyl, C.D., trans., 1987, China illustrata (original by Kircher, A., 1667): Bloomington, Indiana, Indiana University Press, 228 p.

Yamada, T., 2003, Stenonian revolution or Leibnizian revival?: Constructing geo-history in the seventeenth-century: Tokyo, Historia Scientiarum, v. 13, p. 75–100.

Yamada, T., 2004, Descartes versus Gassendi in theories of the Earth with special reference to Nicolaus Steno: Archive for Philosophy and the History of Science (University of Tokyo), no. 6, p. 131–167 (in Japanese).

Yoshimoto, H., 1992, Chemical studies of young Robert Boyle: Their Helmontian phase and the seminal principles: Kagakushi (Journal of the Japanese Society for the History of Chemistry), v. 19, p. 233–246 (in Japanese).

Ziggelaar, A., ed., 1997, Chaos: Niels Stensen's Chaos-manuscript Copenhagen, 1659, Complete edition with introduction, notes and commentary: Copenhagen, Acta historica scientiarum naturalium et medicinalium, v. 44, The Danish National Library of Science and Medicine, 520 p.

MANUSCRIPT ACCEPTED BY THE SOCIETY 17 JANUARY 2006

Geological Society of America
Special Paper 411
2006

Steno, the fossils, the rocks, and the calendar of the Earth

Nicoletta Morello*

Dipartimento di Storia Moderna e Contemporanea, Università degli Studi di Genova, via Balbi, 6, I-16126 Genova, Italy

ABSTRACT

This paper deals with the influence that geological research in Italy during the seventeenth and eighteenth centuries had on the reconstruction of Earth's history. The identification of the true nature (i.e., organic) of fossils by Fabio Colonna in the early seventeenth century and, later in the century, the Stenonian sedimentary geology in agreement with the Genesis and the volcanological studies of Giovanni Alfonso Borelli gave the learned men of the Modern Age important tools in order to establish a numerical dating of Earth's age.

In the seventeenth and eighteenth centuries, two Italian people, Jacopo Grandi and Francesco Bianchini, and the French Charles De Brosses, tried to build a calendar of the past of the world on the basis of natural and historical records. Even if they used the data in a profoundly different way, they reached (in fact, they wanted to reach) the same results: the confirmation of the same calendar of Earth's history elaborated by the most famous Bible chronologists at the middle of the seventeenth century (Lightfoot, 1642; Ussher, 1654). De Brosses rejected the Italian dating, following the steps of his friend Buffon. He enlarged the geological calendar but did not understand, just like the two Italians, that at that time any absolute dating of the age of our planet was impossible.

Keywords: Earth, dating, chronology, fossils, figured stones, glossopetrae, Canis carchariae, Prodromus, sixteenth century, seventeenth century, Ferrante Imperato, Fabio Colonna, Kircher, Descartes, Buffon, Grandi, Bianchini, De Brosses.

INTRODUCTION

When we talk about the geological works of Steno, we usually take into consideration the influence they had, especially on the evolution of geology, paleontology and crystallography (Yamada, 2003). But if we try to reconstruct the period when Steno worked and the success of his conclusions in the decades immediately after the publication of the *Prodromus*, we find a wide range of subjects where Stenonian geology had been applied. In the second half of the seventeenth century, for example, among the numerous geological problems such as orogenesis, lithogenesis, the origin of fossils, which had not yet been resolved despite Steno's writings, there remained the question of the age of the Earth. How old was the planet? How long would it exist? In the context of chronological studies (from the Bible and from nature itself), the Deluge was seen as the "marker" of the end of the ancient world and the beginning of a new one that should be concluded with the last coming of Christ. Dating the Deluge could have been a way to calculate the length of the past and of the future time of the Earth. Could Stenonian geology have provided a method for establishing a numerical dating of the calendar of the Earth?

In *Elementorum myologiae specimen*, a work on muscular anatomy printed in Florence in 1667, Nicolaus Steno (Niels Steensen, 1638–1686) added an appendix of the description of dissections carried out by him on the heads and bodies of two

---

*Nicoletta Morello died unexpectedly on Easter day, 2006, before publication of this volume.

Morello, N., 2006, Steno, the fossils, the rocks, and the calendar of the Earth, *in* Vai, G.B., and Caldwell, W.G.E., The Origins of Geology in Italy: Geological Society of America Special Paper 411, p. 81–93, doi: 10.1130/2006.2411(06). For permission to copy, contact editing@geosociety.org. ©2006 Geological Society of America. All rights reserved.

sharks, a carcharhinid, and a congener of the group of the "sea dogs" (called by him *Porcus Salviani*). These formed the subjects of the *Canis Carchariae dissectum caput* and the *Dissectus piscis ex Canum genere*, respectively. These two anatomical exercises, requested by Ferdinando de' Medici in 1666 while Steno was in Florence, were at the origin of the Dane's notoriety. He was already renowned as a skillful anatomist and as one who could also unravel the intricate problem of the question of the origin of fossils. He would later give a more organic treatment of this topic (in his eyes not a complete discussion) in the *Prodromus*, concluded the year after and published in 1669.

Nowadays, the fact that fossils are *ex-vivi* is well known as is the fact that as a consequence of events linked to the chemical-physical environment of burial, they are preserved intact or have been subjected to diverse transformations which have lithified the remains (traces are also intended as remains) in very different ways, nevertheless maintaining sufficient morphological features to be able to trace, even if with some difficulty, the organism to which they belong.

There was not this certainty at the time when Steno was writing. The word "fossil" (in Latin, *fossilis*, *fossilia*, from the Latin *fodere* = to dig) was still ambiguous, since it indicated, as Georg Agricola (1494–1555) had suggested, a rather vast category of "inanimate subterranean bodies" divided into diverse subgroups as is shown in the table of the *De natura fossilium* (1546) (Agricola, 1546; 1550, p. 185v) (see Figure 6 in Morello's article about Agricola, this volume, Chapter 2). Among these were also included fossils in the modern sense of the term, such as the "stones" of a particular shape which resemble typical forms of the animal and vegetable world. In order to indicate them we will here use the expression "figured stones" except in cases where the term "fossil" may be used correctly in its modern meaning. It must be noted that in the seventeenth century, the recognition of the organic origin of some fossils did not necessarily imply the general recognition of the true origin of all fossils for different reasons; for instance, the identity or living analogue with which to establish a comparison and attempt a determination was not known for some fossils (as in the case of the brachiopods) while others, such as the trilobites, were still completely unknown. There were obviously other reasons also.

In the seventeenth century, the two principal interpretations of fossils—as stones generated by nature or remains of the Deluge—ranged widely due to the diverse cultural matrices from which the various authors still drew on. These matrices were reconnected, for instance, to the Latin and Greek sciences, reviewed also through the Arab influence, to the Medieval tradition of lapidaries, to religious teachings, to the new scientific literature produced by the naturalists of the first Renaissance and so on. There was a need to explain how the figured stones were formed and why they were found enclosed in solid rock in order to understand the therapeutic and magical potential and the mystical meanings concealed in each object (think, for example, of the diversity of the shells, the bones of enormous dimensions, the stones such as the *aquilina*, the bezoar, the belemnite, and so on) but also a need to explain the globality of the phenomenon.

It appears that the petrographic theories which supported the inorganic origin of all the figured stones were the only ones capable of explaining the different types of individual lithification. At the same time, they could explain the general process because they included agents and principles which were mostly metaphysical, of an Aristotelian-Platonic nature, such as the exhalations and the influence of the heavens which were active everywhere and concurrently, protected the specificity of the locally produced effects. Thus, for example, the belemnites could have "generated themselves" in many places (something which did not escape the observers of nature) but different from those places where the shells, the *glossopetra*, and the *aquilina* stones, etc., were "born."

The Deluge instead offered the alternative explanation and allowed an organic nature to be attributed to some figured stones but not to all of them. It served above all to explain the presence of marine animal remains in elevated areas far from the sea and in epicontinental areas, a presence which appeared widespread and chaotic. This reading of the paleontological evidence was a consequence of the modern, wide geographical distribution of fossils of this type, but with regard to the confused mixture of forms, this was affected in part by an epistemic preconception which did not induce a search for proof of an "order" among the fossils. The Deluge had been a rapid and catastrophic event, and its waters had scattered the remains of the living world across the globe in a disorderly manner; in part this could be attributed to the vagueness of the determination of the diverse forms of *conchites* (Bivalves) and *cochlites* (Gastropods) due to the lack of a more refined systematic grid which still did not exist even for the analogous living forms. Therefore, the comparison of the analogous morphology with the forms which actually populated the seas could have been a further element of ambiguity. Confirming the invariability of the species certainly did not put in doubt the diluvial event, but on the other hand, during the seventeenth century, it still did not crack the hypothesis that nature was reproduced in the inorganic parts of the living world.

Due to some (presumed) therapeutic properties, the shells, as well as other animals, whether marine or not, had already formed, even in ancient times, part of the popular pharmacopeia (there are numerous citations in the *Naturalis Historia* by Pliny, I A.D.) as amulets and as ingredients in the preparation of medicines used against pathologies of a diverse nature (such as arteritis, dermatitis, dropsy, jaundice, cancer, amenorrhea, ulcers of the head, foggy vision, parotitis, and as purgatives and antidotes against snake bite). They were renowned for these reasons as well as for alimentary use; for their use by painters as small cups for preparing colors or by pilgrims as drinking cups (*Pecten*); as holy water stoups (*Tridacne*); in the preparation of necklaces; in the adornment of clothes; and in the ornamentation of masonry. They also were widely figured in paintings. The shells were well known throughout Europe, and during the Baroque period they formed an important decorative element perhaps due to their sometimes contorted and spiral form. It is not surprising that they were often collected, given the great variety of species in existence and the remarkable variability in their ornamentation. These collections often included the lithified forms which were similar but at the same time different from the modern species (due to the loss of the ornamentation, the original organic material, cracking, etc.).

Other already renowned figured stones often appeared on the pharmacopoeist's stall, in the jewelry shops, in the cupboards of the nobles and the rich, in collections and in naturalistic museums. The stones started to be discussed from the sixteenth century onward (as it is shown by the documentation): the *glossopetra* (Mercati, 1717, see "Armarium nonum, Lapides idiomorphoi") (Fig. 1). They had also been mentioned by Pliny. Over the centuries, their therapeutic potential had altered very little, linked as they were (due to their trigonal shape, which recalled the teeth or tongue of a serpent) to the ophidians and their venom. They had also been listed in the lapidaries together with the *ceraunie*, stones shaped like the point of a lance, which were thought to have fallen from the sky during thunderstorms (these were often prehistoric stone artifacts). Their greatest diffusion was, however, due to their use against snake bites, as amulets for favoring the cutting of teeth, as tools for stripping flesh from bones when teeth had been lost. When ground down, they were excellent toothpastes. Scattered on the ground throughout a good part of Europe, the *glossopetra* acquired a chosen country, a true denomination with a protected origin when, above all in the sixteenth century, Malta made an Europeanistic advance, presenting itself as the last (in a spatial sense) defense of Christianity against the Turks. The "great christianity" also introduced its *glossopetra*, buried in the Tertiary sediments of Malta and Gozo (*Middle Globigerina Limestone* of Miocene age), into the history of its Christianization, which, begun with the works of St. Paul, started a certain "St. Paul" cult in the archipelago which lives on today.

The *Acts of the Apostles* (27,9–28,10) gives an account of the shipwreck of St. Paul on the Maltese coast (St. Paul Bay) (Fig. 2) (Abela, 1647, p. 225), of his stay on the island, and of the conversion to Christianity of the nobles and the people of Malta convinced of his thaumaturgical powers. The story of the Maltese *glossopetra* is intertwined with the event of the shipwreck: bitten by a viper while preparing a fire to warm himself and the other survivors, the Apostle not only survived, thus showing his virtue as a healing saint, but he also carried out a miracle, depriving the snakes of Malta of the tools of their aggression and bestowing to the ground of the island the capacity to produce stones in the form of the tongues and teeth of snakes that had the power of being an antidote against snake poison. They were also mentioned elsewhere, apart from the episode of the shipwreck of the saint. The *glossopetra* may not be connected to a uniform tradition, and the diverse *lectiones* of the description of their use and their therapeutic properties show some variation that reflects local and profane traditions.

However, the Maltese *glossopetra*, together with the "ground of St. Paul" and the tablets produced from it (Fig. 3), so strictly associated with the St. Paul hagiography, became an excellent medicine for the same spectrum of pathologies due to poisoning as the popular therapies had defined in previous centuries. The grafting of the sacred onto the profane gave new life to the inorganic origin of the *glossopetra* which, during the sixteenth century, had instead already induced some naturalists to discuss it in a contrary light precisely due to their accentuated similarity to the organic parts of living animals. Minerists, pharmacists, and naturalistic collectors as well as physicians, botanists, and "zoographers" from Agricola and his circle of friends to Ulisse Aldrovandi (1522–1605) (Vai and Cavazza, 2003), Ferrante Imperato (1550–1625), and others, and above all those humanists who approached the works of the authors of the classics, were looking at the same objects of the past with a new competence derived from a careful observation of nature in the present. The comparative method which was defined by the *collatio* of the codices for the reconstruction of the original Greek and Latin, became in the hands of the humanist-naturalist the conclusive instrument which from exegesis lead to the living nature, the end terminal of comparison in order to establish the authentic content of the text.

During the sixteenth century, the fact that naturalistic investigations appeared to be almost an imprint of the philological investigation, even if requested by various cultural factors, was not a historiographic analog: the apparent analogy was a consequence in many cases of a true "slipping" of the philological method from the text to nature in as much as, still at the end of the sixteenth century, textual exegesis and descriptive knowledge of nature were deeply intertwined (Fabio Colonna would be a striking example). However, a critical attitude toward tradition and toward the original explanations developed a growing autonomy from the contents of the science of the past which led to a real change in the horizons of the conception of nature and its phenomena. This did not mean, however, a complete rejection of, for example, the assumptions of Aristotelian biology and "meteorology." The effects of this transformation were the recognition

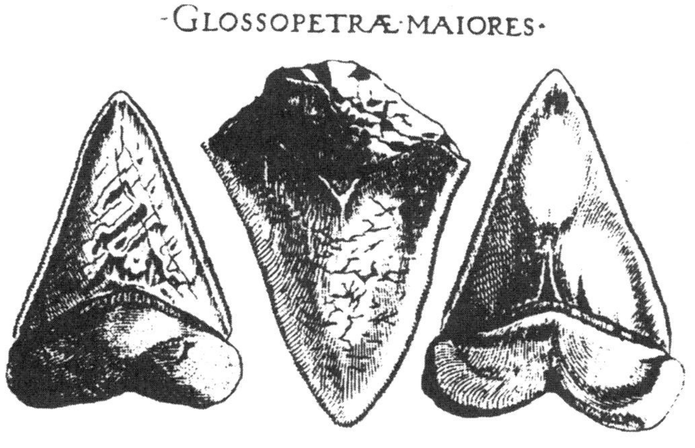

Figure 1. The *glossopetrae maiores*, i.e., the large ones (= teeth of *Carcharodon Carcharias*). From M. Mercati (1717).

Figure 2. Saint Paul Bay (Malta) where the Apostle was shipwrecked and bitten by a snake. From Abela (1647).

Figure 3. One of the famous tablets of St. Paul with a representation of the saint and the snake.

of autonomous principles which regulated the course of natural facts (nature works *juxta propria principia*); a better knowledge of animal, vegetable, and mineral morphology, resulting immediately in more accurate descriptions of natural bodies and allowing new forms to be recognized (not only from the Americas but also from the surrounding world); and a more acute identification of specific morphological characters (lato sensu) with diagnostic value, for more precise determinations that sustained the new attempts of classification. The refinement in observation was translated—for the fossils and above all for those mentioned here, among which the *glossopetra* are a sort of emblem—into a meticulous comparisons, analogies, or identifications with the organic form, and this led to a revision of the ancient interpretations. In the case of an ever-increasing number of figured stones, the similarity with organic morphologies became more precise, and the analogy with living forms or organic parts became ever closer, above all in the case of preservations such as the *glossopetra*: these were shown to be similar to the tongues of birds rather than with the tongues of serpents (Aldrovandi stated this as did some friends of Agricola), but above all to shark teeth, as suggested by Aldrovandi showing a skeletal shark's head (Morello, 2003a, p. 136). However, the question of the true nature of fossils was not resolved by these morphological details, which did not acquire probative value of one origin rather than another; this became a constant of the naturalistic treatises remaining entrapped, so to say, between the Deluge and the petrographic theories. Even if the more widespread explanations in favor of the inorganic hypothesis appeared to some as unsatisfactory, nevertheless the question of the *glossopetra* still remained unsolved.

At least it seemed so to Steno when he decided to tackle the subject again in the *Canis Carchariae* and the *Prodromus*.

FERRANTE IMPERATO AND FABIO COLONNA

In fact, and without having to return to Leonardo—in the sixteenth and seventeenth centuries, his notebooks were not known (Vai, 1986, 1995, 2003)—the question of the shells and the *glossopetra* had already found a solution. At the end of the sixteenth century, the Neapolitan Ferrante Imperato (1550–1625), renowned as an apothecary and for his naturalistic museum (see Figure 8 in Morello's article about Agricola, this volume, Chapter 2), in the *Historia naturale* (Imperato, 1599), returning to the ideas of Agricola on the behavior of the *fossilia* and on subjecting them to empirical tests, had mentioned the difference in reaction to fire of the diverse inorganic bodies. Regarding ambiguous fossils such as the "bucardia" (an internal model of a Bivalve correctly interpreted by him) (Fig. 4) he had briefly described the principal types of fossilization which the organic bodies or their isolated parts were subject. This result was barely considered by Imperato himself— he had not even persisted too much with his conclusions in marine zoology, as original as they were (the nature of corals, the anatomy of the *velella*, etc.)—perhaps due to the main aim of the *Historia*, which was to throw light on the subjects of petrography and mineralogy or perhaps also due to his inability to translate his naturalistic notes into an organic treatise. He was aided by his literary son, Francesco, in the task of writing the work.

It was a friend and pupil of Imperato, Fabio Colonna (1567–1640), a jurist by training and naturalist by passionate curiosity, who exploited and clarified the latter's teachings in botany and marine biology, writing, among others, a brief treatise on the *glossopetra*, the *De glossopetris dissertatio* (1616) (Fig. 5) (Morello, 1979, 1981, 2003a). Summarizing the state of the problem, he demonstrated in this short work the organic origin of these fossils against the petrographic theories and independently from the Deluge. Based on a study of the teeth of a shark and anatomical investigations carried out previously on diverse sharks, he understood that the *glossopetra* were certainly shark teeth. It was from a comparison of the anatomical structure of the fossil and an actual tooth, however, that Colonna understood the

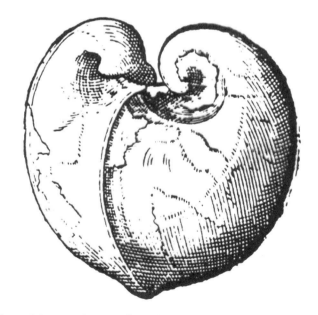

Figure 4. Imperato's Bucardia, an internal model of *Megalodon* sp.(?). From Imperato (1599).

Figure 5. The title page of Colonna's *De glossopetris dissertatio* (1616).

enim dentium usum habere possent, nec testarum fragmenta tegendi, sicuti nec ossa ullum animal fulciendi. Ad hoc affirmandum adducimus aliam observationem quae a Natura inaniter esset elaborata nisi vere dentes non lapides, sed pars animalis demortui fuissent […]. (Dissertatio, p. 3)

If the shape of a stone recalls an animal or vegetable, then that indicates that the stone was an animal or vegetable before having undergone alteration of whatever entity: it should therefore be considered as having an organic nature even if its living analogues cannot be found. This was the case of some fossil shells which Colonna found in the south of Italy whose particular shape consisted of an inequivalve being (they were the Brachiopods—*Terebratula* and *Rhynchonella*—the living analogues of which were not known until the eighteenth century), and which he thus called *Conchae anomiae* (Fig. 6). Regarding these, he commented in the pages of the *Purpura* (1616, p. 2):

nature of the preservation. He deemed the comparison favorable for "demonstrating" the organic origin of the *glossopetra* and for putting an end to the old dilemma. He looked for and found empirically the confirmation of his inferences by subjecting the teeth and fossils to the action of fire and thus after burning them, he obtained identical residues (both carbonized). He therefore concluded definitively in favor of an organic origin of the much discussed *glossopetra*.

In support of the empirical proof, there was also, however, a metaphysical principle which had guided biology for centuries: the finalism or teleology of nature. Nature, which does nothing in vain, certainly did not waste itself in reproducing in a stone a very detailed morphology such as that of the shark teeth: trigonal shapes, serrated, oriented according to the position of the jaw, and endowed with a form which clearly showed that its physiological purpose was for the laceration of prey. The inter-relationship between morphology and physiology, which was attained from this principle of an Aristotelian matrix, lead Colonna (1616) to the conclusion that the *figura* of a stone alone was enough to indicate its true nature.

> E' del tutto falso che le ossa si possano generare in terra come Plinio riferisce da Teofrasto; la natura infatti non fa nulla invano, secondo l'assioma diffuso tra i filosofi, e questi denti sarebbero inutili se non potessero svolgere la funzione di denti, i frammenti delle conchiglie proteggere e le ossa esser di sostegno a qualche animale […]
> Falsum omnino est ossa in terra esse genita, ut Plinius ex Teophrasto refert; non enim natura quid frustra facit, vulgato inter Philosophos axiomate, dentes hi frustra essent, non

Figure 6. The bivalve shells named by Colonna *anomiae* because of their unequal shells. Colonna never saw this kind of animal alive (Brachiopods). From Colonna (1616).

[…] si vede un ammasso di diverse conchiglie e non ce n'è una intera ma tutte sono mescolate e avvinte, rotte non dalla natura ma dall'impeto del mare […] ciò mostra a sufficienza che qualche volta il mare avesse invaso e poi abbandonato diverse regioni […] E' anche degno di nota che queste conchiglie non si trovino adesso vive per qual cosa penso che siano state portate via, lontano dal mare, da un'alluvione e non che la natura abbia smesso di produrle.

[…] maxima variarum testarum congeries conspicitur, nec unam reperies integram sed omnes inter se congestas et implicatas ut non Natura sed maris impulso fractas […] quare mare aliquando variis in regionibus excrevisse et aestuasse satis constat […]. Mirum quidem est huiusmodi testas recentes et vivas hodie non reperiri, quamobrem e longe maris alluvione profectas et advectas censemus, quam Naturam desiisse similes parere […].

Colonna (1616) also clarified why the *glossopetra* and other marine fossils were found within rocks and terrains far from the sea but with arguments which touched only fleetingly the diverse types of fossilization and the formation of sedimentary terrains. Without rejecting the action of the Deluge, he allowed the hypothesis of local variations in the relationship between the land and sea to flourish (*Dissertatio* p. 8): "that in different centuries and different places the changed earth and the sea could alternate each other" ([…] *che in secoli diversi e in luoghi diversi la terra mutata e il mare si fossero vicendevolmente alternati* […]) (*aliis saeculis aliis in locis mutatam tellurem et marem vicissimque alternasse*), thus also the burial of the ancient marine animals, now fossils, could have occurred at different and repeated times.

More than this, however, he did not say. On the other hand, his aim had already been reached: he had resolved definitively the question of the true origin of the *glossopetra* and of all the other fossils using the conceptual instruments available to him and the empirical proof and arguments of a prevalently biological and paleontological nature grafted onto a teleological conception of natural behavior.

STENO VERSUS KIRCHER ?

When Steno, 52 years after Colonna, once again confronted the question of the *glossopetra*, many changes had already taken place in scientific knowledge and also in the investigations of inorganic nature. The relevance of these changes was such that here they are only mentioned in order to emphasize that the studies of the inorganic nature in general and in particular those of Steno had matured within a situation of tension between innovation and conservation whose most radical expression was shown by the Galilean event. The teachings of Galileo and his followers were not to be renounced, for even among the Jesuits, many of those carried out the role of rereading the novelties of science and elaborating upon scientific knowledge—as happened in the *Mundus subterraneus* (1664–65) by Athanasius Kircher (1602–1680) (Morello, 2003b, 2004). He did not leave these completely out of consideration but incorporated them as far as was possible within the perspective of the indoctrination of the young minds, as confirmation of the truth of the sacred pages according to the understanding of the Counter Reformation Church.

It should be noted, however, that in the second half of the seventeenth century the conclusions of science were not so univocal in all the fields of knowledge—in reality, they were not even included within the same astronomy with regard to the motion of the Earth—and this embracing of other solutions which to us appear contradictory represents the cultural substrate from which modern science emerges. The studies of Earth still had links to the meteorological literature and in particular to the *Meteorologica* by Aristotle (IV century B.C.), which included a group of phenomena distinct from those of the heavens, situated between the underground and the internal surface of the sphere of the Moon. René Descartes (Cartesius, 1596–1650) in the *Principia philosophiae* (1644) again dealt with the subjects of the Aristotelian meteorology, inserting it in a theory which, on the basis of physical and metaphysical principles, agreed with the "geogony" and the geological constitution of the planet. Matter and motion together produced a structure of Earth with superimposed and subhorizontal layers that could be explained in mechanistical terms. However, the layers of the globe identified by Descartes were physically undefined (and chronologically indefinable) and were being crossed continually by particles of matter of different provenance. The planet owed many of its phenomena and geological events (volcanoes, earthquakes, subterranean waters etc.) to the *horror vacui* conceived outside of an historical dimension and a progression of Earth.

Meanwhile in the same decade, numerous widely known works on chronology were printed, such as those by the Bishop of Armagh James Ussher (1650 and 1654), the vice chancellor of the University of Cambridge John Lightfoot (1642), and loyal followers of the work of the Parisian theologian Gilberto Genebrardo (1567), testifying to the growing interest on the part of historians and commentators on the sacred texts in the history of humanity and the chronology of the Earth. These works were elaborated on the basis of calculations deduced from a biblical exegesis and profane literature, and they offered a calendar of the history of the planet from the Creation up to the present times. The Deluge was an important element but of difficult collocation within this chronological scanning. It traced a line of demarcation not only in the moral history of humanity but also directly on the planet, it upset the surface, returning it with an altered physiognomy. Inasmuch as it was not univocal, its date was generally collocated as one millennium after the Creation (considered to have occurred in 4004 B.C.).

Connected to the Deluge and the changes it caused to the globe was the orogenetic question which in the seventeenth century dealt with the generation of the mountains as a consequence of the Deluge or Divine Creation or, as Descartes had hypothesized (Fig. 7), due to a passive elevation following the collapse of portions of Earth's crust. Each of these three hypotheses were associated with diverse phenomena such as the presence of relief

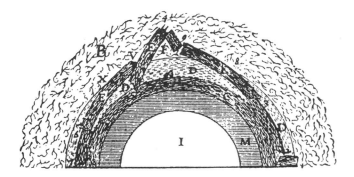

Figure 7. The Cartesian hypothesis of the origin of mountain ranges. Fractures in the upper hard sediments of Earth's original crust caused the collapse of portions of rocks. Not all of them could conserve the original, subhorizontal position, rising up to the level of the surface of the planet. From Descartes (1644).

of different typology (the great mountains, the hills, the volcanoes and the emergent Earth itself), the remains of marine animals far from the sea, cavities full of water or air and the circulation of water (fresh water, thermal springs).

In such a varied panorama, a unitarian explanation of these phenomena of the inorganic world was missing which, as Steno noted in the first pages of the *Prodromus*, had many partial explanations because the phenomena themselves were conceived as being genetically distinct. These divisions produced complications of the individual theories and prevented the recognition of those elements of correlation which would have allowed, within a Stenonian prospective, the resolution of numerous apparently unrelated naturalistic problems. It was for this reason that Steno placed within one category all the inorganic phenomena which could be interpreted as "solid bodies included in other solid bodies" and whose origin could be explained with an inductive-deductive procedure which from the *figura* of the body would allow the way and the place of its production to be obtained and to thus explain its origin. The resulting *Prodromus* was a treatise on the inorganic, but more than that it was the definition of a vast but unitarian conceived field of knowledge which explained with mechanisms that put in action *secundum naturae leges* the formation of solid bodies ranging from Earth's crust to crystals, pearls, and fossils.

In the *Canis carchariae* of 1667, Steno, on the basis of ten pieces of observed data (*historia*) and six hypotheses (*coniecturae*), gave a solution to the problem of the *glossopetra*. Its strong explicative point was that he had connected these fossils to the formation of the sedimentary rocks. He had developed the vague references (also by Colonna) to the ingressions of the sea on the emergent land, hypothesizing that these episodes were repeated for periods long enough to allow the life (and death) of many marine animals, whose remains settled to the bottom of the sediment while material in suspension was being deposited. Upon the retreat of the waters, the residual mud dried and consolidated in the open air, forming superimposed and horizontal strata. The sedimentation was not an isolated episode in time but was repeated many times, enclosing also traces of other natural events (eruptions, floods) which could have occurred before the subsequent ingression of the waters started a new sedimentary cycle.

In effetti, se esaminiamo la loro distribuzione nel terreno, non sembra che si siano potuti ammassare così come si trovano senza che si possa dire che lo hanno fatto lentamente, insieme al sedimento dell'acqua [...] il sedimento che si forma cresce solo con lentezza, per questo gli animali già aderenti al fondo o morti- e quindi resti di cadaveri- o impossibilitati a muoversi sono stati coperti da un nuovo sedimento. Gli altri animali, vivi al di sopra di questo sedimento, riempiono le acque con una prole numerosa prima che anche qui si deponga un altro sedimento.
Sane si illorum in terra situm esaminare libet, non videntur eo modo congeri potuisse nisi cum aquae sedimento sensim dicantur congesta [...] non nisi lente id sedimentum concrescit: unde non nisi quae in fundo iam tum haerent sive mortua animalia, mortuorum spolia sive viva, sed motui inepta, novo sedimento obruuntur; reliqua vero animalia viva, et supra dictum sedimentum eminentia numerosa prole aquas replent antequam novum ibi sedimentum deponatur [...]. (Morello, 1979, p. 130)

Steno included in the *Prodromus* his conclusions on the formation of the rocks expressed in the *Canis Carchariae*, making it the principal lithogenetic mechanism. He distanced himself from Descartes on orogenesis, even adopting the idea of a passive formation of the relief; and, he differed from Kircher, who considered the mountain chains created directly by God, as he intended the orogenesis in general as an effect of natural processes. In particular, he attributed the formation of the great mountains to the sedimentation which occurred in the waters of the primordial oceans, whereas the minor relief could be connected to depositions which occurred during the Deluge. Steno could thus emphasize, without opposing or excluding them, both the constant behavior of nature and the existence of a primigenial ocean as well as the Deluge to which he attributed a second planetary deposition inserted within the previous morphology of the Earth, renewing and rejuvenating it.

The fossils, whether *glossopetra* or shells, had in the Stenonian hypothesis a very clear relationship with the environment in which they were found: they were closely connected only as a result of the sedimentary processes which had entrapped them during the formation of the strata. Like every body in suspension in a liquid, the remains of living organisms were also deposited at the bottom of the subacqueous environment. Because the primordial ocean was devoid of life and because there were only two great global lithogenetic episodes, the fossils could have remained enclosed only in the sediments formed during the Deluge. In the Stenonian geology, therefore, the fossils did not only have their organic origin confirmed but they acquired an ancient significance in a rigorous new way by taking everything into consideration. The new relationship that Steno identified between fossils and the post-Deluge sediments made them *ex vivi*; for him their true nature was no longer in discussion. They became relative chronological indicators being diagnostic elements for distinguishing the post-Deluge rocks from those of the well solidified fine sediment that formed the (apparent) azoic basement of the Earth's crust. At the same time however, they became un-attackable witnesses of the Deluge (or of a second planetary submersion) as they could prove the reality of the happening *without* having to blame the Deluge for "demonstrating" the biological origin of the fossils.

The Deluge was instead "evoked" in the comment to the plate of the well-known six *facies* of the Earth in the appendix to the *Prodromus* in order to underscore the congruence of nature and the Bible (Fig. 8). Despite that, Steno did not accept that the mountain chains could be linked to the creation of the world. It was precisely in the Stenonian orogenesis that the attitude of compromise by Steno between the content of the *Canis Carchariae* and the *Prodromus* was manifested in particular. This was not contradictory but rather the expression of the exigency, difficult to satisfy in geology, of making the conclusions of a global geological theory coincide with that which we may call "geodiversity," transferring terms and concepts from the biological sciences.

The orogenesis was not, however, the only point of disagreement with Kircher; another was the power of the diluvial phenomenon, which, in the rigorously creationist perspective of the Jesuits, could not have made a deep incision into the original structure of the planet. For Steno, the Deluge and its episodes of sedimentation had reconstructed the face of the planet with a global mechanism and unity of deposition which could explain the horizontality and the superimposition of the sediments, and which was repeated after the collapse (reminiscent of Descartes but with rather more immediate evidence from the observations of the geology of the Tuscan basins) that had generated the first morphology of Earth's crust. Deposition and consolidation (diagenesis) unified genetically the azoic primordial rocks, the basement of Earth's crust, and the post-diluvian sediments. These, however, were differentiated by their petrographic nature, by the position of their strata, and by the absence or presence of prevalently marine fossils. These lithological and paleontological characters of the rocks allowed Steno to construct a geological chronology that could be divided into two periods in full agreement with the sacred pages.

On the point of the placement of the mountains, Steno also distanced himself from the Jesuits: he rejected the distribution of the mountain chains according to a divine plan by which they had a precise orthogonal direction, and he emphasized their casual geographic position. In reality, he thus lost the conception of a directional mountain system that Kircher, instead, was able to include by studying the geography of the planet (Fig. 9). Therefore, what Steno rejected was the finalistic justification supporting the Kircherian conception of mountain and volcanic systems and, more generally, the whole structure of the globe. The absolute providence of the divine project did not leave space for the behavior of nature, which, as Steno clearly understood from investigations in the field, had deeply altered the face of Earth's surface more than once. In Kircher, there was no alteration and therefore no space for the history of Earth.

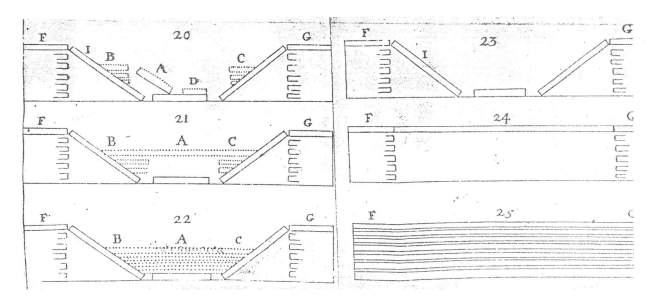

Figure 8. The well-known plate by Steno from the *Prodromus* (1669) showing the different *facies* of Earth before and after the Deluge. The succession of the various moments in the geological history of Earth and the unconformity between pre- and post-diluvial rocks, a mark of their age, is clearly drawn.

Figure 9. The Kircherian hypothesis on the position of mountains on Earth's surface. God gave them an orthogonal frame for keeping the planet safe during Noah's Flood. From Kircher (1678).

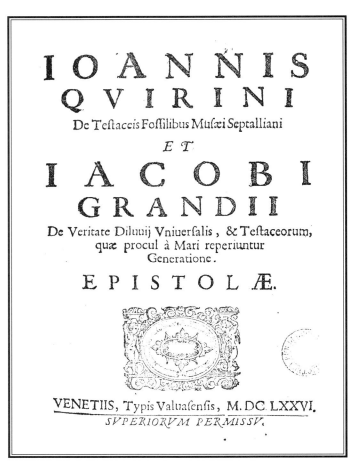

Figure 10. The title page of Grandi's book (1676) on the fossils, the Deluge and the truth of the Bible.

THE STENONIAN INFLUENCE ON THE CALENDAR OF EARTH

A few years after the publication of the *Prodromus*, one of the first applications of the fundamental principles of the Stenonian geology (the horizontality and superimposition of strata) directly exploited their chronostratigraphic valence. Some traditionalist chronologists, followers of the *Vulgata*, intended to confirm the official dating of the age of Earth and of the Deluge and in doing this also intended to confirm the reality of the event and the truth of the Sacred Scriptures. These were in counter tendency with the dating elaborated in the past by other cultures (China, Egypt, and others) which backdated considerably the history of humanity and the world. With this aim they looked for elements external to the sacred or literary traditions which, however, were not in contradiction with them. They thus turned to nature, which in their opinion, as the Stenonian lithogenesis also implied, had conserved objective traces of the principal events which Earth's crust had been subjected to over time. These were the *physicae demonstrationes* supporting the *philologicae rationes*.

An example came from Jacopo Grandi (1646–1691) (Dizionario Biografico degli Italiani [Roma], 2002, vol. 58), a Modenese physician and naturalist, and his *De veritate Diluvii* (1676), written in the form of a letter to Canon Manfredo Settala of Milan (Fig. 10). Grandi used the succession of sediments in the Modenese plain (his native land) to evaluate the time elapsed from the modern soil to the formations which were found lower down in the succession. He identified six strata, the latter of which (i.e., the oldest) was composed of pure material, devoid of organic remains and residues of human activity. These strata seemed to have been deposited over a short time within a Maxima flood, which, according to him, corresponded to the Deluge described by Noah. In support of the chronological work were the traces of a distant past when the plain was a swamp but also of the historical events (the invasions by the barbarians, the successive reconstructions of the city of Modena) documented by the *antiquitates* and by the ruins entrapped in the sediments of diverse ages. Like milestones, these marked for Grandi definite moments of an absolute chronology within which, with the same value, the facts of the geological history were inserted.

L'osservazione del sottosuolo che si possono fare nei pozzi che si cavano nella campagna e nella città di Modena, mia terra natale, si prestano a confermare la verità del Diluvio universale […] Dall'incremento annuo del suolo si può chiaramente individuare la serie di anni che sono trascorsi fin qui, rovesciando così quel vano computo di molte migliaia di anni delle tavole favolose degli Egiziani che hanno confuso gli anni con i mesi o della falsa tradizione dei Cinesi, che alcuni hanno inopinatamente accettato (Grandi, 1676, p. 48, 53)

[…] Asserendae universalis Diluvii veritati maxime conferunt subterraneae observationes, quae passim habentur in puteis, qui fodiuntur in Agro et Civitate Mutinensi, ubi mihi natale solum […] ex annuo incremento soli non obscure colligeretur annorum series qui inde fluxent, ad evertendum vanum illud commentum multorum annorum millium quos ex fabulosis Aegyptiorum tabulis, pro annis varii menses supputantium, aut fictitiis Sinensium traditionibus, nonnulli inepte sectati sunt.

Grandi could therefore date the Deluge in the seventeenth century of the history of the world, endorsing the traditional dating which collocated the Deluge as being 1656 years after the Creation (4004 B.C.) and thus 2348 B.C. He thus reutilized and confirmed the chronology of the Parisian theologian Gilberto Genebrardo (*fl. sec. XVI*), established in his *Chronographia* in 1567, one of the sources of Ussher and Lightfoot, the most renowned chronographers of the seventeenth century.

Grandi's method was again taken up a few years later by Francesco Bianchini (1662–1729), a Veronese physician (Dizionario Biografico degli Italiani [Roma], 1968, vol. 10), in *La Istoria Universale* (1697). It was a chronological work intended to "establish the truth of history" through which he composed a uniform system of the events of humanity, within a global geographic dimension and within a "universal" time and space (i.e., from the Creation up to his own time). With this aim, Bianchini reviewed the chronological and antiquarian works printed from the sixteenth century onward. He differed from these obviously in many respects but mainly in his determination to give historical truth to the Sacred Scriptures through "more universal" documents, such as the changes in Earth's surface, which nature had "imprinted in its parts." The problems which Bianchini dealt with were not, therefore, only problems of dating but rather of the certification of the real occurrence of events, such as the Deluge in the biblical tradition, which were also present in myths, stories, linguistic uses, games, and profane artistic and literary manifestations by civilizations of other places and times.

Potevamo aggiugnere in confermazione delle tradizioni uniformi rimaste tra gli uomini ancora i segni di alcune istorie più universali, che le mutazioni medesime della terra impressero stabilmente nelle sue parti. Ma per non iscostarci in quest'opera dalla professione d'istorici […] reputiamo più conveniente l'apportarne due sole […] la prima, che appartiene alla istoria, si è l'osservazione de'crostacei (così appellano il genere delle conchiglie) soliti a ritrovarsi copiosamente né monti, dove certamente sappiamo che le ostriche, e le conchiglie non figliano, né mantenere si possono in vita; e non pertanto si veggono incastrati dal tempo nel mezo de macigni durissimi […] della qual cosa è difficile allegare altra cagione che un allagamento generale del globo terrestre, accaduto per il diluvio […]. La seconda notizia serve di confronto non isprezzabile per la stabilità cronologica dall'epoca del diluvio fino alla nostra età: e questa similmente ricavasi dalla osservazione, forse non molto nota, mentre non ha più di sette anni, da che un incontro non preveduto suggerì di tentarla sotto il Vesuvio. (Bianchini, 1679, p. 245–246)

With the same chronological aim as Grandi, Bianchini utilized the stratigraphic succession drawn by the Neapolitan architect Francesco Picchetti at the base of Vesuvius within which lava and sediments were alternated. He established as a chronological element of reference the lava flow of 79 A.D. from the eruption that destroyed Pompei (collocated by him in 82 A.D.). The certainty of the exact identification of this lava flow was derived from the historical considerations but also from antiquarian knowledge, which allowed him to date an inscription found in association with the Vesuvian lava to the first century A.D. This inscription remained uncovered during the excavation works, which Bianchini maintained as being remnants of the city of Pompei. The Plinian eruption was followed by the eruption of 472 A.D. and by that of 1631: two recent episodes of great violence (and not three, since the eruption of 79 was also the pivot for the dating of the past) left on the terrain lava flows of adequate dimensions and in the literature a rich testimony of the event and its gravity.

For the past (i.e., the succession of lava and sediments that preceded the lava flow of 79 A.D.), nature testified to three eruptions. The oldest of these was represented by a lava flow of great power that rained on the terrain the characteristic "Stenonian" deposit: subhorizontal, homogenous material, very compact and devoid of any trace of organic life or human activity. Underneath these, there was a great quantity of pure water beyond which there was no trace of sediments and which for Bianchini was the residue of the diluvial flood.

Unlike Grandi's however, Bianchini did not use the sediments and their thickness for dating. Instead, he utilized the lavas emplaced after eruptions of such magnitude and widely documented by history. Bianchini was convinced that nature was rigorously uniform in time and modality of the occurrence of its phenomena. Thus, his calculation could be established on the time intervals that separated the eruptive events (~800 years per eruption) and which would have corresponded to the period nature required to prepare the eruption. If almost 1600 years had passed from 79 A.D. to 1631, ~2400 years would have passed for the three previous eruptions (i.e., from 79 A.D. to the first documented lava flow, which rained on the basement of the crust below which the waters of the Deluge could be found). The Deluge, therefore, had taken place within the same cycle of time as the first Vesuvian eruption, and the date agreed with the one that the traditionalist chronographers had identified on the basis of the exegesis of the sacred texts.

In the first decades of the eighteenth century, this "Stenonian-based" method of dating was again taken up and discussed by Charles De Brosses (1709–1777) in a letter he wrote from Italy to his friend Georges Louis Leclerc Comte de Buffon (1707–1788) regarding a conversation and a previous letter exchanged with Jacques Philippe Fyot de Neuilly (1702–1774) (De Brosses, 1858, p. 286–298) (Fig. 11).

De Brosses, who wanted to have a more precise age of Earth and dating of the Deluge rejected the calculation carried out on the lava flow and the intervals of time between the eruptions, contesting Bianchini's method, which he cited extensively. He sustained instead that a valid calculation should be carried out on what the Veronese physician had expressly rejected, and that was the thickness of the sedimentary strata. The latter were useful because they were interposed between the lava flows of the powerful eruptions, thus facilitating their identification, but also because they were the only means which allowed the calculation of the duration of time. The slow phenomena such as deposition rather than sudden events like the eruptions were for De Brosses the better chronological indicators. The result of this revision of the Grandi-Bianchini method was a calendar which backdated the antiquity of Earth by 42,000 years, in agreement with the length of geological time of which Buffon had spoken in his uniformistic view of the geological phenomena expressed both in the *Théorie de la Terre* (1749) and in the *Époques de la nature* (1780).

> [...] Il nostro globo, per 35000 anni, non è dunque stato che una massa infuocata alla quale nessun essere sensibile poteva avvicinarsi; poi, per 15000 o 20000 anni, la sua superficie è stata completamente coperta dal mare: è percorso un lungo susseguirsi di secoli perché la terra si raffreddasse e le acque si ritirassero, e soltanto alla fine di questo secondo periodo la superficie dei nostri continenti è stata modellata [...] (Buffon, 1960, p. 142).
> [...] Cause il cui effetto è raro, violento e subitaneo [...] non fanno parte dei processi ordinari della natura: ma effetti che si verificano tutti i giorni, movimenti che si succedono e si rinnovano continuamente, operazioni costanti e sempre ripetute: qui sono le cause e la ragioni che cerchiamo[...] (Buffon, 1853–1855, p. 142).

Nell'anno 1689. in una cava fatta nella falda del monte Veſuvio circa un miglio lontana dal mare, nel loco, dove era la villa di Pompeo, ſi oſſervò, che la terra ſoda, e la pietra vetrificata erano collocate con bell' ordine in diverſe regioni; mentre tanto il terreno, che di continuo cala giù del monte alla pianura, e nel mare; quanto il liquore della pietra fuſa, e vetrificata, che in diverſi incendj del Veſuvio hà inondato nella pianura, hanno diſpoſto quelle regioni col ſeguente ordine:

Prima ſi trova da 12. palmi in circa di terra da coltivarſi.
12. *Palmi di terra che ſi coltiva. Appreſſo ſi trova*
4. *Palmi di pietra nera vetrificata, della quale è laſtricata la Città, poi*
3. *Palmi di terra ſoda, poi*
A 6. *Palmi, e mezzo di pietra vetrificata, ſotto della quale ſi ritrovano alcuni carboni, chiavature di porte di fetro, e due iſcrizioni, le quali dimoſtravano quella eſſere ſtata la villa di Pompeo, poi*
10. *Palmi in circa di terra ſoda, poi*
2. *Palmi, e mezzo di pietra vetrificata ſimile à quella di ſopra, poi*
8. *Palmi di terra aſſai più ſoda, poi*
B 4. *Palmi in circa di pietra vetrificata, mà più ſquamoſa, e leggiera della prima.*
25. *Palmi di terra aſſai più ſoda, e di durezza quaſi ſimile al tufo.*
C 16. *Palmi della ſolita pietra vetrificata, & aſſai grave. dopo*
12. *Palmi di pietra di tufo ſi ritrovò l'acqua dolce, e viva, e in gran quantità, nè permiſe il cavare piu oltre.*

Figure 11. The report of the Neapolitan architect Picchetti quoted by Grandi (1676, p. 48, 53), Bianchini (1697, p. 245–246), and De Brosses (1858, p. 286–298). From Bianchini (1697).

The field of application of the sedimentary geology of Steno was obviously not only chronological. It is enough to recall Antonio Vallisneri senior (1661–1730) and the solution he gave to the question of the origin of springs (*Lezione accademica*, 1715 and 1726). This is, however, a field less well known in the successful Stenonian geology perhaps because, different from other cases of application, it did not give persistent results.

CONCLUSIONS

The most important result of this review is that the Stenonian sedimentary geology in the second half of the seventeenth century did not remain isolated in its naturalistic context but became a dynamic instrument of the evaluation of natural events in the attempts to quantify the history of Earth. The failure of repeated dating was not due to the application of the method that was able to provide information on the major and minor length of geological time, but rather to the conceptual shortcoming on which these dating were upheld: the desire to construct, on the basis of natural events, a numeric geological chronology that hinged on the calendar originating from the documentation of the history of the people and the sacred pages in order to demonstrate their contents of truth.

This brief reference to seventeenth-century chronological attempts may also prove useful in understanding how complex the acceptance of the Stenonian lithogenesis was providing fruitful consequences but also stopping lines of research as it traced a pathway leading toward the identification of a geological history separate from the history of the living beings. Approximately two centuries would pass before these "histories" would be reconciled with the new conceptual dimensions within which biostratigraphy would originate.

REFERENCES CITED

Abela, G.F., 1647, Della descrittione di Malta Isola del mare siciliano con le sue antichità et altre notizie libri quattro: Malta.

Agricola, G., 1546, De ortu et causis subterraneorum libri V. De natura eorum quae effluunt ex terra libri IIII. De natura fossilium libri X. De veteribus et novis metallis libri II. Bermannus, sive de re metallica dialogus. Interpretatio germanica vocum rei metallicae: Basileae: per Frobenium et Episcopium (translated in 1955 by Bandy, M.C., and Bandy, J.A., in Geological Society of America Special Paper 63, 240 p.).

Agricola, G., 1550, De la generatione de le cose che sotto la terra sono e de le cause de' loro effetti e nature Libri V. De la natura di quelle cose che da la terra scorrono Libri IIII. De la natura delle cose fossili e che sotto la terra si cavano Libri X. Il Bermanno o de le cose metalliche dialogo, recato tutt'hora dal latino in buona lingua volgare: Venezia, Michele Tramezzino.

Bianchini, F. 1697, La istoria universale provata con monumenti e figurata coi simboli degli antichi: Roma.

Buffon, G.L. Leclerc, Comte de, 1749, Histoire Naturelle Générale et Particulière, v.1: Paris, Imprimerie Royale, [iv] + 612 p. (reprinted in 1884–1885, Oeuvres complètes, nouvelle édition par J.L. de Lanessan Paris, 14 vols.).

Buffon, G.L. Leclerc, Comte de, 1778, Les époques de la nature, 2 vols.: Paris, Imprimerie Royale, (reprinted in 1884–1885, Oeuvres complètes, nouvelle édition par J.L. de Lanessan Paris, 14 vols.).

Buffon, G.L. Leclerc, 1960, Le epoche della natura: Torino. Boringhieri (Italian translation).

Colonna, F., 1616, De glossopetris dissertatio qua ostenditur melitenses linguas serpentinas sive glossopetras dictas, non esse lapideas[…], in Fabii Columnae Lyncei Minus Cognitarum stirpium [...] Ekphrasis [...] Pars altera: Romae apud Jacobum Moscardum. (The Dissertatio is to be found after the Fabii Columnae Lyncei Purpura […]).

De Brosses, C., 1858, Lettres familières écrites d'Italie à quelques amis en 1739 et 1740, edited by Babou, H.: Paris.

Descartes, R. (Cartesius), 1644, Principia philosophiae: Amstelodami, 310 p.

Genebrardo (Genebrardus), G., 1567, Chronographia in duos libros distincta. Prior est de rebus Veteris Populi. Posterior recentes historias praesertim Ecclesiasticas complectitur: Parisiis.

Grandi, J., 1676, Joannis Quirini De testaceis fossilibus Musaei Septaliani et Jacobi Grandi De veritate Diluvii Universalis et testaceorum quae procul a Mari reperiuntur Generatione Epistulae: Venetiis. Valvasensi.

Imperato, F., 1599, Dell'Historia naturale Libri XXVIII. Nella quale ordinatamente si tratta della diversa condition di miniere e pietre. Con alcune historie di piante et animali, sin'hora non date in luce: Napoli.

Kircher, A., 1665, (1678: third ed.) Mundus subterraneus in XII Libros digestus, quo divinum subterrestris Mundi opificium, mira ergasteriorum Naturae in eo distributio, verbo παντομορφον Protei regnum, universae denique Naturae majestas, et divitiae summa rerum varietate exponuntur. Abditorum effectuum causae acri indagine inquisitae demonstrantur; cognitae per artis et naturae coniugium ad humanae vitae necessarium usum vario experimentorum apparatu, necnon novo modo et ratione applicantur: Amstelodami, apud Joannem Janssonium et Elizeum Weyerstraten, v. 1, [xxvi] + 346 + [vi], v. 2, [x] + 487 + [ix].

Lightfoot, J., 1642, A few and new observations on the book of Genesis: the most of them certain, the rest probable, all harmless, strange and rarely heard before: London.

Mercati, M., 1717, Michaelis Mercati Sanminiatensis Metallotheca. Opus posthumum. Auctoritate et Munificentia Clementis Undecimi P. M. e tenebris in lucem eductum; opera autem et studio Joannis Mariae Lancisii Archiatri Pontificii illustratum. Cui accessit Appendix cum XIX recens inventis iconibus: Romae (1719 second ed.).

Morello, N., 1979, La nascita della paleontologia nel Seicento: Colonna, Stenone, Scilla: Milano, Franco Angeli, 265 p.

Morello, N., 1981, De glossopetris Dissertatio: the Demonstration by Fabio Colonna of the true Nature of Fossils: Archives Internationales d'Histoire des Sciences.

Morello, N., 2003a, The question on the nature of fossils in the 16th and 17th centuries (La questione della natura dei fossili nel Cinquecento e Seicento), in Vai, G.B., and Cavazza, W. eds., Four Centuries of the Word Geology: Ulisse Aldrovandi 1603 in Bologna: Bologna, Minerva Edizioni, p. 127–152.

Morello, N., 2003b, The geophysics of Athanasius Kircher, in Schröder, W. ed., Alte und neue Probleme der Physik und Geophysik, Bremen/Postdam: Band IV, v. 1, p. 110–119.

Morello, N., 2004, Il geocosmo di Athanasius Kircher. Lavori in corso su progetto di Dio, in Vai, G.B., ed., A. Kircheri Mundus Subterraneus in XII libros digestus […], Arnaldo Forni Editore, Bologna, p. 11–20 (photostatic reprint).

Morello, N., 2006, this volume, Agricola and the birth of the mineralogical sciences in Italy in the sixteenth century, in Vai, G.B., and Caldwell, W.G.E., The Origins of Geology in Italy: Geological Society of America Special Paper 411, doi: 10.1130/2006.2411(02).

Steno, N., 1667, Elementorum myologiae specimen seu musculi descriptio geometrica, cui accedunt Canis carchariae dissectum caput et Dissectus piscis ex Canum genere: Florentiae, Stella.

Steno, N., 1669, De solido intra solidum naturaliter contento Dissertationis Prodromus, Florentiae, ex Typographia sub signo Stellae, 123 p.

Ussher, J., 1650 and 1654, Annales Veteris et Novis Testamentis e Annalium pars posterior: Londini.

Vai, G.B., 1986, Leonardo, la Romagna e la geologia, in Marabini, C., and Della Monica, W., eds., Romagna: vicende e protagonisti: Bologna, Edison Ed, v. 1, p. 30–52.

Vai, G.B., 1995, Geological priorities in Leonardo da Vinci's notebooks and paintings, in Giglia G., Maccagni, C., and Morello, N., eds., Rocks, Fossils and History, Proceedings of the 13th Inhigeo Symposium (Pisa-Padova, 1987): Firenze, Festina Lente, p. 13–26.

Vai, G.B., 2003, I viaggi di Leonardo lungo le valli romagnole: riflessi di geologia nei quadri, disegni e codici, in Leonardo, Machiavelli, Cesare Borgia Arte, storia e scienza in Romagna: Roma, De Luca, p. 37–47.

Vai, G.B., and Cavazza, W., eds., 2003, Four centuries of the word Geology: Ulisse Aldrovandi 1603 in Bologna: Bologna, Minerva Edizioni, 327 p.

Vallisneri, A., 1715 (1726 second ed.), Lezione accademica intorno l'origine delle fontane: Venezia, appresso Piero Poletti, 407 p.

Yamada, T., 2003, Stenonian Revolution or Leibnizian Revival? Constructing Geo-History in the Seventeenth Century: Historia scientiarum, v. 13-2, p. 75–100.

Manuscript Accepted by the Society 17 January 2006

Isostasy in Luigi Ferdinando Marsili's manuscripts

Gian Battista Vai*

Dipartimento di Scienze della Terra e Geologico-Ambientali, Università di Bologna, via Zamboni 67, I-40127 Bologna, Italy

ABSTRACT

The polymath Luigi Ferdinando Marsili (1658–1730) is generally regarded as the founder of scientific oceanography and marine geology. He also founded the renowned Istituto delle Scienze of Bologna in 1711 (a successor to the Florentine *Accademia del Cimento*). Marsili's major scientific concern and unfulfilled ambition was a *Treatise on the Structure of the Earthy Globe* (*Trattato sulla Struttura del Globo Terreo*), a work-in-progress, containing some 200 sheets with more than 35 water-colored plates and some 50 pen drawings, kept in the Main Library of the Bologna University.

Out of this material, Marsili's manuscript 90, A, 21 (dated 1728) is published for the first time as an appendix to this paper. It is an introduction to the planned treatise, accompanied by a summary and detailed index of its contents. This manuscript is useful in understanding the meaning of some recently published illustrations contained in the same collection. Such illustrations, together with their captions and the text of manuscript 90, A, 21, may be considered the earliest suggestions of the principle of isostasy, as well as of the concept of roots of mountain chains. Also expressed in Marsili's drawings are the concepts of Earth's spheroid-ellipsoid surface and the difference in thickness of "marine" (oceanic) and "mountainous" (continental) crust. His hemiglobal, water-colored section shows a balance of mountain-peak height and seafloor depth compared to sea level. This setting is formulated in the text as a general principle describing an isostatic condition. Different colors indicate three different types of crust beneath the deep seas, the low-elevation continental plains, and the mountain chains, respectively.

Keywords: Earth structure, geotectonics, crustal types, deepest depression of seas, top surface of highest mountains, sea horizon, mountain roots, oceanography, marine geology, stratigraphy, three-fold division of rocks, primary mountains, hills, earthy plains, anti-diluvianism, Descartes, Kircher, Woodward, Burnet, Cassini, Newton, Airy, Pratt, Dutton, Heim.

INTRODUCTION

Count Luigi Ferdinando Marsili (1658–1730) played a leading role in the development of European science through (1) his major publications (Marsili, 1698, 1700, 1711, 1725, 1726) and especially through (2) his founding of the *Istituto delle Scienze e delle Arti* of Bologna, the most renowned Italian scientific institution, after the closure of the *Accademia del Cimento*. Founded in 1711, the *Istituto* was the first publicly funded institution in the world employing scientists to do systematic field geology as full-time professionals (Vai, 2003).

Since the nineteenth century, Marsili has been considered the founding father of oceanography and marine geology. This was initiated by the article *Un des fondateurs de l'océanographie* (1897), by Thoulet, who referred to Marsili's treatise on the *Histoire physique de la mer* published in 1725 (Murray, 1880; Pérès,

---

*vai@geomin.unibo.it

Vai, G.B., 2006, Isostasy in Luigi Ferdinando Marsili's manuscripts, *in* Vai, G.B., and Caldwell, W.G.E., The Origins of Geology in Italy: Geological Society of America Special Paper 411, p. 95–127, doi: 10.1130/2006.2411(07). For permission to copy, contact editing@geosociety.org. ©2006 Geological Society of America. All rights reserved.

1968; McConnell, 1999; Seibold and Seibold, 2001; Sartori, 2003). Marsili was a member of both the Académie des Sciences of Paris and the Royal Society of London. He is recognized as one of the first scientists to have described and drawn stratigraphical columns (Lipparini, 1930; Marabini and Vai, 2003; Sartori, 2003; Vaccari, 2003; Vai, 2003; Vai and Cavazza, 2003) and to have provided cross sections of mountain chains (Gortani, 1930, 1963; Marabini and Vai, 2003; Morello, 2003; Trümpy, 1998; Vaccari, 2003) in the last years of the seventeenth and the first of the eighteenth centuries. In some of these sections, representing both continental and marine settings, angular unconformities are clearly represented (Marabini and Vai, 2003). He is also considered a pioneer in marine and lacustrine sedimentology (Sartori, 2003) and a precursor of the concepts of lateral *facies* change and stratigraphic correlation (Marabini and Vai, 2003).

With such an early stratigraphical approach, it is not surprising that Marsili was granted priority in sketching a small, true, geological map (Lipparini, 1930), which shows narrow folded structures in the sulphurous/evaporitic area of the Northern Apennines (Marabini and Vai, 2003).

Marsili's cartographic genius in potamography (fluvial studies), metallography, and geomorphology is paramount in his second masterpiece, *Danubius Pannonico-Mysicus* (published in 1726), and considered influential on Guettard's and Desmarest's maps (Csiky, 1984, 1987; Ellenberger, 1984; Franceschelli and Marabini, 2004).

An important detail observed in Pl. III of the *Histoire physique de la mer*, where the lowest bathymetric section drawn across the Gulf of Lion and Provence shows the height of the mountain peak (left) equaling in magnitude the depth of the sea bottom (right) (Fig. 1), has not been emphasized until recently. This fact was regarded as a graphical anticipation of the principle of isostasy (Sartori, 2003; Vai, 2003). The observation is not casual because Marsili makes repeated reference in his printed works and manuscript to a volume he intended to write and titled *Treatise on the Structure of the Earthy Globe* (*Trattato sulla Struttura del Globo Terreo*), in which the observation would be recorded.

As a matter of fact, the two major works published by Marsili, *Histoire* and *Danubius*, were almost ready in the first years of the eighteenth century but were not published until decades later. Instead, the work on the structure of Earth was left to Marsili's full maturity, perhaps because it represented his major scientific ambition and concern. For all of the second half of his life, he collected in a special file more than 200 sheets of manuscripts, with over 35 water-colored plates and some 50 additional pen drawings, which are still kept in the Biblioteca Universitaria, Bologna (BUB), Fondo Marsili (FM), manuscript 90.

A few of these illustrations recently were published for the first time (Marabini and Vai, 2003; Vaccari, 2003). They are basically self-explanatory in the sense that they could be considered the earliest embryonic representations of the principle of isostasy and of the concept of roots in the mountain chains. Additionally, these illustrations give geological form to the concept of Earth's spheroid-ellipsoid mathematical and geodetic surface that was under testing by means of the French expeditions planned by the noted astronomer Giandomenico Cassini (1625–1712), a former professor at the University of Bologna, for the French kingdom. Marsili's drawings also show the difference in thickness between "marine" (oceanic) and "mountainous" (continental) crust. This paper aims at providing more information and discussion on the recently published illustrations, comparing them to the related captions and accompanying text, and beginning a step-by-step posthumous publication of the material on planet Earth contained in Marsili's manuscript 90, indexed as *Schedae pro structura orbis terracquei*, i.e., "Files on the structure of the terraqueous globe."

Considering the reputation enjoyed by Marsili during his lifetime and later (Brocchi, 1814; Lyell, 1830–1833; Longhena, 1930; Stoye, 1994), it is somehow surprising to see his work either ignored or mentioned only perfunctorily in recent, general and specialized, otherwise excellent books on the history of geological ideas (e.g., Ellenberger, 1988, 1994; Oldroyd, 1996; Şengör, 2003). Worth noticing is the fact that Marsili developed some of his geotectonic ideas when surveying the Provençal Sea and southern France, the latter being the region that inspired the "pseudo-isostatic" vision of Jean Buridan (1300–1358) (Oldroyd, 1996; Şengör, 2003).

At least two reasons can be suggested why Marsili was largely forgotten during the late nineteenth and early twentieth centuries. (1) Almost two centuries of idealistic bias in the study of the history of science in Italy, together with a persistent self-deprecating attitude in Italian cultural circles, resulted in a progressive underevaluation, even ignorance, of the role Italian scientists played in Europe, especially in the seventeenth and eighteenth centuries. (2) More generally, when geologists began to relinquish their leading role in the philosophical and cultural debate following the professional expansion of their science in the nineteenth century, they were replaced by historians of science of purely humanistic background. These historians were not always trained to appreciate the value of geological ideas, especially those hidden or buried in multiple pages of manuscript and written in unfamiliar languages.

MARSILI'S STRUCTURE OF EARTH IN A COLOR PLATE COMPARED TO OLDER NOTIONS OF EARTH

One of the most spectacular color plates contained in section C of Marsili's manuscript 90 is c. 114, titled *Interna structura terrae*, or "The internal structure of the Earth." The plate contains enlarged detail of one of the three mountain chains punctuating the cross section of a hemiglobe in the upper right corner. It is a north-to-south section across the Swiss Alps with the inscription, "*Tali positu, Strata super Strata, in Montibus Helvetiae jacent*," or "Beds over beds lying with this attitude in the mountains of Switzerland." This is a clear witness of the principle of superposition stated by Steno (1669), who is often quoted in Marsili's manuscripts (Fig. 2).

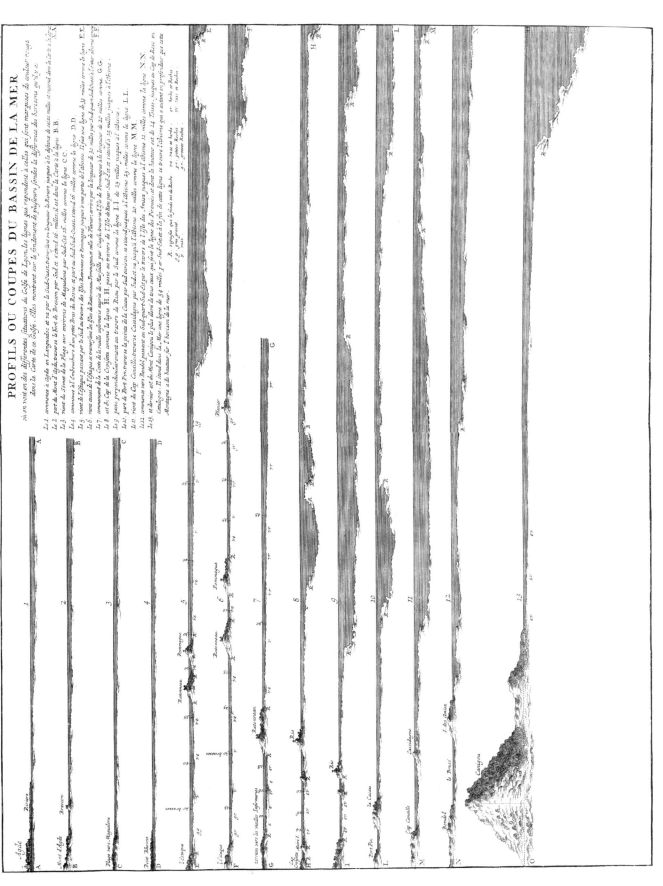

Figure 1. Bathymetric section across the Gulf of Lion where the peak to the left equals the depth in the abyss to the right, thus anticipating elements of the isostatic doctrine (from Marsili, 1725, III Pl. p. 4).

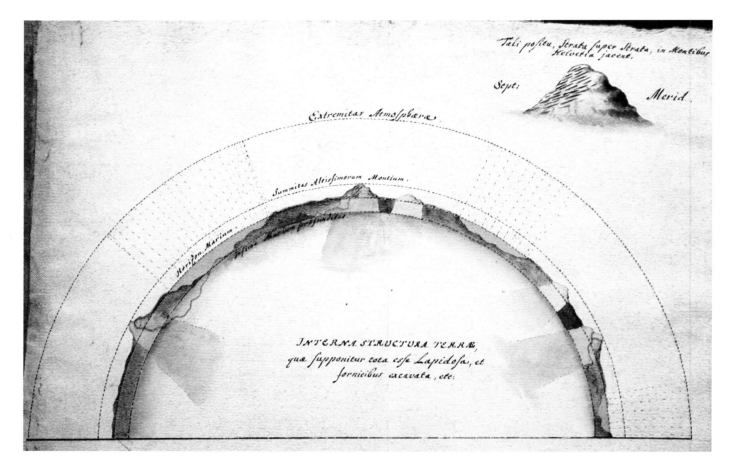

Figure 2. Cross section of the terraqueous globe showing the inner structure of Earth, which is thought to be entirely stony and excavated by arched cavities, etc.; watercolor pen drawing by Marsili (by permission of the Biblioteca Universitaria di Bologna, Fondo Marsili, ms. 90, C, c. 114). Notice that "rainy" springs are clearly distinct from "subterranean" springs, as referred to in the manuscript text (see Appendix).

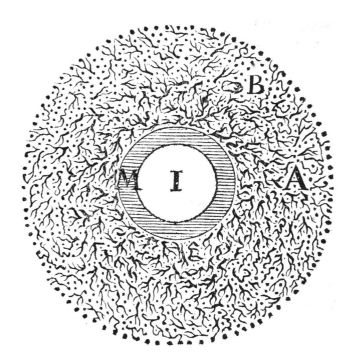

A way of describing and evaluating this posthumously published Marsili figure on the structure of Earth is to compare it to similar figures published earlier by Descartes (1644, 1677, p. 145 to p. 165) and Kircher (1665, 1678). Later, Woodward illustrated his concept in the French and Italian editions (1735, 1739) of his major work. This plate was not in the original edition (Woodward, 1695) nor was it in the Latin version published in 1704, probably slightly before Marsili drew his figure.

Descartes' Earth model in *Principiorum Philosophiae Pars Quarta* shows a four-step evolutionary process that has in common a circular core I ("fiery corpuscles") and a concentric layer M ("matter similar to that of sunspots"), surrounded by an aeriform layer A, rapidly passing into the aeriform layer B (Fig. 3).

Once the four-step evolution had been accomplished, Earth was composed of the following concentric layers from core

Figure 3. An initial step of Earth's evolution. I—fiery corpuscles; M—matter similar to the sunspots; A and B—aeriform (from Descartes, 1677, p. 145; by permission of Biblioteca Universitaria di Bologna).

outwards: I, M (as above), C solid ("*densum, durum & opacum*"; p. 149), D liquid, F aeriform, E solid again ("*prima interior crusta*"), and B aeriform ("*rarum, fluidum & pellucidum*"; p. 149 and figure on p. 153) (Fig. 4).

The body E was permeable and fractured, initially able to remain self-supported in the form of discrete arcs above D and F. As the fractures widened and contraction increased, the individual arcs began gradually dropping to adhere to the top of layer C. Because the surface of layer C is smaller than that of layer E, the remaining arcs of E had to become tilted and to partly overlap each other (figure on p. 156) (Fig. 5). Then, the fluid body D, being lighter than the fragments of E, filled all residual cavities left below and above the fragments of E. At this stage (figure on p. 156) (Fig. 5), one can call: B and F, air; C, the inner crust of the Earth that is the source of the metals; D, water; and E, the outer Earth made of lapideous rocks, clay, sand, and lime. The water lying above fragments 2–3 and 6–7 represents seas, the slightly inclined fragments 8–9 and v-x are plains, and those strongly tilted, 1–2, 9, and 4-v, are mountains (Fig. 5).

Kircher's (1665, 1678) model is quite different, less speculative, and more empirical in the sense that observations play a role in the basic parameters of the model and not only in the eventual details. Kircher's Earth is a solid body punctuated by two internal branching nets of "*pyrophylacia*" and "*hydrophylacia*" (Figs. 6, 7). The two nets are distinct, but intermingled, and form the whole of the Earth's bowels ("*universa Geocosmi viscera*"). *Pyrophylacia* are "fire nests" (a synonym of modern magma chambers), with a major central fire chamber (A) and a series of satellite chambers (B), linked by a net of ascending ("*pyragogi*") channels (C) and related pervasive minor fissures. Channels reaching the surface of the globe originate volcanoes (Fig. 6). *Hydrophylacia* are discrete cavities (corresponding to present-day aquifers and convection cells) filled with water and buried at variable depths from the central fire to the highest mountain chains. Water is formed by condensation of igneous exhalations fed by the central fire and transported through the ascending channels in the form of thermal water and damp,

supplying springs and rivers. Once rich in different minerals, these waters produce metallic bodies by coalescence. Wind and air pressure eject water through subsurface conduits up to the mountains (Fig. 7). Kircher's focus on the internal structure and processes of the globe gives little attention to the proper structure of Earth's crust.

During the first three decades of eighteenth century, in contrast to Descartes and Kircher, Marsili focused on the outer layer of Earth. In its inner part, Earth "is supposed to be entirely solid and excavated by arched cavities, etc." ("*qua supponitur tota esse lapidosa et fornicibus excavata, etc.*") (Fig. 2). The outer layer is laterally segmented into three basic elements: (1) deep seas floored by thin crust (blue), (2) low elevation plains floored by thicker crust (brown), and (3) mountain chains (yellowish) in two different stages of evolution. Both types of mountain chain have roots laterally confined by slightly diverging segments; the higher the mountain chain, the deeper the roots (Fig. 8). Concentric, more-or-less virtual envelopes are drawn, the innermost corresponding to the "*Infima Marium profunditas,*" or "the deepest sea depression," followed upward by the "*Summitas Altissimorum Montium,*" or the "top surface of highest mountains." A thick outer aeriform layer is limited by the upper surface of the atmosphere.

The height of the two crustal layers facing above and below the sea-level surface is the same. As a consequence, sea level once interpolated to extra-marine segments becomes a balanced surface (equipotential surface).

The dashed rays drawn from the top of the atmosphere to sea level are not readily understood. The water cycle is represented as an additional process. This consists of both water flowing down mountains to the oceans and rising from the bottom of the oceans to the top of the mountains through subterranean channels running both below and above the lowest envelope.

Basic inferences from Marsili's figure are: (1) the extrapolated sea-level surface is the equipotential surface of Earth; (2) oceanic crust is different from the crust of flat continental areas, and, again, from the crust of mountain chains; (3) the solid inner

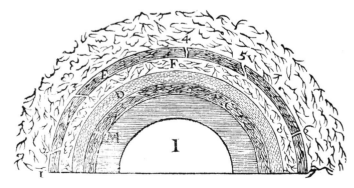

Figure 4. A further step in Earth's evolution. I—fiery corpuscles; M—matter similar to the sunspots; C—solid; D—liquid; F—aeriform; E—solid; B (outermost rim)—aeriform (from Descartes, 1677, p. 153; by permission of Biblioteca Universitaria di Bologna).

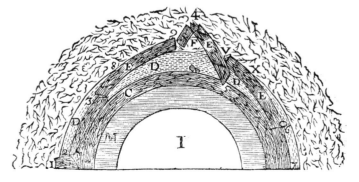

Figure 5. An advanced step in Earth's evolution. I—fiery corpuscles; M—matter similar to the sunspots; C—solid; D—liquid; F—aeriform; E—solid; B (outermost rim)—aeriform (from Descartes, 1677, p. 156; by permission of Biblioteca Universitaria di Bologna).

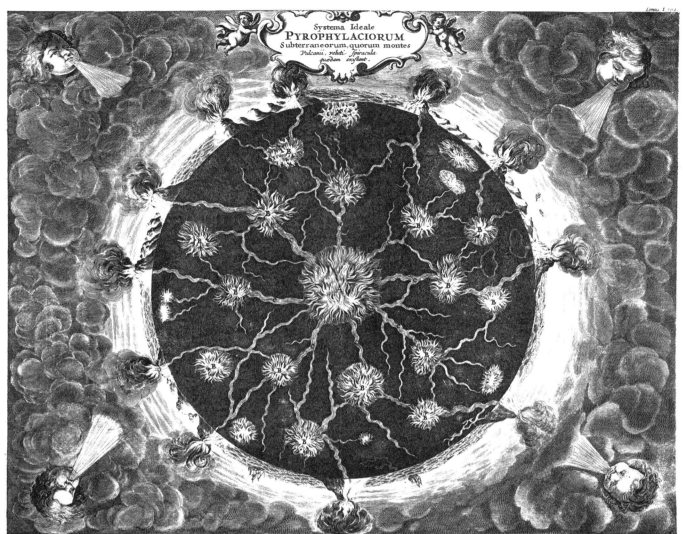

Figure 6. *Pyrophylacia* or fire nests (from Kircher, 1678, plate at p. 104, v. I, by permission of Biblioteca Universitaria di Bologna).

part of Earth contains different amounts of material similar to the crust of mountain chains; (4) the bottom of the oceans is at considerable depth; (5) the mean elevation of the continents is low; and (6) mountainous areas represent a minor part of Earth's surface (Fig. 2). Most of these observations and concepts are nowadays still part of the theory of isostasy.

Even though it does not consider magmatic and volcanic processes, Marsili's model, as seen in this figure, is closer to Kircher's than to Descartes' model.

Woodward's model (1735, 1739) is pictorially attractive, yet more speculative and less observational than any of the other three considered herein (Fig. 9), in spite of continued field activity that allowed him to provide the famous fossil collection still kept at the Gresham College and to establish the first lectureship in geology at the Cambridge University by his bequest (Woodwardian Chair of Geology). Woodward's obsession was to find a reasonable answer to the "Proposition of the Universality of the Deluge."

> I think he [the reader] cannot reasonably doubt of the Proposition: but more especially if hereunto he shall join what I offer concerning the Great Abyss, and thence learns that there is at this day resident, in the huge Conceptacle, Water enough to effect such a Deluge, to drown the whole Globe, and lay all, even the highest Mountains under Water" (Woodward, 1695, Preface).

Figure 7. *Hydrophylacia* or water cavities (from Kircher, 1677, plate at p. 186, v. I, by permission of Biblioteca Universitaria di Bologna).

The water of the great abyss fills the whole interior of Woodward's Earth, as stated in the text (Woodward, 1695) and later illustrated in the French and Italian editions of *Physical Geography* (Woodward, 1735, 1739) (Fig. 9). Even Thomas Burnet, another distinguished British theorist of the diluvian Earth history, did not attempt to assume such a large aqueous interior of Earth in his *Telluris Theoria Sacra* (1699). In Burnet's egg-like *Terra prima* (primeval Earth), which originated and evolved from the Chaos, only layer C represents the Abyss of Water (Burnet, 1699) (Fig. 10).

The two English physico-theologians, Woodward and Burnet, were too concerned about demonstrating a perfect concordance between natural phenomena and Sacred Verse to give more thought to the internal structure of Earth. In contrast, other scientists, notably Marsili and those working in Italy, followed an approach that began first with observational and experimental data.

ISOSTASY

The term and the theory of isostasy were formulated by Clarence Edward Dutton (1889) building upon two quite different, but convergent, models suggested by Georg Biddel Airy (1855) and John Henry Pratt (1855) to fit geodetic-gravimetric observations made in the large mountain chains of the Andes and the Himalayas.

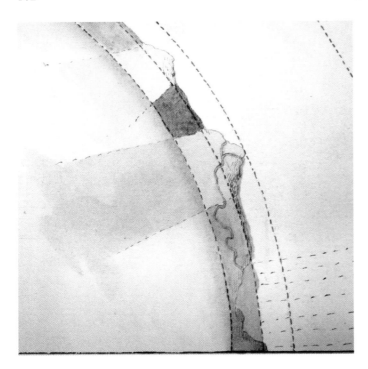

Figure 8. Detail from right-hand side of Figure 2.

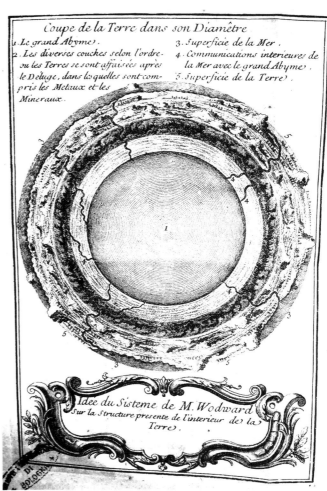

Figure 9. Cross section of the structure of Earth published in the French edition of Woodward's *Physical Geography* (from Woodward, 1735, by permission of Biblioteca Universitaria di Bologna).

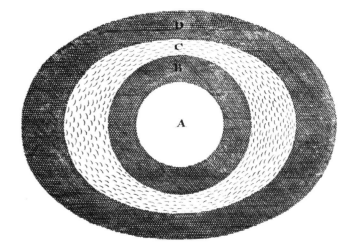

Figure 10. Cross section of the structure of Earth. A—fiery mass at Earth's center; B—solid interior; C—water mass or abyss; D—outer cortex (from Burnet, 1699, by permission of Biblioteca Universitaria di Bologna).

Building on Gian Domenico Cassini's campaign for measuring precisely the length of the degree of latitude on behalf of the French government (1670–1718), it became known, during the eighteenth and nineteenth centuries, that (1) the shape of Earth was ellipsoidal—as theoretically derived by Newton—and not spherical; (2) the Globe could be conceived to be in a state of relative plasticity; (3) gravity measurements often indicated deflection of the vertical expected from an ellipsoidal surface, suggesting heterogeneities within Earth that resulted in the geoid being the figure of equilibrium of Earth; (4) even after correction, gravity anomalies were negative over mountain chains and positive over oceans (this was considered the result of an excess of mass below the oceans and a deficit of mass below the mountains); (5) Earth's crust was thinner under the oceans and thicker under the mountains; (6) an increase in density characterized the crust-mantle transition; and (7) the rocks composing the continents (sial) were of lower density (2.5–2.7) than the basaltic ones (sima) building the mid-oceanic volcanoes (3.3).

Dutton and several of his contemporaries integrated these pieces of evidence and came to the conclusion that (1) Earth is in a kind of hydrostatic equilibrium produced by gravity (isostasy, meaning same or equal standing); (2) once that equilibrium is broken by a disturbance, Earth tends to restore it by slow flow of material below, through the stress produced by gravity; (3) Earth's crust can be viewed as rigid blocks or columns floating on a viscous and denser mantle, with each block, independent of its neighbor, moving vertically; (4) the isostatic equilibrium implies that denudation by unloading the crust allows it to raise, while deposition by loading causes it to sink; (5) at a certain depth,

Figure 11. The origin of gravity anomalies in Earth's crust according to Albert Heim (1892).

there is a compensation level (30–40 km in Airy's model, ~100 km in Pratt's model) where blocks of equal diameter have the same weight; (6) gravity anomalies originate from thickness variation in the rigid and lighter "lithosphere" floating on the heavier "barysphere" (Heim, 1892) (Fig. 11); and (7) there is a condition of gravitational equilibrium established by Earth's lithosphere, which floats on the asthenosphere.

Stated above are different concepts within the doctrine of isostasy, all formulated since the second half of the nineteenth century. In essence, they tell us that the trend for isostatic compensation is a fundamental property of Earth. It then becomes illuminating to compare such formulations, at least some of them and the observations upon which they are based, to those implied in text and illustrations by Marsili.

MARSILI'S ISOSTATIC CONCEPTION

Marsili was a military man by career, as were many of the scientists who contributed to the establishment of the doctrine of isostasy (e.g., Captain Dutton, Colonel Totten, Major Powell, etc.). Like Cassini, formerly a professor at Bologna University, throughout his life Marsili was concerned with measurements and surveys. Thus, Marsili was technically and culturally prepared to give thought to problems of isostasy that were not noted by Descartes and the earlier speculators.

Marsili's insistence on the symmetry between "highest mountain peak(s)" and "lowest marine depth(s)" has already been noted (Sartori, 2003; Vai, 2003) (Figs. 1, 2). The clear graphic illustration of this observation can be viewed as a preliminary statement on the concept of isostatic equilibrium. It is mirrored and confirmed by a sentence found in the contents of the enclosed manuscript (see appendix, c. 165r):

Il Continente Montuoso fu più alto dove è più profondo il Mare

The mountainous continent has been higher where the sea [nearby] is deeper.

To explain the concept further, Marsili planned an introductory chapter dealing with "Proportion of the Earth to the Sea for both extent and depth."

Another point of great interest expressed by Marsili in the materials prepared for his planned treatise is the reference to the different crustal thicknesses below what we now call the cluster oceans–shallow sea–flat cratons on one side and the mountain chains on the other (Fig. 12). In his figure (ms. 90, C, c. 112v), the crust is thin beneath the sea and the flat lowlands (section I in Fig. 12), whereas it is thick beneath the mountain belts (section II

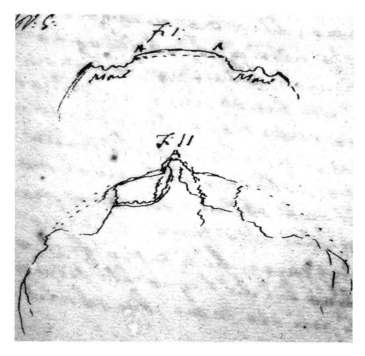

Figure 12. Difference in thickness between "marine" (oceanic) and "mountainous" (continental) crust (from Marsili, Fondo Marsili, ms. 90, C, c. 112v, by permission of Biblioteca Universitaria di Bologna)

in Fig. 12). Such depictions of crustal thickness are amazingly similar to Albert Heim's (1892) illustration explaining the origin of gravity anomalies in Earth's crust (Fig. 11). When referring to his Figure 2 (Fig. 12), Marsili wrote (BUB, FM, ms. 90, C, c. 113r):

Sulla figura II si vedde una sezione del Monte A che con le sue radici
si [s]profonda all'Orizonte quassi, o più del Mare.

Figure 2 shows a section across Mount A, whose roots go down
almost to the sea level or even more.

This appears to be the earliest use of the term mountain root in its modern geological sense.

The two very interesting different sections across the globe (Fig. 12) were meant to illustrate a special discussion of the role mountain chains play in fostering the water cycle (see below). Nevertheless, they are of major importance in understanding Marsili's manuscript 90, A, 21 and especially the most relevant schematic illustration contained in it (Fig. 13). Figures 12 and 13, in fact, complement each other in their meaning.

The caption under Figure 13 states:

Primo sbozzo della maniera con la quale forsse l'Onipotente escavò il mare per rendere la terra arida, e che nello stesso tempo formò li Monti con la <> terra mede[si]ma dell'globo, per

Figure 13. Marsili's early sketch of Earth's surface compared to the ellipsoid-spheroid with embrional hint to the principle of isostasy (from Marsili, Fondo Marsili, ms 90, A, 21, c. 145r, 1728, by permission of Biblioteca Universitaria di Bologna).

<la quantità> formare una profondità da ricoverare aqua destinata al mare. Questa escav[a]tione si fece con la pressione della matteria del globo et facendola ai lati inalzare, <> nella forma dei Monti che sono gl'Argini del Mare; e necessari sono le loro pendenze per tenere adietro(?) la superfizie della terra dall'
acqua, che se non di tale <> divina dispositione non sarebbe statta
che una palude, come si mostrerà.

First sketch shows how perhaps the Almighty excavated the sea to make the dry land, and at the same time formed the mountains using the same earth of the globe to obtain a deep depression where to recover water destined to the sea. This excavation resulted from pressure on the matter of the globe, and the ensuing lateral uplift formed the mountains that became the banks of the sea; their relief is needed to protect the surface of the land from water, which would become a pond without [the use of] such a divine tool, as it will be demonstrated.

As a comment on this statement, there is a preliminary providential approach assigned to the natural process(es) in general. However, the process itself, and the structures originating from it, are described in a smooth, neutral, scientific way, hinting at a natural, operational interpretation of the process. It consists of a pressure or vertical push on Earth matter, displacing masses from the sea downwards and outwards (arrows in Fig. 13) to form the rising mountain chains with their roots reaching well under the sea level (dotted circle) (Fig. 8). Thus, God is always present in the back, while nature is operating in the front following its internal rules.

Even more plastic is the metaphoric example provided in the text to explain the same concept (c. 152r):

Quello che si vedde in un globo di pasta nel quale si vorrebbe alogare un pugno, che, premendo la pasta nel mezzo dalle parti laterali si vedde alle parieti quell'innalzamento di pasta, che era uguale alla mole del pugno.

It is similar to what one sees happening in a globular mass of pasta, where if you insert a fist into the middle of the pasta it produces a lateral squeezing and a rim rise equal to the size of the fist.

Such a clear vision of Earth's mass balance is closely consistent with the concept of isostasy.

That above being indeed a Baroque example, it seems to me unsuitable, if not insulting in Marsili's religious perspective, to see the hand of God pushing down the seas, instead of a natural vertical centripetal and/or centrifugal force (see dashed force lines in Fig. 2). Given the friendly relations Marsili shared with

Isaac Newton (1643–1727), it is reasonable to see this force as Earth's gravity. In fact, Marsili planned a dedication to the Royal Society of a work on "The probable organic structure of the Earth" and had various meetings with Newton (and Woodward) during his long stay in London in 1721 (Cavazza, 2002, p. 8–9).

From Figures 2 and 8, it is clear that Marsili's Earth is "supposed to be entirely solid and excavated by arched cavities." The outer layer is laterally segmented into three basic components: (1) deep seas (blue), floored by thin crust (orange); (2) low-elevation plains, floored by thicker crust (orange and brown); (3) mountain chains (yellowish), with downward-protruding roots laterally confined by diverging segments. Four concentric zones partly coincident with virtual envelopes are shown to limit the different layers. The thickness of the two layers facing the sea-level surface is the same. Thus, sea level, once it is interpolated to extra-marine areas, becomes an equipotential surface. It is interesting to remark that this same surface, the "*Horizon Marium*" line of Figure 2, actually relates to the point where surface waters enter the sea (water cycle). No direct mention is made of the varying densities of the different components of Marsili's Earth crust. This makes the rudimentary, but very significant, isostatic observations of Marsili different from the late concept of isostasy (see above). One should not forget, however, that the manuscript under discussion is only a summary and a punctuated list of contents of a treatise that was never completed.

The basic inferences from Marsili's Earth figure are: (1) the extrapolated sea-level surface is the equipotential surface of Earth; (2) the oceanic crust is different from the crust of the flat continental areas and from the crust of mountain chains; (3) the solid inner part of Earth contains different amounts of material similar to that of the crust of mountain chains; (4) the bottom of the oceans is at considerable depth; (5) the mean elevation of the continents is low; and (6) the mountain areas represent a minor part of Earth surface (Fig. 2). Remarkably, most of these observations and concepts are still part of the theory of isostasy. Moreover, the observations made clearly separate Marsili from the pseudo-isostatic view of Jean Buridan (Şengör 2003, p. 49–53).

Figure 14. Cover of Marsili's manuscript 90 (right) and back of the volume contained in it and titled "Treatise on the structure of the earthy globe" (by permission of Biblioteca Universitaria di Bologna).

ADDITIONAL REMARKS

Marsili's manuscript 90, A, 21, dated 1728, is published for the first time as an appendix to this paper because it provides explanatory text for the remarkable Figures 2, 12, and 13, all included under the general cover of manuscript 90 (Fig. 14). There is also curiosity about the contents and the leading ideas Marsili planned to develop in his treatise. Manuscript 90, A, 21 was probably a draft introductory chapter to the treatise and was accompanied by a summary of some other chapters and by a detailed list of the items to be developed.

This manuscript provides consistent evidence that Marsili was fully aware of a dynamic equilibrium between seas and mountain belts. He explained the equilibrium through vertical pressure over the seas inducing depression of their floors and ensuing displacement and uplift of the mountain belts. The uplift was then followed by the erosional cycle of the stony strata of Earth (c. 151r). Thus, he can be considered the first to have introduced and illustrated, albeit in embryonic fashion, the concept of isostasy.

Additionally and consistently, Marsili was mainly concerned with detailed demonstration in the field and on maps that mountain belts are "organic" structures in the sense that they are continuing across and beneath plains, lakes, and seas, and that they do this not fortuitously, but following a system (Fig. 15A, 15B, 15C).

Marsili strove to test his hypothesis by means of observations and experiments made in the subsurface (mines) and on the sea bottom (dredging).

In three different sentences he opposed "the method of those speculators sitting comfortably in their cabinets to filling

A

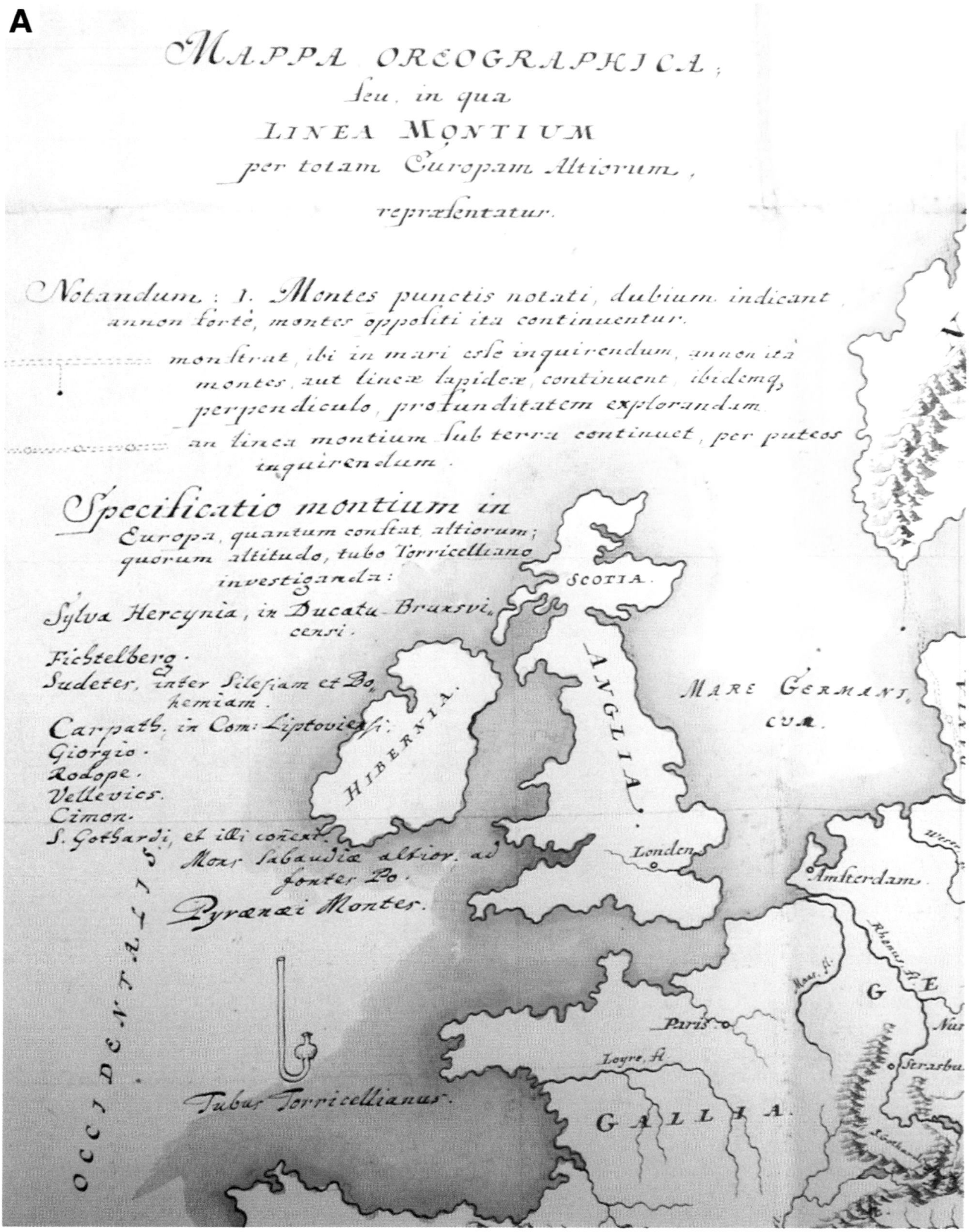
Mappa Oreographica;
seu, in qua
LINEA MONTIUM
per totam Europam Altiorum,
repræsentatur.
Notandum: 1. Montes punctis notati, dubium indicant annon forte, montes oppositi ita continuentur.
monstrat, ibi in mari esse inquirendum, annon ita montes, aut lineæ lapideæ, continuent, ibidemq, perpendiculo, profunditatem explorandam.
an linea montium sub terra continuet, per puteos inquirendum.
Specificatio montium in Europa, quantum constat altiorum; quorum altitudo, tubo Torricelliano investiganda:
Sylva Hercynia, in Ducatu Brunsvicensi.
Fichtelberg.
Sudeter, inter Silesiam et Bohemiam.
Carpath. in Com: Liptoviensi.
Giorgio.
Rodope.
Vellevics.
Cimon.
S. Gothardi, et illi conexi.
Mons Sabaudiæ altior, ad fontes Po.
Pyrenæi Montes.
Tubus Torricellianus.
SCOTIA.
ANGLIA.
HIBERNIA.
MARE GERMANICUM.
Londen
Amsterdam.
Paris
Loyre, fl.
GALLIA.
OCCIDENTALIS
Rhenus fl.
Maas. fl.
Strasbu

Figure 15. (A) Title and legend of Marsili's "Orographic map of major mountain trends in Europe" (from Marsili, Fondo Marsili, ms. 90, A, 21, c. 148/155, before 1728, by permission of Biblioteca Universitaria di Bologna [BUB]).

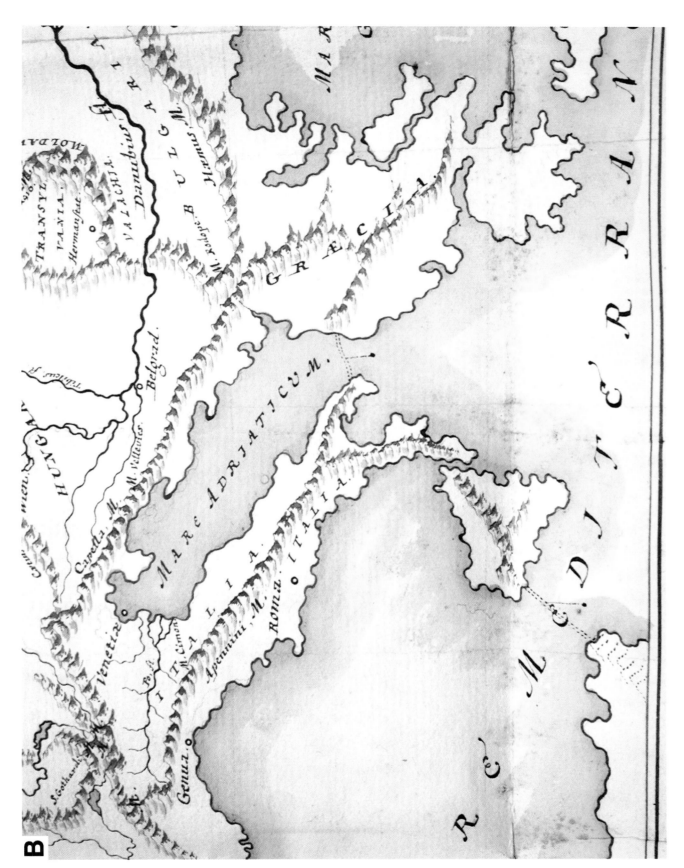

Figure 15. (B) Detail of the central Mediterranean area showing the Apennine chain in Sicily crossing the Sicily Channel and continuing into the Maghrebian chain in northern Africa. Also, the outer Southern Apennine chain is shown to cross the northern Ionian Sea and continue into the Hellenic chain (same, by permission of BUB)

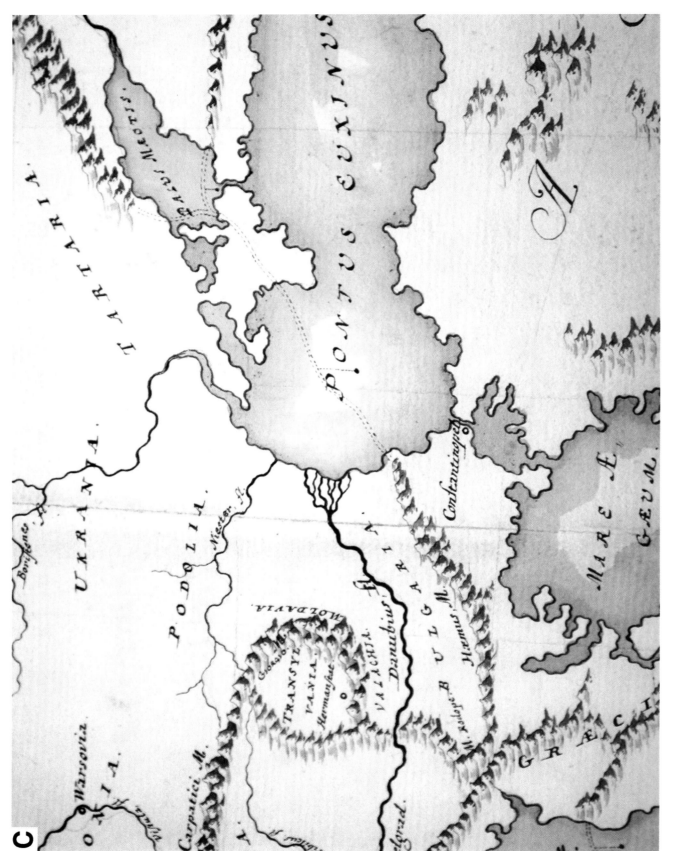

Figure 15. (C) Detail of the eastern European area showing the Balkan chain continuing across the Black Sea into Crimea and bifurcating toward the Caucasus and the Scythian chains (same, by permission of BUB).

up paper sheets with chimerical configurations." Among them, it is possible to glimpse Descartes, while Marsili acknowledged the role played by Steno's *De solido* (1669) in understanding the organic structure of Earth (c. 148r).

As a Roman Catholic believer, Marsili shared a finalistic (providential) view of the globe created by God. As a scientist, however, he studied the body of Earth by the method of observing nature, without taking into consideration "God's miraculous intervention that cannot be understood by us mortals."

Marsili also conceived of a solid Earth interior with conduits allowing for the water cycle running from the inner to the outer part of Earth and vice versa, as well as for the exhalation of many vents that, once stopped, may ferment and burn "producing the fire called volcanoes and earthquakes," which he viewed as "organic disorders."

Worth mentioning, too, is the emphasis Marsili gives to the formulation of Steno's principle of superposition (1669), both in Figure 2 and in the text (c. 151r, 165v), by repeating and explaining the sentence "*strata super strata.*" This is illustrated by an example from the Swiss Alps (Marabini and Vai, 2003; Sartori, 2003; Vaccari, 2003)—a locality very convenient for Marsili's purpose, where more than 20 km of Earth's crustal thickness is exposed.

Under the heading "*Capi dell'Oppera,*" three pages of the manuscript (c. 165r, 165v, and 166r) provide a detailed list of the contents of the planned treatise in the form of 27 mainly synthetic sentences as titles or brief explanations of different chapters. A clear summary statement of Marsili's views on geodesy, physical geology, lithology, tectonics, geodynamics, stratigraphy, and Earth's history comes out, which is in sharp contrast with the complex contorted syntax of the rest of the manuscript and other Marsili manuscripts.

The first chapters were designed to deal with the geometry, size, and relative proportion of the land to the sea in both depth and extent. Some chapters were destined to describe the type, setting, and mutual order of superposition of the materials building the terraqueous globe in the early creation, subsequent growth, and decrease by erosion with time. Other chapters were supposed to deal with mountain belts, especially showing that the highest mountainous areas occur where the sea is deeper. Such mountain belts were primarily created, and additional new hills originated only from alluvial sediments or through earthquakes and bituminous deposits in the presence of volcanoes.

The role of rain and springs in producing mountainous morphology through the action of water running in streams was introduced. Beach profiles were used to recognize the daily effects of sea tides.

Of special interest is the announced intention to describe new, ephemeral land areas consisting of grass and leaves bound by roots of ditch reeds, which create floating islands in the brackish seas. Other chapters were devoted to "stone breakdown" through alteration and "stone growth" by mineral formation when mineral vents had found the necessary structure within the mountains.

The last chapters were planned to discuss some difficult questions. How thick was the layer of soft earth that was originally superimposed by God onto the stony layer? Can ideal submarine conduits connecting surface-land water with marine water be tentatively mapped when only submarine springs are known? In fact, only the "rainy springs," not the "submarine springs," disappeared during the last severe droughts (see Figure 2 for the distinction between "rainy" and "submarine" springs).

Marsili also raises doubts concerning the concept of a stable subterranean fire. If this fire lasted over 5000 yr, it would have calcined the stony mass of the globe. Even at volcanoes, which perhaps are generated by internal vents that burn after fermenting, the fire is noted to have discontinuous activity.

As for stratigraphy, the manuscript emphasizes the basic bipartition into two superposed layers consisting of hard-stony and soft-earthy material, the latter deriving from and supported by the destruction of the former. This couplet is not static, but dynamic, passing to a triplet consisting of (1) "primary Mountains," (2) "Alluvial, Earthquake, bituminous, and volcanic hills," and (3) "earthy plains." This appears as an original development of Steno's principles of geology and an anticipation, by three decades, of Lehmann's "*Gang-Gebürgen, Flötz-Gebürgen*, and surface bodies" as well as Arduino's "four orders" of terrains (Vaccari, this volume, chapter 10).

Marsili's position on Noah's deluge, as expressed in the appended manuscript, will be considered now. It is relevant to the history of geological thought, regardless of its limited relation to the title of this paper. At first, the two sentences below (c.148r and c.165v) will come as a surprise even to those who have considered Marsili a nonradical and liberal diluvianist (e.g., Vai, 2003). Marsili basically writes:

"I have compared the many marine bodies [fossils] found in the mountains that have been explained as one of the effects of the deluge in so many printed reports with the many observations I have made on the living marine organisms. My recognition of the order and the setting such marine bodies have in the mountains of Italy, Germany, and France is a reason to ascertain the groundlessness of that idea" (c.148r).

"[They] show that the deluge has not produced the alleged bad effects to the solidity of the Earth except for the slaughter of the living beings" (c.165v).

However, this late, but explicit withdrawal of Marsili from diluvianism (especially from the physico-theological Woodwardian version well introduced into the Marsilian *Istituto delle Scienze di Bologna*) is entirely consistent with the completely different view of Earth's structure that Marsili had in comparison to those of Woodward and Burnet (as outlined above).

DISCUSSION AND CONCLUSIONS

Marsili was well renowned among European savants during his lifetime for having founded in 1711 the *Istituto delle Scienze e delle Arti of Bologna*, the most famous Italian scientific institution after closure of the *Accademia del Cimento* in Florence. Marsili has also been regarded, since the end of the nineteenth century, as the founding father of scientific oceanography and marine geology. However, his role in the development of other geological ideas is still underestimated, probably because many of his geological manuscripts remain unpublished. His manuscripts published by Gortani (1930) and Lipparini (1930) have revealed Marsili's priority in drawing stratigraphical columns, describing general types of tectonic structures, producing regional cross sections of mountain chains, and sketching a proto-geological map. This last one is not a surprise because Marsili's cartographic genius in fluvial, hydrogeological, geomorphological, and mining studies was always widely appreciated.

Unfortunately, Marsili's *magnum opus*—a work on the structure of Earth—remained unfinished, although he had prepared a number of figures and written hundreds of pages, which are still preserved in the Bologna University Library. Some of these recently published figures show the earliest embryonic representation of the principle of isostasy and of the concept of roots in mountain chains. Moreover, they give a geological meaning to the idea of Earth's spheroid-ellipsoid mathematical and geodetic surface. During the late nineteenth and early twentieth centuries, Marsili's works were neglected by idealistically biased Italian historians of science and his manuscripts were largely not explored. Italian historians were poorly attracted by works and manuscripts following an experimental approach and addressing only scientific and mechanistic matters.

The recent revival of interest is expected to shed new light on Marsili's still-hidden geological views as shown by this paper based on the study of manuscript 90, A, 21, which also demonstrates how Marsili's figures are closely related to manuscript texts.

Although he did not use the formal term "isostasy," which was introduced only at the end of the nineteenth century, Marsili gives a preliminary definition of isostatic equilibrium when he writes: "the Mountainous Continent has been higher where the Sea [nearby] is deeper." As for the concept of roots in mountain belts, he expressly says: "Figure II shows a section across Mount A, whose roots go down almost to the sea level or even more."

Manuscript 90, A, 21 goes further by giving a possible interpretation of how seas and mountains were formed and by providing a series of observations and concepts that are still part of the theory of isostasy.

Two additional ideas contained in Marsili's manuscript 90, A, 21 deserve attention. The first is Marsili's concept—already expressed in his major printed works—that mountain chains are not stopped when crossing seas, lakes, and plains.

Secondly, and quite surprisingly, Marsili says that, "the idea that the order and setting of many marine bodies found in the mountains of Italy, Germany, and France are a result of the effects of the deluge is groundless, based on the many observations I have made on the living marine organisms." To emphasize the concept, Marsili adds, "The deluge has not produced the alleged bad effects to the solidity of Earth except for the slaughter of the living beings." Although late, Marsili's clear withdrawal from diluvianism is consistent with both his different view of Earth's structure and on the question of fossils Marsili had as compared with the radical Woodwardian diluvianists and the moderate ones of the *Istituto delle Scienze di Bologna*.

Our knowledge of Marsili's diverse personality may be expected to increase considerably as more of his manuscripts receive careful scrutiny, analysis, and interpretation.

ACKNOWLEDGMENTS

Biancastella Antonino, director, and the staff of the Biblioteca Universitaria di Bologna are gratefully thanked for providing facilities and access to the manuscript section.

This paper was considerably improved by careful reviews of Victor R. Baker, Ellis L. Yochelson, and W. Glen E. Caldwell.

APPENDIX

This appendix contains the first posthumous publication of Luigi Ferdinando Marsili's manuscript 90, A, 21, c. 144r–166r, kept in the Biblioteca Universitaria di Bologna, Fondo Marsili (BUB, FM).

Legend

<>—text crossed out.

Primo Sbozzo dell' Organica Strotura della Terra
Adi 4 Novem. 1728

[c. 144r]

Leopoldo Cesare, che con la liberatione
di Viena da Turchi per una serie di
17 <anni> campagne si vidde vittorioso
in guisa tale da duvere sperare <qual>
una Pace che gl'assicurasse un aumen-
to considerabile al di lui Imperio <>
dentro di sicuri limiti più della
natura, che per l'Arte, non erano scor-
si quasi due secoli, che tante Provin-
cie che facevano il teatro della guer-
ra per esere poi anche quello della
pace <> si trovavano senza comerzio di
giente capace <> di darne notizie
e per guerra, e pace <>, e ne tampoco
v'erano mappe, ne
descriptione impresse [stampate], che amaestrassero
al bisogno dell'uno, ed altro uso, anzi
che molte di queste [Province] dalle frequenti
mandrie de Tartari si trovavano all'ora senza

abitanti, ed affatto deserte per ne meno
pottere essere provisti di guide
fidate, ed alla fine si potteva dire, che
con l'esercito s'andava a scoprire
nove terre, che per tagliarle con <limi-
ti bisognano> proietti [progetti] di limiti
più sicuri, facceva d'oppo [uopo] di riconoscerle.
A ciò dal mede[si]mo Cesare fui io scielto, et
instrutto da lui, e suo ministero per stabilire le frontiere sicure contro de Turchi <>,
e con riflesso <>
alle convenienze, che pottevano impor-
tarli a riguardo del presente, e futu-
ro con gl'Aliati [Alleati] confinanti con quelle
vaste Provincie, che sono atorno
no del Danubio, e che vedevamo di prossima conquista. L'obbedienza mia
fu eguale alla fede, ed applicazione con
ché intrapresi una così vasta oppera
che durò per <> dieci campagne, nelle
quali essendoli varietà de' successi
dell'armi anche <> diverse erano le pro-
posizioni de' limiti, che con la quiete
dell'inverno

[c. 144v]
si dibatevano in gabinetto per regolare
l'opperazio-
ne della prossima futura campagna, ed anche
essere pronti ad una pace, che si solleci-
tava alla Porta dagl'Ambasiatori d'Inghil-
tera, ed Olanda.
Da questo confidentiale comando di Cesare
<> unito alli viaggi, che avevo benché
in età giovanile per bona parte di Turchia
fatti, e poi da tutti gl'altri <>
posteriori sino all'ora fatti <>, e <>
sino adesso per altre parti d'Europa naque
l'idea mia di tentare
[left-hand column inserted here]
<> appunto
una spetie d'Anatomia della organica
strotura della Terra con quelli lumi che
mi venivano dati dall'esecutione delle mie
incumbenze e riconoscendo le linee
de Monti, e corso
de fiumi, e spiagie maritime,
notitie, che unite a quelle interne intie-
re de mede[si]mi monti esaminando <>
le tante miniere dell'Ungaria
e d'altre parti dell'Europa
per la loro strotura <> como-
da alla generazione de minerali, e mezzi minerali, comin-

ciai a formarmi un sistema,
che non ardivo <> di stabilirlo
senza che di prima <>
mi fossi assicurato severamente
l'Alveo del mare in tutte le
sue parti fosse composto della
continuazione del continente
della Terra.
Le mie aventure lasciandomi
libero da tutte l'occupazioni
mi scielsi per tale esame li due littorali di
Provenza e Linguadoca in
Franzia il primo, che [è] tutto
montuoso e batuto dal mare,
ed il secondo che è piano
quasi sino alle falde orientali de Pirinei ed a setentrio-
ne contermine fra diverse
larghezze alla linea montuosa fra delle (?) differenti
di littorali, che n'abisognavano
<> dall'Onipoten-
te la strotura della Terra coi fondamenti della continuata linea de monti
non solo a traverso de fiumi, ma
anche de mari, e come che questi siano
nel loro interno <> construtti, e con
quali matteriali composti, e riflettere
se le moderne pianure
fossero alla creazione <> della Terra, e se <>
in quelle fossero mari tentando sopra
questi miei pensamenti <> osservazioni, esperimenti, e mostrare
con il fatto, che anderò dicendo quasi
con ordine Anatomico di questa mole
terra aquea, e slontanarmi dal metodo di quelli, che in un comodo Gabinetto meditarono speculazioni, che ese-
guirono sopra piani fogli patienti a
riccevere qualunque chimerica con-
figurazione.
Io non esponerò se non
quello, che ho toccato con mani tanto
nella superfizie <> montuosa, che
piana, e dentro de Monti, che Mari,
e finché per angusto tratto, <>
respetivamente all'ampiezza di
tutta la mole terra aquea, ad ogni
modo mi lusingo, che possa dare
lumi per una sola quasi universale Ipotesi sino a quella profondità, che mi è statto possibile con vari
tentativi di penetrare, e per quello

[continued on c.148r]

[c. 145r]

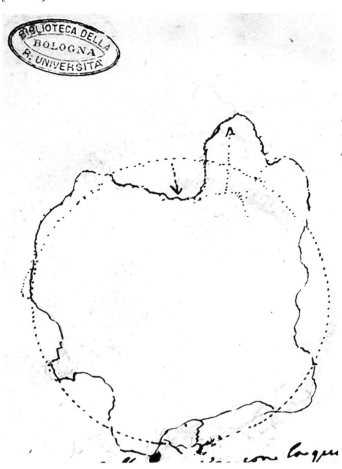

Figure A1. No title in the original (c. 145r). Note that this figure is the same as Figure 13 in text.

*Primo sbozzo della maniera con la quale forsse l'Onipoten-
te escavò il mare per rendere la terra arida, e che nello stesso tempo formò li Monti con la <> terra mede[si]ma
dell'globo, per
<la quantità> formare una profondità da ricoverare aqua
destinata al mare. Questa escav[a]tione si fece con la pressione
della matteria del globo et facendola ai lati inalzare, <> nella
forma dei Monti che sono gl'Argini del Mare; e necessari sono
le loro pendenze per tenere adietro[?] la superfizie della
terra dall'
acqua, che se non di tale <> divina dispositione non
sarebbe statta
che una palude, come si mostrerà.*

[c. 146r]

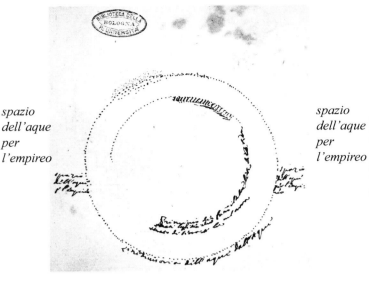

spazio dell'aque per l'empireo

spazio dell'aque per l'empireo

Figure A2. No title in the original (c. 146r)

linea divisoria dell'aque dall'Aque

[Writing inside the drawing:]

*Principio dei
Strati lapidei che fanno il t........della
terra
Strato di terra .... la mal ....*
[continued on c.147r without writings]

[c. 147r]

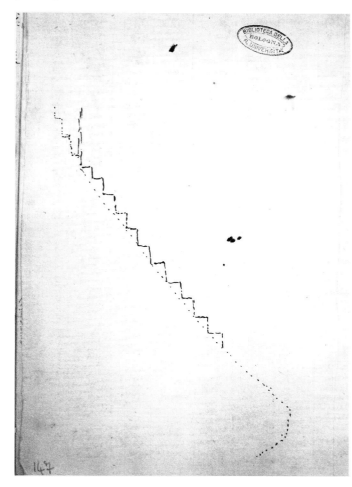

Figure A3. No title in the original (c. 147r).

[Contains only a staircase diagram without text; see reference to the diagram on c.166r.]

[c. 148r]
che non ho veduto, o che le mie osservazio-
ni non lo possono comprovare, lasciarlo
ai felicissimi ingieni speculatori sopra di
com(o)da sedia senza fatica, ne di corpo,
ne di forza, perché vaddino più avanti
nei loro indovinelli.
Da omo grande l'eruditissimo Stenone il
primo nel suo tratato de solido intra
solidum *<> acienò qualche cosa*
che può influire all'Organica Strotu-
ra della Terra, per alcuni altri poi
lascierò, che il publico ne giudichi lui
e massime doppo delle mie esposizioni.

La necessità, che hebbi di veramente chiarirmi
se li monti continuassero dentro del mare
o sopra l'orizonte della di lui aqua per Isole,
o pure sotto in diversità, m'obligò al sogior-
no dei lidi di Linguad'oca faccendo quelli
tanti sperimenti nei loro mari, che com-
ponendo un Historia fisica maritima
scritta alla Regia società di Parigi, fu d'
oppo per renderla più chiara, che <>
dimostrasse molto <> di quello
che riguarda alla strotura della Terra,
e che in ora a suo loco doverò ripiglar-
lo colorandolo <> dove doverò
dimostrare la continuazione dei Strati
del continente dentro del mare, dove
gl'agiungerò molto di quello, che osser-
vai nel lago di Garda, del quale avven-
done uno fisico tratato per la stampa,
e che cossì bene non solo corrisponde, ma anzi
con maggiore chiarezza mi serve a dimostrare
tutto correlativo col come do di non essere
la di lui estensione fra monti cossì vasta
come quella del mare.
Non è possibile d'avanzare questo mio
ardito tentativo con le sole mie osservazioni
senza d'avvere per base la terza gior-
nata della divina Onipotenza
per la creazione del Mondo alla quale
si devve essere somerso per quel tan-
to molto, che non siamo capaci di
intendere, anzi conosco per
mio duvere d'applicare le mie
fisiche osservazioni
[continued on c. 148v]

[The fine text at the left of the main column is placed below.]
Li paesi fatti d'Olanda tutti piani pare
manufatti da un'arte incredibile
a volte pare anatomizate con somma
spesa, e fatica, come alta Pianura
Lombarda atorno del Po in Italia
e con aplicatione anche al solievo
delle campagne di Bologna
anegate da quella gran copia di
aque, che <> confluiscono lentamente
nell'Adriatico Mare. La <> re-
cognizione, che presi per li monti
d'Italia, Giermania e Francia
dell'ordine con il quale sono
disposti li tanti corpi marini
già decantati con tante relazioni
stampate per uno di quelli effetti del
diluvio, e paragonando tante
mie osservazioni fatte nei viventi

marini ho avuto motivo di
riconoscere l'insusistenza di
sì fatto pensamento, e la compu
tazione de' miei prencipi nel presente mio assunto

Lago di Garda <>

In fine osservazioni di poco meno di 31 (?)
anni <> per le <> mentovate <>
parti dell'Europa, ed ancora dell'
Asia minore, e navigazione a mano
della costa di Tartaria e particolare esame del
Bosphero Trace, ...scrissi a R ... saranno
[writing not clear]
il <> fondamento di questa mia
intrapresa forssi [forse] tutta nova, e che
ordendola mi serve molto a miei
studi naturali.
<Benché già pubblicato>
Dove bisognerà valersi di figure per spie-
gare con brevità, e chiarezza le mie
esposizioni, non <> sono rispar-
miati, e principalmente la mappa <>
orografica del corso de monti osservato nel me ...rano (?) del
continente, e d'atorno dei littorali
d'Europa, ed Asia, ed Africa da
me veduti, e notati in essa.
Li <> siti nei quali <>
sono <> quelle osservazioni, che avve
ranno bisogno d'essere dissegnati
in grande.
Vi sono distinte le linee di monti da me non
veduti, ma riferitimi da altri, e
massime nell'Asia, che hanno li
monti fatti di punti, <> per
non essere debitori di quello ch'io
non riconobbi.

[A canceled section follows.]

Per <> darò principio al mio Tratato con
l'infallibile frase del sacro Testo della
Genesi, che è quello della terza
giornata dell'Onipotente siano<>

[c. 148v]
alla dimostrazione
di quello, che con qualche
oscurità è scritto in quelle sacre carte. Alcuni moderni han
no sopra il sugetto della terra ins-
tituire [istituiti] sistemi più per far pompa
di chimerico ingenio, che di solidità somessa a quello, che dobbia
mo venerare per debito di Reli-
gione, cercando <> che mettere più in chiaro
con qual ordine l'Onipotente
faccesse <> con le sue divine
mani la grande oppera della
terra, dalla quale construsse l'omo,
e stabilì l'alimento ad esso
disposto in tanti <> vegietabili, ed animali.
#
[Between the above section and the next, a # marks where to insert the block "Lago di Garda."]

Per meglio assicurarmi dell'osservato
nei mari, volsi pure veddere, se in
un lago, <> vi fosse corispondenza
nella sua strottura con quella
del mare, e scielsi quello di Garda,
che per la sua angustia <>
mi mostrava ambe le ripe, e le profondità sue misurate con l'altezze littorali, <> ed in guisa tale, che ne com-
posi una disertazione fisica naturale
scritta al mar.se Poleni professore
matema[tico] nell'Università di Padova perché pronta per le stampe, e col mio sommo
piacere, che con più chiarezza viddi
corrispondere tutto con l'osservato
nel mare.

[c. 149r]
[This page appears to have been crossed out by Marsili in the
original, probably to avoid redundancy.]
Oppera dell'Onipotente mano di Dio fu la
creation del Mondo, che lui stesso divise in
sei giornate secondo, che insegnano le Sacre
Carte, che descrivono la Gienesi.
Nella terza <>, che fu quella della divisione
dell'aque dall'aque, <> cioè più più pure
pure dalle meno pure. Quelle colocò altrove
per altri usi noti alla
di lui Onipotenza, e che non è del mio assunto
esaminare, e queste <> relegò in quel loco,
che è il mare con le sue sovrane parole da
Creatore <u>Congregentur aque, qui sub celo
sunt in locum unum, ut appareat arida</u>. Questo <loro> recipiente <indubitabilmente>
probabilmente nelle due
prime giornate non era stato da Dio creato
perché se no l'aque col loro peso si sarebbero
<> precipitate in esso naturalmente
parlando, perché se poi ricoreremo a mira-

colo, che <> sopra del recipiente marino
tenesse sospesa in aria cossì gran mole dal
primo giorno sino al terzo nel quale gl'intimò
le sopra memorate parole, non vi è da parlare
giaché tutta la Gienesi in quelli sei giorni non
fu, che un'<> oppera sopranaturale di fami-
gliare a cossì grande Artefice, che appunto
per tenere sospese l'aque <> lo fece
comprendere nel Mare Rosso
apprendo in esso un'arida strada argina-
ta dall'istesse aque levate da <> quella, e
fisatessi assieme come per continuarla arida sino a tanto, che l'esercito eletto avves-
se sopra de asciuto varcato quel Mare.
Mi conviene di restare sospeso a <>
legiere le tante dispute, che si fanno, come la
mole aquea avvese potuto <> capire
fra la superfizie del Terra
all'ora senza li posteriori recipienti, ed il
firmamento, quando veruno pottia nemeno
mai <> comprendere, che da un pezzo
di creta si faccesse un omo animato e orga-
nizato come l'Anatomia ci mostra, e da
una costa d'esso creare una femina
egualmente in tutte le sue competenti
parti perfetta, e cossì di tutti li viventi <>, e
piante terrestri, ed aquei

[c. 149v]
[This page also was crossed out by Marsili, probably to avoid duplication.]
come non era impossibile, che al tutto
di lui pottere non le condensasse per ridur-
le a quella perfetta fluidità, che avve-
vano determinata per farne la divisione della
pura dal impura <>.
Alle meno pura destinò il Mare,
ed i fiumi per servitio [ripetizione] della <>
terra che scoprì da essa, per-
ché <> la rendesse fertile con
quella ordinata circulazione, che
le moderne osservazioni insegnano ne Viventi, e diante
qualunque osservato degl'effetti provenu-
ti dai Miracoli di Dio può sperare
d'intenderne qualche cosa, ma non
già di comprendere in qual guisa
gesso lui opperasse in essi, non essendone
noi capaci. A questa cognizione<li pottiamo
giungere per mezzo> libera all'
omo in molte cose alle volte giun-
giamo con metodiche, e longhe osservazioni dell'esistenti miracolosi
opperazioni del Creatore Autore

della natura, e n'abiamo l'esempio
nell'Astronomia, nella quale cognitione siamo giunti per l'osservationi de steo-
li, nell'Anotomia dell'omo, e tanti
altri viventi per <le tante sezio[ni]> li tanti
tagli d'essi, per li quali avvendo vedute
le loro stroture, habbiamo dovuto venerare il Creatore, <che dobbiamo far>
sotto nome di natura, che avvendo
le ben comprese fra la loro reciproca
unione habbiamo potuto parlare dell'
omo con magiori dimostrazioni dello
<> di lui complesso esterno, ed inter-
no, ed uso d'ogni di lui parte che lo
compone. <Egualmente> con eguale
ordine mi sono impegnato nel non
esaminare per quanto mi è stato possibile il
corpo terreo aqueo col metodo d'osservatore della natura senza più parlare
di quella parte miracolosa di Dio
incapace d'essere da noi mortali
compresa.

[c. 150r]
Nelle narate congiunture esaminando
le tante parti componenti la Terra, non
mancai d'avvere avanti di me la terza
giornata della Gienesi, che quella di sepa-
rare l'aque dall'aque, <> le più pesanti
relegarle nell'ampla carcere del Mare
e l'altre più pure assignarle altrove con
auttorità da Onipotente creatore, perché la Terra restasse arida nella di lei
superfitie cir[c]uendo <> il Mare per
reciproco loro comodo a quelli tanti usi
che lui predeterminò <> con la sua divina
mente, e non già per la pretensio-
ne temeraria di farmi giudice delle diverse questioni fra li Santi Padri in uno
ed altro successo di questa giornata, ma
unicamente di tenermi dentro dei limiti d'osservatore delle di Lui grandi Opperazioni sino a quel grado che la mia
<più debole> mente più debole di tutti gl'omini
habbia potuto pervenire fra l'oscurità
che in se tiene fra la di lei ampiezza, e
dificultà d'avvere sotto degl'ochi le parti
componenti, <> e pratica
re la somistione ai modi miracolo-
si dell'Onipotente, <> senza <> altro pensiero,
che d'adorarli, e solamente essere contento
di mostrare, che tentai per l'Anatomia
d'un corpo, dal quale <>
noi omini fossimo formati, e che ne

*viviamo, e ne viveremo sino al di lei
anichilamento.
Giaché il Creatore unì all'elemento
della terra quello dell'aqua incorporandoli
assieme, perché reciprocamente s'assistesse-
ro dentro d'un solo corpo negl'usi, che
gl'assignò a benefizio dell'omo, fa d'
oppo di parlare dell'aqua, che distin-
guerò nelle di lei spetie, e motti, che
fa per la Terra, la di cui strotura
fu formata anche con un'organizatione*

[c. 150v]
*perché questo fluido elemento
la potesse irigare per diverse strade.
Nel mio tratato del Mare già stampato parlai
molto dell'aque marine, <e dolci>
e terrestri, <> e che bisognarà reppli-
care con l'aditione <> d'altre posteriori osservationi in questa <>
mia Anotomia della Terra, nella
quale poco di sicuro preciso pottrò mostrare quali siano li condoti, che danno
comodo al aque di corere per le parti
solide interne della mede([si]ma, <> come per
irigare la di lei superficie <>.
Io dimostrerò più osservazioni senza
volermi impegnare a fare figure <sezioni>
dell'interno della Terra figure
ideali, che servino più a <dimostrare>
comprobare <> ideali sistemi che per
lo più <> sono opposti all'esenza
del fatto insegnando più tosto imagi-
nazioni, che la vera esistenza
<del fatto>, e se questi non corrisponde-
ranno ai loro sistemi
non sarò io che gl'impugni, ma solamente l'esposizione dell'occorsomi
ne miei tentativi, sopra de quali
stentarei di pottere essere come altri
facile di determinarmi sopra d'un
assunto, li fondamenti del quale
non veddo imediatamente, ma solo
per il mezzo d'osservazioni, le quali
pure <> sino ad ora non sono
in quella ordinata serie, che ci
bisognerebbe per pottere stabilire
qualche cosa di positivo.
In fine la mia intrapresa è ardita
e se fosse bene dimostrata sarebbe
<> utile, ma con questo
mio sbozzo altri doppo di mè potranno
<> mettere più in chiaro l'Anotomia
del Globo terreo, aqueo.*

Anotomia
[Marsili commonly writes *Anotomia*(!), and in
a couple of cases only *Anatomia*]

*tentata per dimostrare l'
Organica Strotura del <la Terra>
Globo Terreo Aqueo*

[151r]
Cap.1°
Della Figura, e
Grandezza della Mole
Terrea, ed Aquea
del Mondo

*Non hanno mancato certi <> uni di voler
provare, che l'intiero di questo complesso-
so fosse <> di figura piana, altri delitica
ed altri sferica, che è la corrispondente
alle tante dimostrazioni astronomiche
e che è quella, che io seguito senza volermi punto intricarmi in cossì fatte dispute
che al mio assunto <> d'Anatomia del
Corpo Terreo Aqueo poco importa, essendo
a mè lo stesso di mostrare le parti <>
che lo compongono sotto di qualunque figura che <vogliano> possino figurarsi.
Io seguitando <con la corrente> dei più savi
astronomi <e do conto e divido il siste.>
il sistema, che il
Globo terreo sia sferico, e diviso in 360 gradi e <> che ogn'uno d'essi contiene miglia
<> 60, che fanno la circonferenza di milia
<> 21600, che ha un diametro ....*
[The dots are present in the original.]
*spatio che contiene le Moli Terree, ed Aquee
e che forma il corpo, che intraprendo d'
anatomizare col fondamento di <> quelle
osservazioni, che ne' miei impieghi, e viagi
<> mi è statto permesso di <fare> racogliere.*

[Marsili's left-side note is placed below.]
*L'ampiezza del Mare
respetivamente a quella
della Terra e ....*

Cap.2°
Dei Matteriali con li quali
il Globo Terreo Aqueo fu fatto

*Questi furono Pietra viva, Terra <fruttifera
Ed atta alla Vegetazione dell'Erbe, e Piante>
mol[l]e <> e manegiabile con la quale Dio
creò-l'omo, e come tale anche atta alla*

vegietatio-
ne di tutte le <> piante e l'aqua doppo
che gl'ebbe fatto il vaso ampio del Mare
perché in esso congregate <> discopersero
la Terra, e che per benefizio de' viventi
e vegietabili servis[s]e alla Terra.
La pietra viva, e dura, che troviamo egualmente continuata nel continente, che nella
catena del Mare ed in certi siti <>
degli alvei dei fiumi magiori fra le parti
montuose col nome generico di
cataracte è insegna, che questa serve
di sostegno della terra mole [molle], e fructifera
che senza di cossì forte folcro [fulcro] <>
non acciperebbe susistito contro la curusione
de' fiumi, ed atratrione solare, e pressione
de' viventi sopra d'essa, <> e del rivolgimento
d'essa dall'arte della coltura d'essa, e dallo
sbatimento dell'onde del Mare <> nei

[151v]
vicini ad esso.
Ambi questi materiali sono disposti con ordine di strati sopra strati. Quello della terra come[?]
dirsi mole, e capace d'essere con ordegni ò di fer[r]o
ò di legnio manegiato di sotto è un suolo d'una natura da pottere essere irigata dalle piogge, ò
dalle rugiade, e d'atraere in sé <> per la
coltura, e per la sua natura parmuosa [?] le
parti nitrose dell'aria, <> alimentando
tutte le piante di qualunque sorte, che Dio
creò per il benefizio del uomo. L'altro della
pietra, ò roc[ci]a viva è di quella durezza,
e solidità, che esigie il bisogno di sostenere
il grave peso della terra vegetabile, e di
tutta mole aquea, che de <> laghi, e mari che con la consistenza dei loro fondi
devvono sostenere.

Cap. 3°
**Qual proportione sia fra
La Mole Terrea, e <la Pietra viva> altra lapidea.**

Sarebbe tanta necessaria, che impossie di
sapere che proportione vi sia fra la
parte terrea, che <> pietra viva, perché
in quella altezza, che dicevo, non si
trovano più per la continua curusione delle
acque, che la fa sdrucciolare, e massime da
monti <dentro del Mare> che sono coltivati
con il rivolgimento della terra, che negli
alti Monti che vanno sotto la catagoria
d'Alpi <hanno levato> e ha bisognato, che
il solaro lapideo rimanga scoperto, ed

esposto all'ingiurie de venti g[h]iaci, del sole
che in diversi modi macerandolo a grani
che chiamiamo arena in diverse grandez-
ze <>, ed a gran pezzi <li fa>
gl'anichila, come nel capitolo <>
di ciò mostrerò a suo loco.
Solamente dobbiamo essere contenti di pottere comprendere che la mole pura terea
quando anche era di nova creatione, era d'una pochissima mole respet-
tivamente al diametro <dell'intiero globo aqueo-terreo> del rimanente del corpo
lapideo, perché in quello bastava tanta altezza che li più grandi arbori
potessero aradicare, e nutrirsi, quando
tutto il rimanente lapideo dovveva
essere impiegato per il fondo del Mare,
e per tutte quelle tante organizazioni
massime a comodo dell'aque delle
piogge, e marine, e dare essito ai
tanti aliti, che <> mancando
di questi causano quegli incendi, che
chiamano Volcani, e teremotti, effetti
tutti <d'una disordinata> d'un qualche
disordine organico in quelli tali

[152r]
tali condoti, che ivi <> rista-
gnandossi (veddasi la Favor)
[Note in the left-side column, see c. 165v.]
causano fermentazioni, che
s'accendono, e fanno quelli disordini, come a
suo loco in un capi[tolo] si parlerà. Io non mi posso
cossì prontamente come altri determinare
che nelle viscere della Terra, <> vi sia un foro
che <di continuo ... tante ragioni> che chiamano sotterraneo, che dalla creazione del Mondo sino al di lui nichilamento andar detto
non ho ne il genio, ne la capacità di farmi
col lapis divisioni nella vasta ampiezza della
mole lapidea senza avverne una minima
preliminare dimostrazione, ma solo mi conten-
tarò di narare a suo loco, che ho <> veduto
dentro de monti alla profondità di quasi 200
tese,
[tesa: a unit = 1.9490366 m]
e nelli diversi fondi <profondità> delle linee
<> di Mare col scandaglio <>
come nel Danubio alcuni vortici, e nell'
ultimo tante correnti.
(al dotor Monti, che vedda nel tomo de'
Minerali la profondità della miniera di
Semnia)
[Note in the left-side column.]

Cap. 4°
Della disposizione di questi materiali

Come dissi secondo le tante magiori <> osservazioni degl'Astronomi, e Geografi dispose questi in forma rotonda, e con una superfizie ineguale, la quale inugualianza è quella che chiamiamo monti, oppera che Dio si devve creddere, che la facesse la terza nata quando volse asiugare la tera <congregando l'acque nel Mare> che secondo il più probabile <> sino a quel punto avveva tutta la superfizie equale, che sosteneva l'aqua fra essa, e firmamento. Per fare la gran Vasca del Mare ordinò che tutta la mole lapidea con la terra sovra postoli per quanto volse amplo il Mare che si rompesse <alzandosi d'ogni parte li strati lapidei con la loro terra sia sopra> gonfiandossi da ogni parte tanto quanto <fosse> era di bisogno <> alla mole dell'aqua che copriva tutta la Terra, e per più chiaramente spiegarmi ne succedesse <come> quello che si vedde in un globo di pasta nel quale si vorebbe alogare [inserire] un pugno, che, premendo la pasta nel mezzo dalle parti latterali si vedde <> alle parieti quell'alzamento di pasta, che era eguale alle mole del pugno. Cossì è da creddere che seguisse nella fabrica del Mare, che nello stesso tempo <> dispose il continente <tutto montuoso> come tera <come li declivi ai Mari, e senza se> più alto del pelo

[152v]
[o] superfitie dell'aqua <del Mare> marina, e con egualità di altezza magiore, e minore, che fu più, e meno in certi siti fatta magiore pressione ne materiali terrei, e lapidei, come dimostrerò al suo loco. Con questa grande opperazione il Creatore rinchiuse li mari, pose il continente più elevato dell'universale recetacolo dell'aque perchè le terre fertili avvessero li loro scoli, che essendo tutta la superficie della Terra piana non sarebbe stata, che una Palude.
[The intervening pages are blank.]

[165r]
Capi dell'Oppera

Figura della Terra piana, elitica, rotonda

Grandezza nella circonferenza, e diametro

Proportione della Terra col Mare tanto per l'estensione, che profondità

Materiali che compongono <> il Globo Terreo acqueo nella prima Creatione

Materiali cresciuti, e diminuiti nel corso dell'esistenza della Terra

Disposizione de materiali, ed ordine loro

Profili, e tagli del Globo terreo aqueo per diverse plaghe

Mostrare, che quando Dio rese asciuta la terra, che pare non vi fossero Pianure, ma che tutta fosse montuosa

Mostrare, che le pianure si sono fatte dal ritiramento dal Mare per ricevvere il disfacimento terreo

Mostrare, che li monti si disfano nella parte tenera, e Pie[t]rosa

Mostrare, che non si gienarano novi monti, ma solo colli d'Aluvioni <> o per Teremoti, e deposizioni bitumanose massime dove sono volcani, e che mai si legono con la primaria Creazione de Monti

Il Continente Montuoso fu più alto dove è più profondo il Mare

Non sarebbe il corso montuoso cossì destinto senza li tanti scoli che <hanno fatto li> si sono fatti dalle piogie, e sorgienti

Il corso de Monti non solo fu necessario per il scolo dell'aque <>, ma ancora per servire il Mare

Il Mare indispensabilmente di giorno in giorno si ritira, ed alza, e li profili lo mostrano

Terre nove accidentali fatte da erbe, foglie d'arbori legate da radici di canne palustri, che succedono in aque palustri formando isole natanti

[165v]

<Le Pietre>

La Parte pietrosa per più infirmità si diminuisce, <ed anche cresce vegetando>, e per l'iruzione dell'aque, ed anche cresce vegietando

Li Minerali Metalici, e mezzi minerali nella creazione del Mondo duvettero essere creati ma poi multiplicati quando gl'aliti minerali trovano dentro de' monti la necessaria strotura, finché nell'ambiente <creschino> si formino in Ungaria

Mostrare, che il diluvio non ha fatto nella solidità della Terra il male preteso fuori dell'eccidio de viventi

Non pottiamo con certezza mostrare quanto fosse alta la terra, che Dio pose sopra la sustanza pie(t)rosa

Quali siano <> gl'aqui dotti, che possino essere dentro del Mare per una reciproca comunicazione delle aque fra lo[ro] se ne possa su la carta fare proietti ideali non avvendo fondamento se non di fonti tamarini, [possibly *sottomarini*] *che nelle grandi secità ultime habbiamo veduto perdersi le Vene pluviateli, ma mai quelle dei Fonti Tamarindi*

Per un foco soteraneo <> stabile, e che dia moto a tante opperazioni della natura ne lascio la determinazione ai facili fabricatori di sistemi, avvendo per opposizione a ciò, che un foco di 5000 anni avverebbe calcinato la massa lapidea del nostro Globo, oltreche nei Wulcani veddiamo il foco non continuato, come forsi nati dall' incontro d[e]le aginosi(?), che assieme fermentando <> s'accendono. <u>esperimento la Favor</u>

L'Unione della Terra con l'aqua vole che si parli d'essa nei di lei moti, natura salsa amara che è già fatta nella mia storia, duvvendoseli agiungere più esperimenti per l'atrazione, che ne fa il sole paragonandoli con altri dell'acqua dolce

Nel parlare della terra manufatta e massime d'Olanda si parlerà dell'arti per fare la turba [torba] *e dei Cuur* [Dutch term?], *o isole natanti mostrando, che la canna con le sue radici n'è l'orditura La strotura de' monti e dei strati, sopra strati, e con un interspazio, che è*

[166r]

una terra mo[l]le [molle], e glutinosa

Il corso di questi strati va da Settentrione a mezzodì, e se alle volte si trovano questi o troppo obliqui, o perpen-[di]colari <> non è la causa l'ordine con <> il quale li fiumi tagliano li monti per scolare l'aque

Ho osservato, che nei passagi a traverso delle grandi Alpi, che dividono l'Italia, <> che il corso della pendenza è fatto a scala cioè un pezzo che discende, un altro piano, e poi un altro <> che di novo descende, e con tale ordine sino alla tottale pianura, altrimente se la pendenza fosse tutta d'un tratto ne seguirebbero gran sconcerti per la decadenza dell'aqua

Li fiumi reali <nella loro> nei siti, che si congiungano [ai] monti hanno cataratte, come nel Reno nel Danubio ed altri, che comprovano l'unione della catena de monti continuata [nel]le cataratte.

Free Translation

Capitalization of names follows strictly the original Italian text. Some changes have been needed in punctuation and text order for ease in reading.

Legend

◇— text crossed out.
()—alternative or synonymic translations.

First draft of the Organic Structure of the Earth
November 4, 1728

[c. 144r]

King Leopold, after a series of 17 campaigns, liberated Vienna from the Turks. He triumphed so successfully that he hoped to create peace and reign wholly within the area's natural borders, which were surer than the artificial ones. After having been the theater of war for almost two centuries, many provinces were lacking people that could give reliable information about the landscape; neither printed maps nor descriptions were available. Many of the provinces were devoid of inhabitants and completely deserted since the Tartars' invasion. Reliable guides could not be found and one could say that following the army involved discovering new lands. Segmenting these lands and recognizing the safe boundaries within them became a much-needed project.

I was selected (or entrusted) by the King to do this job, and I got instructions from him and his ministry to establish safe borders against the Turks. Additionally, I had to establish the present and future advantages of his interest concerning the Allied countries bordering those large Provinces around the Danube, which we considered to be next for colonization. My obedience was equal to the belief and devotion I exercised in such a major work, which lasted for ten campaigns. Because the army's successes were different, the suggested borders were various and often discussed during the quiet winter times

[c. 144v]

in the cabinet, planning the operation of the next campaign and also preparing peace talks requested by the English and Dutch Ambassadors.

My idea to attempt a sort of Anatomy of the organic structure of Earth started with the opportunities provided to me by the fulfillment of my duties born from this confidential request by the King, together with the experience made during the voyages I had made through a large part of Turkey in my youth and all of the other later trips made through other parts of Europe.

I had previously recognized the trends of the Mountains, and the course of the rivers, and the maritime beaches and put that information together with what I had collected from the interior of the same mountains, which came from examining the structure of the many mines of Hungary and other parts of Europe and from my understanding of the generation of minerals and half minerals. Thus, I started building for myself a system that I did not attempt to establish before having been assured that the seabed in all its parts consisted of the continuing of Earth's emerged surface.

As my adventures left me free from all occupations, I chose the two littoral areas in Provence and Languedoc to check my theories. The first, totally mountainous area is located in France directly in front of the sea; the second is a plain almost up to the eastern parts of the Pyrenees. Toward the north, this littoral area is almost parallel to the mountain trend between different parts of the littoral area. Different littoral areas were needed by the Almighty to align the structure of the Earth with the foundations of the mountain trends, not only across the rivers but also across the seas. How are these mountains made in their interior, and by which materials are they composed? Have the modern plains existed since Earth's creation, and did they replace former seas? I have attempted to make observations, to run experiments [to prove or disprove] my thoughts, and to answer these questions with facts, which I will report in an almost Anatomical way about this terraqueous mass [Earth]. I have departed from the method of those who, staying inside a comfortable cabinet, pondered speculations on plane table sheets, able to support any chimerical configuration.

I am not going to expound on anything except that which I have touched by my own hands on both the mountainous and plane surfaces and inside both Mountains and Seas. Although [my observations have only been made] in narrow sections of the Earth, as compared to the extent of the terraqueous mass [Earth], nevertheless, I flatter myself that I can shed light on one single, almost universal, hypothesis extending down to the depth in the subsurface that it was possible for me to penetrate by various attempts.

[continued on c. 148r]

[c. 145r]

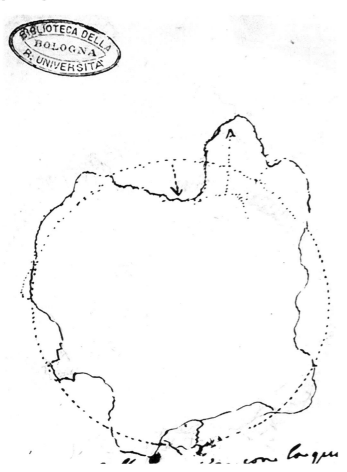

Figure A1. No title in the original (c. 145r). Note that this figure is the same as Figure 13 in text.

[c. 146r]

space for the waters to the empyrean

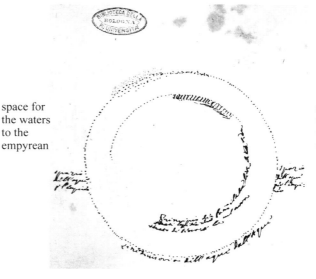

space for the waters to the empyrean [heaven]

Figure A2. No title in the original (c. 146r)

line dividing the waters from the Waters

[Writing inside the drawing]

Beginning of the
Lapideous strata building the top (?) of Earth
Bed of earth … the bad …
[continued on c. 147r without writings]

The first sketch shows how perhaps the Almighty excavated the sea to make the dry land, and to form at the same time the mountains, using the same earth of the globe, to obtain a deep depression whereby He could recover the water destined to the sea. The excavation resulted from pressure on the globe matter and the ensuing lateral uplift formed the mountains that are the banks of the sea; their relief is needed to protect the surface of land from water, which would become a pond without the use of such a divine tool, as will be demonstrated.

[c. 147r]
[Contains only a staircase diagram without text; see reference to the diagram on c. 166r.]

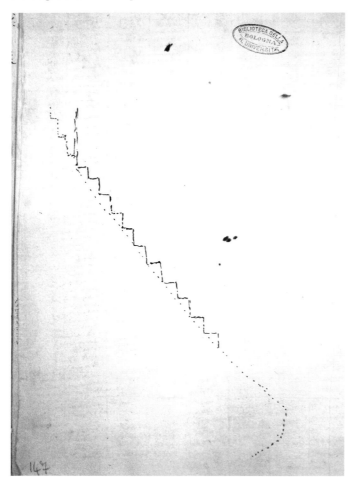

Figure A3. No title in the original (c. 147r).

[c. 148r]
What I could not see, or what could not be proven by my observations, I left to the happiest wizards (or great minds), speculating over a comfortable chair without bodily and heartily pain, so that they can advance more in their riddles.

As a pioneer, and great man, the polymath Steno mentioned something that may be influential to the Organic Structure of the Earth in his treatise *De solido intra solidum*. For some other authors, I shall leave to the public its own judgment and mainly offer my exposition.

The need to make it really clear to me whether the mountains were continuing inside the sea or over the sea-water level through Islands, or even beneath in a different way, forced me to stay (or live) in the Languedoc beaches to make those many experiments in their seas. They were described in a maritime physical History for the Royal Society of Paris. To make it [the history of the sea] clearer, it [my book] had to demonstrate much of what the structure of Earth consists of. I shall now reconsider this matter at its proper place using colors where I shall demonstrate the continuation of the continental Strata into the sea, adding much of what I had observed in the Garda Lake. In fact, I had prepared a physical treatise on the Garda Lake to be printed, which fits quite well and is also useful for me to demonstrate more clearly that all strata on both parts are correlative because the extent of the lake between the mountains is not as large as that of the sea.

It is not possible for me to begin my audacious attempt, based only on my observations, without starting from the third day of the divine Almightiness creating the World. We owe to it the submersion up to the top that we are not able to understand. Rather, I know to be my duty to pursue my physical observations.
[continued on c. 148v]

[The fine text at the left of the main column is placed below.]

The Netherlands' plains are seemingly handmade by an incredible craft (or skills), sometimes seemingly anatomized with great expenses and labor, as in the high Lombardy Plain around the Po [River] in Italy, and with the aim to bring relief to the Bologna countryside drowned by that great amount of water, which slowly flows into the Adriatic Sea. The reconnaissance survey I undertook through the mountains of Italy, Germany, and France aimed to understand the order by which the many marine bodies are arranged. They [the marine bodies] were reported to be one of the effects of the deluge in many printed works. After comparing my observations [made] on living marine organisms, I had reason to recognize the groundlessness of such a thought and the evaluation of my principles in my present assumption.

Garda Lake

Together with detailed observations made slightly less than 31 years ago through the mentioned parts of Europe, and also of lesser Asia, and after manual sailing along the Tartarian coast and the special survey of the Thracian Bosphorus, … I wrote to R …[writing not clear] the ground of this my enterprise perhaps totally new, the warp of which is very useful for my natural studies.

<Although already published>

I shall use illustrations to explain and clarify my position, not limited, and mainly the orographic map of the trend of the mountains observed in the…[not clear] of the continent and around the European, Asiatic, and African beaches I have seen and mapped.

[Report] the sites where those observations have been made and that will need to be drawn in detail. I have not seen some of the mountain trends, especially those in Asia, but they have been reported to me by other people and are drawn in the map. They have the mountains represented by dots, to avoid confusion about what I did not personally recognize (or observe).

[A crossed out section follows.]

So, I will begin my Treatise with the sentence of the sacred Text of Genesis, which is that on the third day of the Almighty …

[c. 148v]
to demonstrate what is written in those sacred scriptures with some obscurity. Some modern authors have established systems about Earth more in an attempt to show off their imaginative minds, rather than to show respect to our Religious debt; they are trying to uncover the ways in which the Almighty made the great Earth with his divine hands, from which he built man and established his nourishment with the many vegetables and animals.
#
[Between the section above and the next, a # marks where to insert the block "Lago di Garda."]

To be surer of the observations I made on the seas, I also wanted to check whether lakes have a structural correspondence like the sea. For this, I chose Garda Lake, which because of its narrowness, showed me both slopes, and its depth measured together with the littoral altitudes in a way as to write a natural physical dissertation to the marquise Poleni professor of mathematics at the University of Padua when it was ready for printing. I was pleased to see that everything corresponded with the observations made at sea.

[c. 149r]
[This page appears to have been crossed out by Marsili in the original, probably to avoid redundancy.]

The creation of the World was a work of the Almighty hand of God, who subdivided it in six days, as taught by the Sacred Script described in Genesis.
In the third [day], the more pure waters were separated from the less pure. The former were located elsewhere for other uses known to his Almightiness, which I am not going to address here, and the latter were directed to the sea by his sovereign words as Creator, "*Let the water under the sky be gathered into one place, so that the dry land may appear.*" This container <undoubtedly> probably was not created by God during the first two days, otherwise the waters would have rushed into it by their own weight, naturally speaking. Instead, when we look for a miracle that would suspend such a great mass of water in air from the first to the third day, when he commanded to them the mentioned words, there is nothing to speak about because the whole Genesis during those six days was only a supernatural work familiar to such a great Almighty, who indeed demonstrated how to hold water when he divided the Red Sea and let his chosen army cross the dry land.

It is not worthwhile for me to read the many controversies that are made on how the water mass could have been contained between the Earth's surface, which had not yet been provided with the future containers, and the firmament. Moreover, no one could even understand that an animated and organized man as shown by the Anatomy could have been made from a piece of clay, and a female, equally perfect in all her components, be created from one of his ribs, and all terrestrial and aqueous living beings and plants in the same way.

[c. 149v]
[This page also was crossed out by Marsili, probably to avoid duplication.]

Similarly, it was not impossible for his supreme power to condensate waters to reduce them to the perfect fluidity achieved for dividing the pure from the impure water.
He destined the less pure to the Sea and let rivers to serve [redundancy] the earth that was uncovered from the water, so as to make the earth fertile through the tidy circulation that the modern observations show in the Living Beings. In observing the effect of God's Miracles, we can only hope to clarify something, not to understand the ways in which he worked, because we are not able to do so. Sometimes we reach this knowledge of many things through methodical and long observations of existing miraculous works made by the Creator Author of nature, for example, in Astronomy, the knowledge of which was obtained through the observations of the stars, and in Anatomy, where the knowledge of man and many other living beings was made through their many cross sections. Having seen their structures, we are required to worship the Creator, under the name of nature, who, having created all of the parts that compose man, gave us the power to speak of him and, thus, demonstrate the body's complex outer and inner structure.

I have committed myself equally to examining as far as possible the terraqueous body by observing nature and without speaking more of God's miraculous involvement, which is unable to be understood by us mortals.

[c. 150r]
When examining the many parts composing Earth in the circumstances I have discussed, I did not fail to have before me the third day of the Genesis, that when the pure waters were divided from the impure waters, the heavier ones were sent into the wide jail of the Sea and the other, purer water assigned elsewhere by the Almighty creator, such that the Earth surface surrounding the sea remained dry for the many uses he had predetermined with his divine mind. I have no rash pretension to make a judgment on the different questions discussed among the Saints Fathers concerning one or another event of that day, but instead, I intend to keep myself within the limits of an observer of his great Works up to the level my mind, the weakest of all men's, could have arrived at in the darkness existing between the Almighty's lofty creation and my understanding of its component parts. My aim is to practice submissiveness to the miraculous ways of the Almighty without any thought but to worship them, and to be pleased only that I attempted to show through the Anatomy of a body from which we men were formed and live, and will be living until it [referred to the Earth] is annealing.

Since the Creator joined together the elements earth and water, so that they could intermingle inside a single body in the uses he assigned them for the benefit of man, it is needed to speak about water, which I shall distinguish in the forms and motions it makes through Earth. Earth's structure was also formed in a setting

[c. 150v]
such that this fluid element could irrigate the Earth through different ways.

In my treatise on the Sea, which is already printed, I spoke a lot about marine and terrestrial waters. I shall duplicate this information with the addition of other later observations in this, my Anatomy of the Earth, in which I will be able to show not so much surely and precisely which are the conduits allowing the waters to run through the solid inner parts of the same [Earth] as well as to irrigate its surface.

I shall demonstrate many observations without being required to make illustrations of the interior of Earth, which would only be used to prove ideal systems rather than the true existence of facts. They are commonly opposed to the essence of facts, and prefer to teach speculations. If the facts do not fit with their systems, I will not contest them; they will be contested, instead, only by the statement of what happened during my attempts. It is not easy for me—as it is for others instead—to decide upon one assumption starting from such attempts. I do not immediately see the foundations of an assumption. My observations, which until now have not been ordered in such a way as to yield anything conclusive, have only hinted at an assumption.

My enterprise is daring and could be useful if well demonstrated. This, my draft, will allow others after me to reach an improved knowledge of the Anatomy of the terraqueous Globe.

Drafted Anatomy to Demonstrate the
Organic Structure of the Terraqueous Globe

[151r]

Chapter 1
On the Outline and Size of the Terraqueous Mass of the World

Some have tried to prove that Earth is a plane figure (or outline), others have suggested an elliptic or spherical shape. I subscribe to this last figure, which corresponds with many astronomical demonstrations (or evidence), but I do not wish to get entangled in such disputes that have poor relevance to my aim of describing the Anatomy of the Terraqueous Body. My demonstration of the component parts of Earth is not dependant on its possible figure. I am following the most learned astronomers and the system that the earthy Globe be spherical and divided into 360 degrees, each of which contain 60 miles, which makes a circumference of 21,600 miles, and has a diameter of … [not filled] a space containing the Earthy and Aqueous Masses and building the body, I am going to anatomize, based on the observations I was allowed to collect during my duties and voyages.
[Marsili's left-side note is placed below.]
The amplitude of the Sea in respect to that of the Earth is …

Chapter 2
On the Materials by Which the Terraqueous Globe was made

These materials were hard Stone, soft and handy earth <fruitful and suitable to the Vegetation of Herbs, and Plants>, by which God created man, and as such suitable for the vegetation of all plants and water, after he had made the large pot of the Sea so that once the waters were gathered into it they uncovered the Earth, and served the Earth for the benefit of the living organisms and vegetables.

The hard stone, which we found continuing equally in the continent as in the Sea chains and in certain sites of the major river beds at the intersection with mountainous parts with the generic name of cataract [cataracts of the River Nile and other rivers like the Rhine and Danube], serves as a support for the soft and fertile earth. Such earth could not have resisted against the river erosion, and the attraction of the sun, and the pressure of living organisms on it, and its upheaval by cultivation, and the crashing Sea waves without the strong support of the hard stone.

[151r]
Both of these materials, hard stone and soft earth, are arranged in an order of strata over strata. The one of the earth called soft, suitable to be turned out with iron or wood tools, is a soil capable of being irrigated by the rain or dew and to attract the nitrous particles of the air through its permeable nature as to cultivate and feed all plants of any type created by God for the benefit of humankind. The other, stony material, or hard rock, is of such a hardness and solidity as required from the need to support the heavy weight of the land and the whole aqueous mass, which lakes and seas have to support with the strength of their floor.

Chapter 3
What is the Proportion between Earthy and Lapideous Mass?

It would be helpful to know the proportion (or ratio) between the earthy part and the hard stone, because at that elevation I spoke about, [earth] is no longer found because of the continuous erosion by the water that makes it slip <into the Sea>, especially from the mountains that are cultivated through upheaval of the earth. The lapideous bedrock of the high Mountains, which is included in the categories of the Alps, had to remain uncovered and exposed to the ravages of the cold winds and Sun, which, soaking it in different ways, broke it down into the grains that we call sand and also destroyed it in great blocks, as I shall show later.

We have to be happy to be able to understand that the pure earthy mass, even when it was newly created, was of very low mass in respect to the diameter of the remaining lapideous body. The earthy mass, in fact, required such a height as to allow the tallest trees to root and feed, whereas all remaining lapideous (or stony) mass had to be used for the floor of the Sea and for the many structures designed for the rain and marine waters, and to provide an escape for the many breaths (or vents) that, if they didn't exist, would cause fires, called Volcanoes, and earthquakes. All of these are effects of some organic tangle in such conduits.

[152r]

If the vents become stagnant, (see under *la Favor*) [note in the left-side column, see c. 165v] they cause fermentation that can light up, and make those tangles that will be treated in a chapter to follow.

I cannot as promptly as others establish here that a hole called subterranean is present in the interior of the Earth from the creation of the World until its annealing. It does not appeal to me, nor have I the skill to create with the pencil, subdivisions in the large width of the stony mass without having a slightest preliminary demonstration. I shall simply be pleased to tell, in its proper place, what I have seen inside the mountains at a depth of almost 400 m, and what I have learned by sounding the different depths of Sea lines such as was done for some whirlpools in the Danube, and finally for many currents.

[note in the left-side column]

([Remember to ask] doctor Monti to look for the depth of the Semnia mine in the tome of Minerals)

Chapter 4
On the Setting of These Materials

As I said, according to many observations of the Astronomers and Geographers, such materials were arranged in rounded outline and with an uneven surface. This unevenness is what we call mountains. One should believe that this work was made by God on the third day to dry up the earth <gathering the waters into the Sea>, which until then most probably had an equal surface supporting the water between itself and the firmament. To make the large Sea Tank, he commanded the whole stony mass with the overlying soft earth to break <raising all around the stony strata with the earth overlying them> swelling the surrounding stony strata as much as was needed until the mass of water covering all of Earth ran into the Sea Tank. For me to explain clearer what happened, consider what one sees happening in a globular mass of pasta, where if you insert a fist into the middle of the pasta it produces a lateral squeezing and a rim rise equal to the size of the fist. It is to believe that he followed [this way] in building the Sea, and at the same time he arranged the <mountainous> continent as earth higher than the surface of marine water, and with upper

[152v]

or lower altitude depending on the major or minor pressure made on earthy and stony materials, as I will show at its place. With this great work, the Creator shut up the seas, placed the continent higher than the universal receptacle of the waters so that the fertile earths would be drained; otherwise, all of the Earth surface would be plane, it would have been but a Marsh.

[165r]
Headings of the Work

Figure of the Earth plane, elliptical, round

Size in circumference and diameter

Ratio of the Earth to the Sea as for surface and depth

Materials composing the Terraqueous Globe in the first Creation

Materials grown and diminished during the existence of the Earth

Arrangement of the materials, and their order; Profile, and cross sections of the terraqueous Globe through different areas

Demonstration, that when God let the earth dry up, it was all mountainous, and seemingly there were no Plains

Demonstration, that the plains originated after the Sea retreated to receive the earthy destruction

Demonstration, that the mountains are destroyed in both the soft and Stony parts

Demonstration, that no new mountains were generated, but only Alluvial and Earthquake hills, and bituminous deposits especially where volcanoes are present, which never are connected with the primary Mountains Creation

The Mountainous Continent was higher where the Sea is deeper

The course (or trend) of the mountains would not be so finalized without the many drainages which have been produced by the rains and springs

The course (or trend) of the mountains was necessary not only for draining waters, but also to serve for the Sea

The Sea retreats and rises necessarily day after day, and the profiles show it

New accidental earths made by herbs, tree leafs bound by palustrine reed-roots, forming floating islands

[165v]
The stony Part decreases by many sicknesses [erodes or degrades], and by water flooding, and also increases through vegetation

The Metallic Minerals and half minerals must have been created during the creation of the World; however, they are multiplied when the mineral breaths find inside the mountains the structure needed to form them in the environment as in Hungary

Demonstration that the deluge has not hurt the Earth solidity, as claimed, except for the carnage of the living organisms

We cannot show with certitude how thick was the layer of earth placed by God upon the stony matter

Demonstration which water conduits can be found inside the Sea allowing for a link of the waters to each other; whether one can make on the map ideal projections of them based only on submarine springs, because we show that the pluvial Springs vanish during the last droughts, but not the submarine ones

I leave to the easy builders of systems the statement concerning a stable subterranean fire driving many natural processes; I am opposing that a fire lasting for 5000 years should have calcined (or ignited) the stony mass of our Globe; moreover, we see a discontinuous fire in the Volcanoes possibly born from the gathering of breaths which fermenting together light up. [Place here the] experiment *la Favor*

The unity of Earth with the water requires one to speak of it in its motions, of its bitter salt nature I have already made in my history. One should add more experiments concerning the attraction made by the Sun on it [the water] and to compare them with other experiments about the fresh water

When speaking about the handmade earth and especially that of the Netherlands, one will report on the techniques to make the peat and the Cuur, or floating islands, showing that they can make its framework with their roots

The structure of the mountains and the strata superposed to strata, and with an interspace (or a parting), which is

[166r]
a soft glutinous earth

The course (or strike, or trend) of these strata goes from North to south, and when sometimes they are found too oblique, or perpendicular, this is not depending on the order the rivers cut the mountains to drain the waters

When crossing the great Alps dividing Italy, I have observed that the pattern (or course) of the dip is staircase, that is a descending piece, a plane piece, and then another newly descending, and following the same order down to the full plane; otherwise, if the dip would be continuous there would be a great perturbation for the water falling down

Where the true rivers meet the mountains they form cataracts, such as for the Rhine, the Danube, and others, thus proving the continuity of the mountain chain through the cataracts.

REFERENCES CITED

Editions of ancient works having publication dates different from the edition examined and quoted in this paper are recorded with their date of publication under square brackets.

Airy, G.B., 1855, On the computation of the effect of the attraction of mountain-masses, as disturbing the apparent astronomical latitude of stations in geodetic surveys: Philosophical Transactions of the Royal Society of London, v. 145, p. 101–104.

Brocchi, G.B., 1814, Discorso sui progressi dello studio della conchiologia fossile in Italia, *in* Conchiologia fossile subapennina: Milano, dalla Stamperia Reale, I-LXXX p.

Burnet, T., [1684] 1699, Telluris Teoria Sacra: Amstelaedami, apud Joannem Wolters, [iv] + 558 p.

Cavazza, M., 2002, The Institute of Science of Bologna and the Royal Society in the eighteenth century: Notes and Records of the Royal Society of London, v. 56, no. 2, p. 3–25, doi: 10.1098/rsnr.2002.0164.

Csiky, G., 1984, Forerunners of mining-geological mapping in Hungary in the 18th century / L. Ferdinando Marsigli, Ignac Born, Johann E. Fichtel and Janos Fridvaldsky, *in* Dudich, E., ed., Contributions to the history of geological mapping, INHIGEO 1982: Budapest, Akademy Chiadó, p. 399–410.

Csiky, G., 1987, Luigi Ferdinando Marsigli, an Italian discoverer of Hungary, *in* Hala, J., ed. Rocks, fossils and history. Italian Hungarian relations in the field of geology: Budapest, Hungarian Geological Society, p. 327–341.

Descartes, R., [1644], 1677, Principia philosphiae: Amsterdam, Apud Ludovicum Elzeverium, [xxxiv] + 222 p.

Dutton, C.E., [1889], 1925, On some of the greater problems of physical geology: reprinted in Journal of the Washington Academy of Sciences, v. 15, p. 259–369.

Ellenberger, F., 1984, Early French geological maps: trends and purposes, *in* Dudich, E., ed., Contributions to the history of geological mapping, INHIGEO 1982: Budapest, Akademy Chiadó, p. 73–82.

Ellenberger, F., 1988, Histoire de la géologie: Paris, Lavoisier Tec Doc, v. 1, viii + 352 p.

Ellenberger, F., 1994, Histoire de la géologie: Paris, Lavoisier Tec Doc, v. 2, xiv + 383 p.

Franceschelli, C., and Marabini, S., 2004, Luigi Ferdinando Marsili: A geomorphological and archaeological approach to the earth science [abs.]: 32nd International Geological Congress, Florence, 2004, abs. T20.01, Origin of modern geology in Italy, p. 1122.

Gortani, M., 1930, Idee precorritrici di Luigi Ferdinando Marsili su la struttura dei monti, *in* Corbelli, A., and Marsiliano, C., eds., Memorie intorno a Luigi Ferdinando Marsili: Bologna, Nicola Zanichelli, p. 257–275.

Gortani, M., 1963, Italian pionieris in geology and mineralogy: Journal of World History, v. 7, no. 2, p. 503–519.

Heim, A., 1892, Geologische Nachlese: Vierteljahresschrifte der naturforschenden Gesellschaft: Zürich, Nr. 1, Oktober.

Kircher, A., S.I., 1665, Mundus subterraneus in XII libros digestus; ...: Amstelodami, Apud Joannem Janssonium et Elizeum Weyerstraten, 2 volumes, [xxvi] + 346 + 6; [x] + 487 + 9 p.

Kircher, A., S.I., 1678, Mundus subterraneus in XII libros digestus; ...: Editio tertia: Amstelodami, Apud Joannem Janssonium à Waesberge & Filios, 2 volumes, [xviii] + 366 + 6 p.; [x] + 507 + 9 p.

Lipparini, T., ed., 1930, Storia naturale de' gessi e solfi delle miniere di Romagna, *in* Lovarini, E., and Marsiliano, C., eds., Scritti inediti di Luigi Ferdinando Marsili: Bologna, Nicola Zanichelli, p. 189–211.

Longhena, M., 1930, Il conte Luigi Ferdinando Marsili. Un uomo d'armi e di scienza: Milano, Alpes, 346 p.

Lyell, C., 1830–33, Principles of geology, being an attempt to explain the former changes of the Earth's surface, by referring to causes now in operation: London, J. Murray, 3 volumes, xv + 511 p.; xii + 330 p.; xxxii + 398 + 109 p.

Marabini, S., and Vai, G.B., 2003, Marsili's and Aldrovandi's early studies on the gypsum geology of the Apennines, in Vai, G.B., and Cavazza, W., eds., Four centuries of the word geology: Ulisse Aldrovandi 1603 in Bologna: Bologna, Minerva Edizioni, p. 187–203.

Marsili, L.F., 1698, Dissertazione epistolare del fosforo minerale ò sia della pietra illuminabile Bolognese: Acta Eruditorum di Lipsia, A Lipsia, 32 p.

Marsili, L.F., 1700, Danubialis operis prodromus, ad regiam societatem anglicana: Norimbergae, Apud J.A. Endteri filios, 60 p.

Marsili, L.F., 1711, Brieve ristretto del saggio fisico intorno alla storia del mare scritta alla regia Accademia delle Scienze di Parigi: Venezia, A. Poletti, 52 p.

Marsili, L.F., 1725, Histoire physique de la mer. Ouvrage enrichi de figures dessinées d'apres le naturel: Amsterdam, Aux dépens de la Compagnie, [vi] + xi + 173 p., 40 tabs., 12 pls.

Marsili, L.F., 1726, Danubius Pannonico-mysicus, observationibus geographicis, astronomicis, hydrographicis, historicis, physicis perlustratus, Hagae Comitum, Apud P. Gosse et al.; Amstelodami, Apud Herm. Uytwerf et Franç. Changuion, 6 vols., [ix] + 96 + 4, 46 pls.; [iii] + 147 + 6, 66 pls.; 137 + 4, 35 pls.; [ii] + 92, 33 pls.; 154 + 6, 74 pls.; 128, 28 pls.

McConnell, A., 1999, Introduction, in Dragoni, G. ed., Natural history of the sea by Luigi Ferdinando Marsigli (translated by A. Mc Connell): Bologna, Li.Pe., photostatic edition, ix + 570 p.

Morello, N., 2003, The birth of stratigraphy in Italy and Europe, in Vai, G.B. and Cavazza, W., eds., Four centuries of the word geology: Ulisse Aldrovandi 1603 in Bologna: Bologna, Minerva Edizioni, p. 251–263.

Murray, J., 1880, Historical introduction, in report on the scientific results of the voyage of H.M.S. 'Challenger' during the years 1872–76: A summary of the scientific results, v. 1, p. 60.

Oldroyd, D.R., 1996, Thinking about the Earth: a history of ideas in geology: London, Athlone, and Cambridge, Massachusetts, Harvard University Press, xxx + 410 p.

Pérès, J.-M., 1968, Un précurseur de l'étude du benthos de la Méditerranée: Louis-Ferdinand, comte de Marsilli: Bulletin de l'Institut océanographique, Monaco, special issue 2, v. 2, p. 369–376.

Pratt, J.H., 1855, On the attraction of the Himalaya Mountains, and of the elevated regions beyond them, upon the plumb-line in India: Philosophical Transactions of the Royal Society of London, v. 145, p. 53–100.

Sartori, R., 2003, Luigi Ferdinando Marsili, founding father of oceanography, in Vai, G.B., and Cavazza, W., eds., Four centuries of the word geology: Ulisse Aldrovandi 1603 in Bologna: Bologna, Minerva Edizioni, p. 169–177.

Şengör, A.M.C., 2003, The large wave-length deformations of the lithosphere: Geological Society of America Memoir 196, xvii + 347 p.

Seibold, E., and Seibold, I., 2001, Antonio Vallisnieri—Ein moderner geologe vor 300 Jahren: Max Pfannenstiel zum Gedächtnis (1902–1976): International Journal of Earth Sciences, v. 90, p. 903–910, doi: 10.1007/s005310100204.

Steno, N., 1669, De solido intra solidum naturaliter contento dissertationis prodromus: Florentiae, Ex Typografia sub signo Stellae, 80 p.

Stoye, J., 1994, Marsigli's Europe 1680–1730. The life and times of Luigi Ferdinando Marsigli, soldier and virtuoso: New Haven and London, Yale University Press, 356 p.

Thoulet, J., 1897, Un des fondateurs de l'océanographie. Marsigli: Revue Scientifique, y. 34, s. 4, v. 8, p. 801–805.

Trümpy, R., 1998, Tectonic units of central Switzerland. Their interpretation from AD 1708 to the present day: Bulletin of Applied Geology, v. 3, no. 2, p. 163–182.

Vaccari, E., 2003, Luigi Ferdinando Marsili geologist: from the Hungarian mines to the Swiss Alps, in Vai, G.B., and Cavazza, W., eds., Four centuries of the word geology: Ulisse Aldrovandi 1603 in Bologna: Bologna, Minerva Edizioni, p. 179–185.

Vaccari, E., 2006, this volume, The "classification" of mountains in eighteenth century Italy and the lithostratigraphic theory of Giovanni Arduino (1714–1795), in Vai, G.B., and Caldwell, W.G.E., The origins of geology in Italy: Geological Society of America Special Paper 411, doi: 10.1130/2006.2411(10).

Vai, G.B., 2003, A liberal diluvianism, in Vai, G.B., and Cavazza, W., eds., Four centuries of the word geology: Ulisse Aldrovandi 1603 in Bologna: Bologna, Minerva Edizioni, p. 221–249.

Vai, G.B., and Cavazza, W., eds., 2003, Four centuries of the word geology: Ulisse Aldrovandi 1603 in Bologna: Bologna, Minerva Edizioni, 327 p.

Woodward, J., 1695, An essay toward a natural history of the Earth and terrestrial bodies especially minerals as also the seas, rivers and springs. With an account of the universal deluge and of the effects it had upon the Earth: London, R. Wilkins, [xii] + 277 p.

Woodward, J., 1704, Specimen geographiae physicae quo agitur de terra, et corporibus terrestris speciatim mineralibus: nec non mari, fluminibus, et fontibus. Accedit Diluvii universalis effectuumque eius in Terra descriptio, Tiguri, Typis D. Gessneri, (translator J.J. Scheuchzerus), 305 p.

Woodward, J., 1735, Géographie physique, ou essay sur l'histoire naturelle de la terre: A Paris, Chez Briasson.

Woodward, J., 1739, Geografia fisica ovvero saggio intorno alla storia naturale della terra: Venezia, presso Giambatista Pasquali, 421 p.

MANUSCRIPT ACCEPTED BY THE SOCIETY 17 JANUARY 2006

Geological Society of America
Special Paper 411
2006

Luigi Ferdinando Marsili (1658–1730): A pioneer in geomorphological and archaeological surveying

Carlotta Franceschelli*
Dipartimento di Archeologia, Università di Bologna, piazza S. Giovanni in Monte 2, 40127 Bologna, Italy

Stefano Marabini*
Dipartimento di Scienze della Terra e Geologico-Ambientali, Università di Bologna, via Zamboni 67, 40127 Bologna, Italy

ABSTRACT

The relevance of Luigi Ferdinando Marsili to the history of geology is related to his publications, to the design and foundation of the Academy of Science of Bologna, and to his unpublished *Trattato de' monti* (*Treatise on the Mountains*). Additionally, his name is well known among scholars of antiquities because he assembled during his life a rich collection of archaeological findings that today forms an important part of the Civic Museum of Bologna.

Beyond his achievements in many different fields, from the earth sciences to antiquities studies, one of the most original aspects of Marsili's work lies in the methodology he developed. He assigned basic importance to field research and to direct visual checks (nowadays known as field surveys), which were for him the real instruments of knowledge. His scientific approach was probably spontaneous, deriving from his long military career, during which he was engaged in reconnaissance exploration and cartography. However, it is not unreasonable to acknowledge that he was also influenced by the work of Philip Cluver, one of the founding fathers of ancient topography, who, nearly 100 years earlier, wrote *Germania Antiqua*, which stressed the importance of autopic sensibility as the best way to validate the results of field research.

In practice, Marsili did not accept a sharp break between the natural and human-caused aspects of land science, which show many reciprocal influences; he studied and represented them cartographically together, with a multidisciplinary and modern approach.

Keywords: ancient topography, *Canstatter Travertin*, cartography, Diluvianism, fossil bones, geoarchaeology, history of geology, Marsili, Scheutzer, survey.

INTRODUCTION

The Italian Count Luigi Ferdinando Marsili (1658–1730), who in 1711 founded the Academy of Sciences of Bologna, is internationally known for having played a significant role in the field of cartography, geography, oceanography (of which he is considered to be the founding father), zoology, botany, and geology (Thoulet, 1897; Gortani, 1930; Lipparini, 1930; Csiky, 1987; Stoye, 1994; Seibold and Seibold, 2001; Marabini and Vai, 2003; Vaccari, 2003; Vai, 2003) (Fig. 1).

Marsili was very popular among his contemporaries (Marsili, 1725, 1726). In 1691, he was admitted to the Royal Society

---

*E-mails: cafrances@racine.ra.it; stemarabini@libero.it

Franceschelli, C., and Marabini, S., 2006, Luigi Ferdinando Marsili (1658–1730): A pioneer in geomorphological and archaeological surveying, *in* Vai, G.B., and Caldwell, W.G.E., The origins of geology in Italy: Geological Society of America Special Paper 411, p. 129–139, doi: 10.1130/2006.2411(08). For permission to copy, contact editing@geosociety.org. ©2006 Geological Society of America. All rights reserved.

Figure 1. Portrait of the young Luigi Ferdinando Marsili. By permission of the Biblioteca Universitaria di Bologna.

of London and in 1715 to the Academy of Sciences of Paris; in 1722 he visited the Royal Society, where he was welcomed by an 80-year-old Isaac Newton. Notwithstanding, as a scientific pioneer, he has not been fully recognized, perhaps due to the fact that most of his works are partly still unpublished.

However, Marsili was rediscovered during the celebrations of the bicentenary of his death (1930), when several of his works were published for the first time and a cataloging of his vast personal archive took place (Frati, 1928; Marsili, 1930a). The definition of Marsili as "*uomo d'arme e scienza*" ("man of army and science") given by one of his biographers, the socialist writer Mario Longhena (1876–1967), dates back to that time (Longhena, 1930).

Until 1703, Marsili's main activity was his military service for the Emperor Leopold of Austria, for whom he fought against the Turks until the treaty of Carlowitz in 1699 (Marsili, 1930b). This experience was the start of the particular approach to the geological matters that later led him to become a scientist. As a soldier, he put a lot of energy into the exploration and cartographic survey of vast regions along the Danube, from Dalmatia to Transylvania. These surveys were used for the planning of military campaigns, fortifications, streets, and bridges, and for drawing borders (Marsili, 1930b).

In order to better understand Marsili, and to restore him to the history of geology, it is necessary to publish the manuscripts and sketches taken from the *Trattato de' Monti*, which, unfortunately, he did not finish (Vai, this volume, Chapter 7), and to understand his influence on the so-called Diluvianistic school of Bologna in the 1700s (Sarti, 1988, 2003), with its essential concepts for the development of geology (Gortani, 1930; Vaccari, 2003; Vai, 2003).

Marsili was among the first geologists to use stratigraphic columns, sketches of outcrops and landscapes, and, above all, geological maps (Marabini and Vai, 2003; Vaccari, 2003) to represent his on-site geological observations. Moreover, in practicing geomorphology, he also investigated land use and other anthropic aspects of the landscape, carrying out observations normally considered unusual for a mere naturalist; for example, those concerning archaeology.

This sound and pragmatic multidisciplinary approach to land study led Marsili to investigate archaeology and ancient topography. For this reason, he can be considered one of the precursors of the well known Italian school of geologists, which in the following century, with men like Giambattista Brocchi (1772–1826) and Giuseppe Scarabelli (1820–1905), played a significant role in developing stratigraphic archaeology (Brocchi, 1819; Scarabelli, 1850).

MARSILI'S CARTOGRAPHY OF THE DANUBE AREA (1681–1704)

Marsili was well aware of the scientific importance of the information he had collected in the central Danube area. He spent most of his military career there, improving his cartographic method and adding to his naturalistic knowledge through experience. In fact, after the treaty of Carlowitz, he worked on an extensive treatise on the Danube, *Danubius Pannonico-Mysicus*, which was published 25 years later in five illustrated volumes dedicated to history, geography, archaeology, geology, botany, zoology, and other fields (Marsili, 1700, 1726) (Fig. 2). According to his autobiography (Marsili, 1930b), his interest in the Balkan-Danube area began in his youth when, after a sentimental disappointment, he joined the journey by sea to Constantinople undertaken by the new Venetian ambassador (1679–1680). Eleven months later, he came back by land, passing through Sofia, Belgrade, Serbia, and Dalmatia. A few months after returning, his father died, and Marsili decided to make the most of his knowledge of the Turkish language and culture. He joined the army of the Emperor Leopold of Austria in the war against the Ottoman Empire, on the Balkan front.

Marsili's long military career was characterized by frequent disagreements with his senior officers. He also experienced a number of adventures, some of them unfortunate. For example, he did not take part in the defense of Vienna during the Turkish

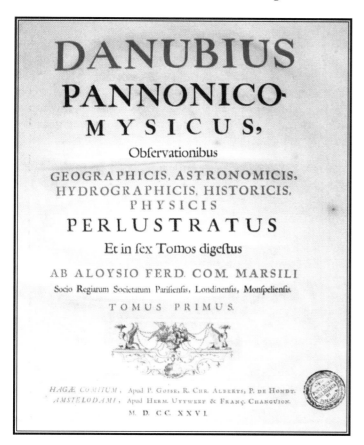

Figure 2. Title page of *Danubius Pannonico-Mysicus* (1726). By permission of the Biblioteca Universitaria di Bologna.

Figure 3. Marsili's drawing of a partially marshy valley bottom plain, bordered by hills and fluvial slopes (Danube plain). By permission of the Biblioteca Universitaria di Bologna, Fondo Marsili.

siege of 1683 because he was a prisoner in the besieging camp. During his long imprisonment in the remote valleys of Hungary and Bosnia, which ended in 1684, he developed a passion for exploration and cartography.

Marsili showed from the beginning a particular versatility for scientific matters when surveying and mapping the Danube area. He was certainly helped by what he had learned during his early attendance at the Universities of Bologna and Padova (Marsili, 1930b). He became an expert on mining sites and, above all, on the wide marshes around the bottom of the Danube and Tisza valley, which he portrayed in a number of maps, showing innovative approaches both in terms of features observed and of the techniques of representation (Fig. 3). His interest in the marshes led him later to study the marshes around the Po delta (Marsili, 1930b) and to visit the low countries of Holland (Biblioteca Universitaria di Bologna [BUB], Fondo Marsili [FM], ms. 99, A, 1–7 and ms. 87, F; Longhena, 1930).

Although his education in humanistic subjects was not thorough, he spent a considerable amount of time identifying and cataloging various Roman archaeological sites along the Danube *limes* (border), focusing on defense walls and ancient routes, which he also found interesting from a military point of view (Brizzi, 1983). He also drew a number of sketches and maps, more or less refined, in order to localize them in their geomorphological context (Fig. 4).

In his autobiography, Marsili describes when, after a victory at Nissa in 1690, he moved with the imperial troops south to Vidun (today's Bulgaria):

In questo intervallo di tempo, con l'aiuto di più di cento uomini, m'applicai ad osservare tutte le antichità romane, in quelle vicinanze esistenti, ed in particolare le vestigia del ponte di Traiano.
Nel luogo in cui era stato il ponte suddetto, assai famoso, pretendeasi di fabbricarne un altro sul Danubio, anzi in più comodo sito per le marcie; ed io, che avevo la cura di questa opera, ne sentivo particolar passione, per rispetto d'un'antichità rinomata. (Marsili, 1930b, p. 119)

In this period of time, with the help of over 100 men, I observed all the Roman antiquities around that area, most of all the remains of the bridge of Trajan.
In the place where this very popular bridge was, another bridge on the Danube was going to be built, in a better position for the marches; being in charge of this work I was particularly fond of it, because of my respect for a prestigious past.

Marsili's most outstanding maps and landscape views were those he made to define the new borders between the Hapsburgs and the Ottoman Empire after Carlowitz peace (1699), when he drew, on a daily basis, some hundreds of them (Longhena, 1933). The original surveys, made during 1700 and 1701, are kept in the Marsili Fund at the Library of the University of Bologna (BUB, FM, ms. 66), where Marsili classified them as *Diaria Geographica in itinere limitaneo collecta*.

In these sketches from life, Marsili accurately represented the morphology of mountains, ridges, slopes, river beds, and

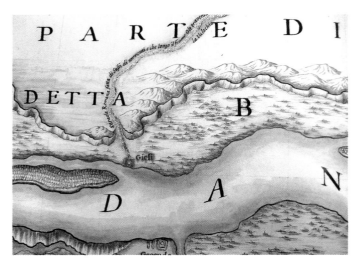

Figure 4. Watercolor of a Roman road made from pebbles (*glareata*) near the river Danube. By permission of the Biblioteca Universitaria di Bologna, Fondo Marsili.

valley plains. In many cases, his sketches are framed as if they were pictures, and therefore the corresponding sites can probably be recognized today (Fig. 5).

It would be reasonable to assume, given his early interest in geology (Marabini and Vai, 2003), that one of Marsili's main aims was to understand the internal composition of Earth through the analysis of its surface. Marsili himself talks about this in some unpublished works, such as *Trattato de'monti*, where he defines the subject of his studies as "*organica structura del globo terreo acqueo*" (the organic structure of the terraqueous globe), in which he gives equal importance to the rocks, earth, and water that make our globe (Vai, this volume, Chapter 7). In fact, he noticed that the globe was made by "*uno substratto di pietra viva, e dura*" (a substrate of alive and hard rock), covered by a skin of "*terra mole*" (i.e., the soils and the soft detrital covers), and that the geologist's task was equal to that of an anatomist who works with human bodies and takes off their skin to investigate what is inside them (Fig. 6).

Looking at all these sketches and maps of the early 1700s, it can be said that Marsili's approach to cartography, with all the limits of his knowledge and possibilities, was similar to that of a modern geomorphologist or geologist.

SURVEYING IN THE STUTTGART AREA (MARCH 1704)

The turning point in Marsili's life, transforming the "*uomo d'arme*" (man of army) into a "*uomo di scienzia*" (man of science), was the French takeover of the Hapsburg stronghold of Breisach, a small town on the Rhine near today's Freiburg, in 1703. As commandant of the imperial garrison, Marsili was involved in the defeat and the following rumors and trial (details in Longhena, 1930; Stoye, 1994).

Having been condemned and reduced in rank, he reacted by focusing more on his scientific interests. He went on a number of study trips to Italy and Switzerland (Vaccari, 2003), clearly to extend his "geological survey" to areas other than the Balkans, and to collect the information necessary to write the *Trattato de' Monti*. During 1704, he was assisted by Johann Scheuchzer, brother of the famous naturalist Johann Jakob Scheuchzer, who had a good knowledge of Greek and Latin (Marsili, 1930b; Longhena, 1930; Vaccari, 2003).

Marsili's activities in these years can be seen in an unpublished manuscript of March 1704 (BUB, FM, ms. 23, cc. 2–3), which describes an ossiferous deposit near *Canstadium*, today Bad-Cannstatt, in the northeast of Stuttgart, in the plain of the river Neckar, tributary of the river Rhine (Fig. 7).

Figure 5. Pen-and-ink sketch from life from *Diaria Geographica*, November 1700. By permission of the Biblioteca Universitaria di Bologna, Fondo Marsili, ms. 66.

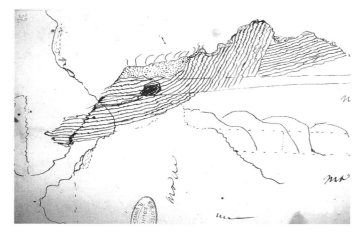

Figure 6. Pen-and-ink sketch of a spring (black spot) from a stratified substratum; it is fed by the meteoric water that passes through a permeable detrital cover (dotted). By permission of the Biblioteca Universitaria di Bologna, Fondo Marsili, ms. 90.

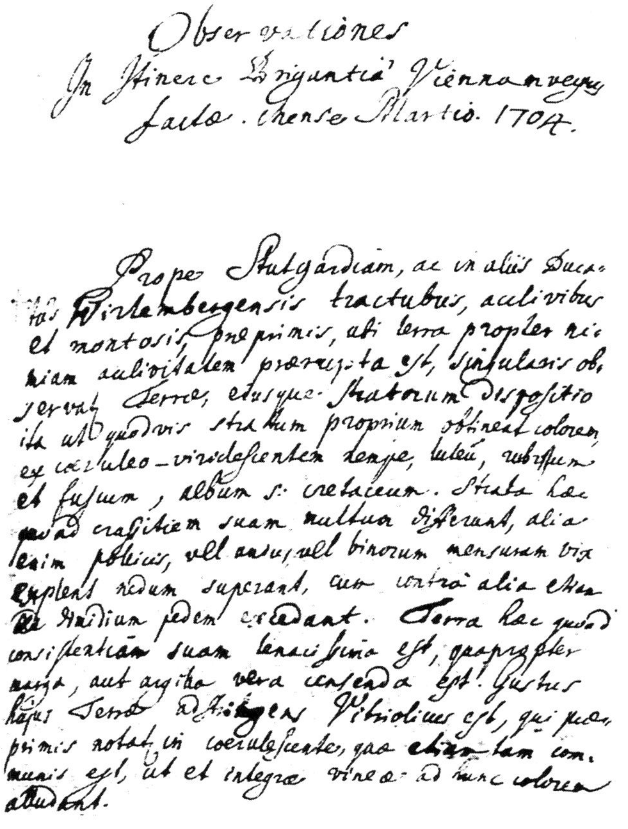

Figure 7. *Incipit* of the manuscript *Observationes In Itinere Brigantia Viennam versus factae. Mense Martis 1704*. By permission of the Biblioteca Universitaria di Bologna, Fondo Marsili.

The authorship of the manuscript is a problem. It belongs to one of the volumes of the Marsili Fund with the title *Osservazioni naturali sopra varie miniere fatte dal generale Co. Marsili e sopra il lago di Costanza* (Natural surveys in several mines carried out by General Commandant Marsili and in the Lake of Constance), but has a headline, on page 1, reading: *Observationes naturales factae per decursum anni 1704 a Johanne Scheuchzero* (Natural surveys made in 1704 by Johann Scheuchzer), which seems to suggest that what comes after was written by Scheuchzer. However, comparison between this text and other works surely written by Marsili—his 1680 journal of the return journey from Constantinople and several comments to the geological sketches that he drew on the field (BUB, FM, ms. 51; ms. 90, A, 21, c. 145n. a m.; ms. 90, C, c. 112v)—and the fact that Marsili's writing style was peculiar and therefore recognizable, seems to suggest that this text was also written by him. The date seems to back up this hypothesis, as in March 1704, when he was reduced in rank, Marsili was in Bregenz, by the Lake of Constance (Longhena, 1930). Moreover, there is a pen-and-ink sketch of the site beside the text, accompanied by a short but comprehensive explanation, typical of Marsili's style for his early works (Marabini and Vai, 2003).

Before examining the text, we quote a vast extract here (see Fig. 8).

Observationes
In Itinere Brigantia Viennam versus factae. Mense Martis 1704

[c. 2]
[…] *Locus prope Canstadium, ubi ossa et unicornua fossilia fuēre reperta ita se habet. Petrae s(ive) lapides hanc fodinam nunc undique circumdantes, varissimi sunt, et quasi ex stalactita compositi quod testantur crebrae ibi reperiundae concretiones veniformes nigricantes, ex mera crustarum invicem supernatarum figura constantes, lapidibus istis etiam in tantam duritiem compactis immista est crebro Ferri Minera, satis, ut pondus indicat, dives, constant hae petiae ex stratis ad Horizontem et quidem versus se invicem inclinatis, ita ut hac inclinatione verum fornicem efformatum fuisse colligam (ut figura monstrare poterit.) patietur Lector, et ingenuitati rationum quas adiligare animus nobis est, ignoscet, si probaverimus haec ossa revera an inantium olim demortuorum, vel occisorum esse reliquias. Quoad loci sitam, in perfecta planitie fodina est prope Coemeterium Canstadiense extra urbem, ut ex hac sola consideratione sua sponte concidat erronea quorundam opinio, quod nempe haec ossa genereantur, aut producta sint.*

[c. 3]
ex marga quadam subtili quae fluida tam varias assumpserit formas, unde quaeso affluxus talis margae, cum nec in petris appareant talia ductuum lineamenta, nedum maiora interstitia pro tam copioso materiae affluxu satis ampla, nedum etiam inter ipsa petrarum strata ubi et ossicula vario et inordinato situ iacentia inter argillam luteam recondita cernuntur, cum e contra nulla ubique reperiatur margae tam subtilis albae vestigium. Relinquimus hanc opinionem, properamus si non ad veram, saltem probabilem. Historiarum Romano(rum) Antiquarum monumentis traditu(m) est, quod exercitus Romanus haud quidem semel sed saepius usque ad Necarim fluvium venerit qui locus noster s(ive) Canstadium Necari immediate appositus est,
novimus etiam iisdem Romanae Historiae tradentibus pandectis, quod Romani tamquam populus magno reru(m) apparatu instructus praesertim ubi in castris vixerunt, quod inquam Romani secum in acie duxerint non solum equos, boves et caetera eiusmodi animalia victui saltem, aut equitationi dicata. Habuerunt insuper Elephantes, et alia, maioris molis et extensionis bestias ipsi assaltui praeprimis accomodas. Tam quid quaeso paradoxum sta-

tuesco si dicam ibi locorum ut in exacta planitie, ad egregium fluvium, summatim, ubi pro militibus omnis erat commoditas, fuisse Romanas Castrametationes, quod etiam ibi locorum aut fuerit grassataque sit communis inter bestias etiam homines lues, aut ibi locorum facta pugna tales bestiae occubuerint, et per consequens bestiarum cadavera subter hisce fornicatis rupibus fuere Abscondita, sicque tractu temporis fuere quasi calcinata, affluente per petrarum eamque stratorum rimas aqua particulis ferri metallicis facta, quod multum conducit ad calcinationem corporum, oleaginosas partes absorbendo, et absumendo, Salinas item particulas obstipando, earumque quasi vim constringendo, quare hisce factis calcinatio iam dum factam esse censeo, hisce enim sublatis principiis iam ossa friabiliora magis sunt ac porosiora, enim hypotheseos argumentum haud exiguu(m) est quod ossa haec non solum superficie tenus, sed etiam in ipsa substantia ostendant nigra punctula et lineamenta dendritarum s(ive) arbuscularu(m) effigies repraesentantia, oleosarum et salinaru(m) partium dissolutarum, et hin(c) inde fixaru(m) indicia. […]

This text, which comes after a general geological outline of the Stuttgart area, reports on a visit to a small quarry near *Canstadium*, which had probably been visited by other scholars after several large bones were found there. The description of the stalactites and the fact that the quarry is defined as a cave ("*cripta*") in the watercolor that Marsili reproduced twenty years later while he was still working on the *Trattato de' monti* (Vai, Chapter 7, this volume) seem to suggest that the bones were found mostly inside the muddy cracks or, perhaps, inside real karstic holes (Fig. 9).

Figure 8. Pen-and-ink sketch from life of the quarry near Bad-Canstatt in the manuscript of 1704. By permission of the Biblioteca Universitaria di Bologna, Fondo Marsili, ms. 90.

It is worth noting that this area, and the fossil vegetal rests of the *Canstadium* travertine in particular, had been mentioned in the script *De Spongite lapide*, by Johann Mattheus Faber, published in 1694 in Leipzig in the *Ephemeridum Medico-Physicarum Germanicarum Academiae Caesareo-Leopoldinae Naturae-Curiosorum Decuriae III, Annus Primus, Anni M.DC. XCIV*. As this was a scientific institution of the Emperor Leopold, whom Marsili had been serving for many years, he certainly knew this study.

In particular, Marsili's text recognizes the organic origin of these bones, clearly disagreeing with the opinion, still widespread at this time, that fossil bones were mainly inorganic, the result of a spontaneous production process of rock (Sarti, 1988; Vai, 2003). What really shows Marsili's originality in interpreting the evolution of Earth's surface is the fact that, although he was the founder of the Academy of Sciences of Bologna—later known as the "Temple of the Diluvianism"—he does not mention the theory of the Deluge according to which fossils were the victims of that Flood (Sarti, 2003). Even in his *Trattato de' monti*, Marsili dismisses the Deluge theory, as explained by Vai (this volume, Chapter 7).

The Bologna cultural and scientific environment between the seventeenth and eighteenth centuries broadly accepted the organic nature of fossils and was indeed characterized by a pragmatic "*liberal Diluvianism*," which saw in the Deluge merely an easy explanation of the large presence of fossils in several areas (Vai, 2003).

Marsili used his knowledge of the history of the region, acquired from the classics, to suppose a link between the fossil bones and the presence of Roman camps along the right side of the river Neckar. In his opinion, the large quantity of bones mentioned would have resulted from the carcasses of horses, cattle, or elephants used by the Romans, which would have died during pestilence or after a battle and which had been piled up in caves. The link between fossil bones and the Romans had also been suggested by Marsili a few years before, when he had found some elephant bones and teeth in the Danube marshes (Marsili, 1726).

In fact, these bones were even more ancient, as they were probably linked to the fossil fauna of the *Canstatter Travertin*, presumably dating back to the middle-upper Pleistocene (Adam et al., 1986). Still, although the link between these bones and the Romans is not correct—as there are no records of elephants being used by the Romans in that area—Marsili's hypothesis can rely on some historical facts, widely recognized today. The region was in fact known during the Roman age as *Agri Decumates* and belonged to the province of *Germania Superior*, whose eastern border, marked along the river Neckar between the age of *Flavii* and that of *Antoninus Pius*, was in those years the eastern *limes* of the Roman Empire. Further, there is record of a fortress along this defense line in the area around Bad-Canstatt (Baatz, 1974; Schalles, 1998; Rinaldi Tufi, 2001).

Marsili's hypothesis is additionally backed by taxonomic studies; only fifteen specimens out of the original Marsili

Figure 9. Watercolor of the site near Bad-Canstatt, which Marsili visited in 1704 and reproduced twenty years later for the *Trattato de'monti*. By permission of the Biblioteca Universitaria di Bologna, Fondo Marsili, ms. 90.

collection are still preserved, and fourteen have been classified as Proboscidea, species *Palaeoloxodon antiquus* (Falconer and Cautley) and *Mammuthus primigenius* (Blumenbach) (Sarti, 1984).

MARSILI AND THE ANTIQUITIES STUDIES

Marsili's early and continuing interest in archaeology and antiquities is well known, in particular from a collector's point of view. In fact, he played a fundamental role in the preparation of the so-called "*stanza delle antichità*" (room of the antiquities) in the Academy of Sciences in Bologna, which today is an important part of the Civic Museum (Morigi Govi, 1982; Brizzolara, 1986).

Looking at Marsili's works, especially the unpublished ones, it is evident how his interest in the ancient world ranged over a wide spectrum. For instance, modern antiquity topographers today study Roman units of measurement, which Marsili also studied, using a method still in use today that observes real models of measurement instruments and their ancient representations. This he reported in the *Manoscritto del Generale Marsili delle Misure, Linee, dei Pesi di pietra e metallo ricavati ò dà Bassi Rilievi esistenti in Roma, ò dagli Originali che sono nell'Istituto, ed in altri Musei, con ogni diligenza e proporzione figurati* (Manuscript of the General Marsili on measurements, lines, stone and metal weights obtained from bass-relief in Rome or from the originals in the Institute and in other museums, represented thoroughly and in their proportions) (BUB, FM, ms. 92). Marsili's goal was, in fact, the realization of synoptic tables of ancient weights and measures based on comparisons with those used at his time.

Marsili also drew archaeologically thematic maps. All the archaeological sites found or just visited by Marsili are precisely indicated, described in detail, and always graphically documented in *De Antiquitatibus Romanorum ad ripas Danubii*, the first part of the second volume of the *Danubius Pannonico-Mysicus* (1726). This can be considered a treatise on Roman archaeology of the Danube area. Among these sites, the most important is the Bridge of Trajan on the Danube (103–104 A.D.), near the present town of Turnu-Severin in Romania, corresponding to the ancient *Drobeta* (Barbulescu, 1998; Sacchetti, 2001), of which Marsili, from a scientific point of view, can be considered the real discoverer (Figs. 10 and 11). In fact, in 1689, with the help of 100 men, he completed a topographic relief of the visible structures of the bridge, probably after some target excavations (Longhena, 1930; Marsili, 1930b, p. 119) (Fig. 12).

This investigation is described in detail in the second volume of *Danubius*, on pages 25–32, and in a letter sent in 1715 to

Figure 10. Map of a segment of the river Danube, with the remains of the Bridge of Trajan on both sides. Also noted are other Roman remains, such as fortifications ("*castrametationes*") and roads ("*viae*"). By permission of the Biblioteca Universitaria di Bologna, Fondo Marsili, ms. 1.

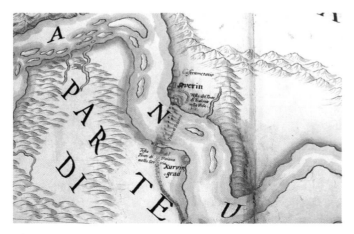

Figure 11. Watercolor representing the river Danube, with detail of the Bridge of Trajan. By permission of the Biblioteca Universitaria di Bologna, Fondo Marsili.

Montfaucon, a French Benedictine monk of the congregation of St. Mauro (BUB, FM ms. 102 D, c. 17–22; *Lettera del Co. Marsili intorno al ponte di Traiano sul Danubio a Padre Montfaucon Benedettino della Congregazione di S. Mauro. Parigi*) (Figs. 9 and 10). From this letter, we can understand how Marsili, as a precursor of ancient topography, was interested in the reconstruction of the Roman road network, to favor the movements of the armies in the border regions:

> *Non mi fu dificile col benefizio delle tante Marchie de li Eserciti Cesarei per le Misie, e Dacie, e di molte informazioni che per altri riguardi dovetti pigliare di tali Paesi ubbidienti all'Impero Ottomano di rinvenire le vestigie delle strade che avevano la comunicazione di quel Ponte, e per mezzo di esse intendere le marchie de gli eserciti da Roma fino all'ultima estremità delle stesse Dacie, ed unirle coll'altre che pure scopersi per le Pannonie, e tutte corrispondenti al sovrano centro di Roma* [...] (BUB, FM, ms. 192 D, c. 21).

It was not difficult for me to find the remains of the roads that linked to that bridge, thanks to the many marches with the

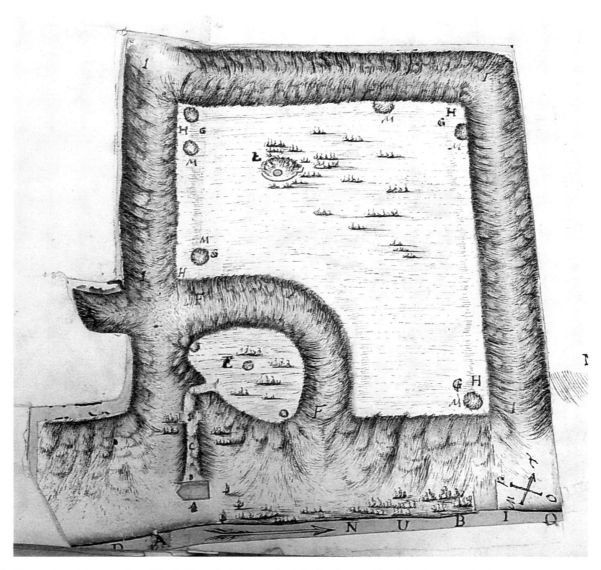

Figure 12. Watercolor of the remains of the Bridge of Trajan on the left side (Rumania) of the river Danube (*Diaria Geographica*, November 1700). By permission of the Biblioteca Universitaria di Bologna, Fondo Marsili, ms. 66.

Emperor's army across Misia and Dacia region, and from the information that I collected in other ways, and through them I could trace the ways walked by the Roman armies until the bottom of the Dacia and to put them together with the others that I found in Pannonia, all of them belonging to Rome [...].

It is evident how the importance of Marsili's work in the field of antiquities is mainly due to his methodology and above all to the role he attached to field research. His skills as a geologic surveyor, especially in Switzerland, made him particularly receptive to the method, known for about a century as the "topographic method." In fact, some of the key words of this method, such as multidisciplinarity and verification on the ground of the land information, were clearly anticipated by Marsili.

To this extent, it can be said that Marsili knew *Germania antiqua* (Lugduni Batavorum, 1613), the work of the 33-year-old Philip Cluver, born in Danzica in 1580 and one of the strongest supporters of direct visual check ("*autopsía*") as the best method to validate the results of a territorial survey. Marsili does not mention Cluver, but this is not a surprise, as it was his habit not to quote reference works in detail. Thus, in a letter about the Bridge of Trajan, in which he writes about sending the second volume of *Danubius* to his friend Monsignor Fontanini:

[...] *il quale dovrà avere il peso di fargli quelle note che sono più proprie di uno consumato al tavolino com'egli è, che ad un soldato che non ebbe per oggetto altro che dimostrare la verità dell'esistenza delle reliquie dell'antico*

Impero Romano, fra nazioni che furono, sono, e saranno sempre barbare. (BUB, FM, ms. 102 D, c. 21).

[…] who will have to add those notes that better suit a scholar like him, used to working at his desk, than a soldier who could only demonstrate the existence of the ruins of the ancient Roman Empire, among countries that are, and always will be, barbarian.

It would not be unreasonable to think that Marsili's archaeological work had its roots in Cluver's speculations, according to which it was important to put the remains of the past in the actual landscape and to try, where possible, to understand the relationships between them and their contemporary environment (Uggeri, 1993). All these requirements had been adopted by Marsili in a natural way, as the aforementioned letter contains a number of references to the direct visual check: "[…] *osservazioni mentre io stava coll'esercito Cesareo accampato ad ambe le ripe di quel fiume nell'anno 1689* […]" (BUB, FM, ms. 102 D, c. 18), ("observations while I was with the Imperial army on a camp on both sides of that river in 1689"), where he says: "*Questa ricognizione che dovetti fare con ogni diligenza* […]" (c. 20), ("This survey that I had to carry out thoroughly"), or where he talks about Roman relics that remained: "[…] *neglette sino alla mia ricognizione* […]" (c. 21v) ("neglected until my survey").

CONCLUSIONS

The significant role that Luigi Ferdinando Marsili played in the development of geological sciences between the seventeenth and the eighteenth centuries is now well recognized. Furthermore, the huge quantity of unpublished manuscripts and drawings of the Marsili Fund at the University Library of Bologna shows how he developed a personal method of field survey, characterized by its remarkable multidisciplinary approach, since the time he was a young soldier in the army of Emperor Leopold of Austria in the Danube area.

One of the most innovative aspects of Marsili's work is exactly this method, used in today's geological and archaeological field surveys. In his manuscripts, it is evident how, from a conceptual point of view, Marsili saw no separation between the different aspects of geomorphological land study, aiming, on the contrary, at a comprehensive knowledge of it, well aware of a cause-effect relationship between human activity and environmental evolution, with mutual retroactive influences. In this sense, Marsili's work can be considered a precursor of the modern discipline that lies between the geology of the Quaternary and archaeology, known today as geoarchaeology.

REFERENCES CITED

Adam, K.D., Reiff, W., and Wagner, E., 1986, Zeugnisse des Urmenschen aus den Cannstatter Sauerwasserkalken: Stuttgart, 100 p.

Baatz, D., 1974, Der Römische Limes, Archäologische Ausflüge zwischen Rhein und Donau: Berlin, Mann, 364 p.

Barbulescu, M., 1998, Un ponte per l'eternità, *in* Traiano ai confini dell'Impero: Milano, Electa, p. 132–133.

Brizzi, G., 1983, Studi Militari Romani: Bologna, CLUEB (Cooperativa Libraria Universitaria Editrice Bologna), 154 p.

Brizzolara, A.M., 1986, Le sculture del Museo Civico Archeologico di Bologna, La Collezione Marsili: Casalecchio di Reno (Bologna), Grafis Edizioni, 305 p.

Brocchi, G., 1819, La carta fisica del suolo di Roma, *in* Brocchi, G., 1820, Dello stato fisico del suolo di Roma—Memoria per servire d'illustrazione alla carta geognostica di questa città, Con due tavole in rame: Roma, Nella Stamperia De Romanis, 282 p.

Csiky, G., 1987, Luigi Ferdinando Marsigli, an Italian discoverer of Hungary, *in* Hala, J., ed., Rocks, fossils and history: Italian-Hungarian relations in the field of geology: Budapest, Hungarian Geological Society, p. 327–341.

Frati, L., 1928, Catalogo dei manoscritti di Luigi Ferdinando Marsili conservati nella Biblioteca Universitaria di Bologna–Firenze, Leo S. Olschki, 162 p.

Gortani, M., 1930, Idee precorritrici di Luigi Ferdinando Marsili su la struttura dei monti, *in* Corbelli, A., and Comitato Marsiliano, eds., Memorie intorno a Luigi Ferdinando Marsili, pubblicate nel II° Centenario della morte: Bologna, Zanichelli, p. 257–275.

Lipparini, T., 1930, Storia naturale de' gessi e solfi delle miniere di Romagna, *in* Lovarini, E. and Comitato Marsiliano, eds., Scritti inediti di Luigi Ferdinando Marsili: Bologna, Nicola Zanichelli, p. 189–211.

Longhena, M., 1930, Il conte Luigi Ferdinando Marsili, Un uomo d'armi e di scienza: Milano, Edizioni Alpes, 346 p.

Longhena, M., 1933, L'opera cartografica di L.F. Marsili: Roma, Società Anonima Tipografica Leonardo da Vinci, 84 p.

Marabini, S., and Vai, G.B., 2003, Marsili's and Aldrovandi's early studies on the gypsum geology of the Apennines, *in* Vai, G.B., and Cavazza, W., eds., Four centuries of the word "geology," Ulisse Aldrovandi 1603 in Bologna: Bologna, Minerva Edizioni, p. 187–203.

Marsili, L.F., 1700, Danubialis Operis Prodromus, Ad Regiam Societatem Anglicanam: Norimbergae, Apud. J.A. Endteri filios, 60 p.

Marsili, L.F., 1725, Histoire physique de la mer, Ouvrage enrichi de figures dessinées d'après le naturel & c.: Amsterdam, Aux dépens de la Compagnie, 173 p.

Marsili, L.F., 1726, Danubius Pannonicus-Mysicus, Observationibus geographicis, astronomicis, hydrographicis, historicis, physicis perlustratus et in sex tomus digestus: Hagae Comitum, Apud P. Gosse et al., Amstelodami, Apud Herm. Uytwerf et Franç. Changuion, 6 vols.

Marsili, L.F., 1930a, Scritti inediti di Luigi Ferdinando Marsili, raccolti e pubblicati nel II centenario della morte, a cura del Comitato Marsiliano: Bologna, Nicola Zanichelli, 273 p.

Marsili, L.F., 1930b, Autobiografia di Luigi Ferdinando Marsili: Bologna, Nicola Zanichelli, 263 p.

Morigi Govi, C., 1982, Per la storia del Museo Civico Archeologico di Bologna: Atti e Memorie della Deputazione di Storia Patria per le Province di Romagna, v. 33, p. 3–32.

Rinaldi Tufi, S., 2001, Archeologia delle province romane: Roma, Carocci, 442 p.

Sacchetti, F., 2001, Il ponte di Traiano sul Danubio nella testimonianza di Luigi Ferdinando Marsili (1658–1730): Atti e Memorie della Deputazione di Storia Patria per le Province di Romagna, v. 52, p. 317–386.

Sarti, C., 1984, Paleoloxodon e Mammuthus della collezione paleontologica di L.F. Marsigli: Museologia Scientifica, v. 1, no. 1-2, p. 103–113.

Sarti, C., 1988, I fossili e il Diluvio Universale: Bologna, Pitagora Editrice, 189 p.

Sarti, C., 2003, The Istituto delle Scienze in Bologna and its geological and paleontological collections, *in* Vai, G.B., and Cavazza W., eds., Four centuries of the word "geology," Ulisse Aldrovandi 1603 in Bologna: Bologna, Minerva Edizioni, p. 205–219.

Scarabelli, G., 1850, Intorno alle armi antiche di pietra dura che sono state raccolte nell'Imolese: Nuovi Annali delle Scienze Naturali, Bologna, v. 2, no. 3, p. 258–266.

Schalles, H.J., 1998, Il limes renano da Domiziano a Traiano e la presenza romana nelle due Germanie, *in* Traiano ai confini dell'Impero: Milano, Electa, p. 33–38.

Seibold, E., and Seibold, I., 2001, Antonio Vallisnieri—Ein moderner Geologe vor 300 Jahren: Max Pfannenstiel zum Gedächtnis (1902–1976): International Journal of Earth Sciences, v. 90, p. 903–910, doi: 10.1007/s005310100204.

Stoye, J., 1994, Marsigli's Europe 1680–1730: The life and times of Luigi Ferdinando Marsigli, soldier and virtuoso: New Haven–London, Yale University Press, 356 p.

Thoulet, J., 1897, Un des fondateurs de l'océanographie: Marsigli: Revue Scientifique, v. 34, s. 4, v. 8, p. 801–805.

Uggeri, G., 1993, Filippo Cluverio e il metodo topografico, in Les Archéologues et l'Archéologie: Caesarodunum, v. 27, p. 342–354.

Vaccari, E., 2003, Luigi Ferdinando Marsili, geologist: From the Hungarian mines to the Swiss Alps, in Vai, G.B., and Cavazza W., eds., Four centuries of the word "geology," Ulisse Aldrovandi 1603 in Bologna: Bologna, Minerva Edizioni, p. 179–185.

Vai, G.B., 2003, A liberal Diluvianism, in Vai, G.B., and Cavazza W., eds., Four centuries of the word "geology," Ulisse Aldrovandi 1603 in Bologna: Bologna, Minerva Edizioni, p. 221–249.

Vai, G.B., 2006, this volume, Isostasy in Luigi Ferdinando Marsili's manuscripts, in Vai, G.B., and Caldwell, W.G.E., The Origins of Geology in Italy: Geological Society of America Special Paper 411, doi: 10.1130/2006.2411(07).

MANUSCRIPT ACCEPTED BY THE SOCIETY 17 JANUARY 2006

Mattia Damiani (1705–1776), poet and scientist in eighteenth century Tuscany

Giancarlo Scalera*

Istituto Nazionale di Geofisica e Vulcanologia (INGV), Via di Vigna Murata 605, I-00143 Roma, Italy

ABSTRACT

Mattia Damiani da Volterra (1705–1776), "renowned Doctor," was the author in 1754 of a collection of scientific poems, Le Muse Fisiche (*The Physical Muses*) on two subjects: Newtonian physics and the plurality of the worlds. Damiani's interest in science was precocious, but even at that, it was superimposed on his studies in jurisprudence completed in Pisa in 1726. In 2003, Damiani's lost text, *De Hygrometris et eorum defectibus disputatio* (*Disputation about hygrometers and their defects*), which was printed in 1726 in Pisa, was brought to light. It characterizes him as a young scientist who reflected upon the properties and limits of laboratory instruments and on nascent aspects of climatology. In this *Disputation*, a delightful amalgamation of scientific and humanistic literature is pursued. A discussion of the properties and limits of contemporary hygrometers and a comparison of the Cartesian and Newtonian hypotheses about cloud formations are interspersed with quotations of verses on natural phenomena, mostly from poems of the classic age—a prelude to the author's future involvement in writing scientific verses. The poetry of Damiani, which often shows a musicality comparable to that of the poet Giacomo Leopardi (1798–1837), deserves to be recognized and saved from oblivion. Especially remarkable is the implicit "multimedia" project of a union among science, poetry, theater, and music. The rediscovered *Disputation about hygrometers* opens a new window on the personages involved and on the evolution of meteorological concepts in Europe in the context of the then-new Galilean and Newtonian physics.

Keywords: geophysics, meteorology, climatology, Earth's evolution, hygrometers, Newton, Descartes, Galilei, Metastasio, Jansenism, Enlightenment.

INTRODUCTION

My encounter with the works of Mattia Damiani (1705–1776) was fortuitous. Some years ago, I found a book of scientific poems, *Le Muse Fisiche* (*The Physical Muses*), dated 1754, in poor condition, on the bookstall of a street vendor. On first inspection and before buying it, I judged the text to be a work of Metastasio, because of the large-size type of the Pietro Metastasio name printed on the frontispiece. Moreover, Damiani's name was absent from the frontispiece. Only more careful inspection led to recognition of the actual author, who had signed a short introductory poem and was the addressee of a foreword by Anton Francesco Gori (1754), the well know, learned archaeologist who founded a museum in Florence.

Preliminary investigation on the World Wide Web led to confirmation of Damiani's identity as a friend of Pietro Metastasio (alias Pietro Trapassi, Italian poet, Imperial poet in Wien, and author of many libretti, 1698–1782) (Astaldi, 2001). A volume of the complete edition of Metastasio's works was dedicated to a collection of the 54 letters from the imperial poet to Damiani

---

*E-mail: scalera@ingv.it

from September 1734 to March 1776 (Metastasio, 1847). At least one wrong classification of the book was made by an Italian library, resulting in Damiani's poems being attributed to Metastasio. Notwithstanding my attempts to find the name of Damiani either in a modern or old history of Italian literature, I was not able to find any reference to him, and only the indication of a referee of this paper allowed me to become aware of a shortest mention of Damiani in a book by Giulio Natali (*Il Settecento*, 1964). Girolamo Tiraboschi's (1731–1794) large panorama of Italian literature (1829) stops at the first years of the eighteenth century, and in the continuation to the next century, which was written by his pupil, Antonio Lombardi (1768–1847), no mention of Damiani's books is found (Lombardi, 1827–1830).

Figure 1. The frontispiece of Damiani's most famous book *Le Muse Fisiche (The Physical Muses)* published in 1754.

BIOGRAPHICAL SKETCH

The known modern accounts of the Damiani's life are the short notes of Losavio (1925), Bertini (1965), and Marrucci (1997), all based on an old manuscript of Persio Benedetto Falconcini (1729–1809), *De claris Viris Volaterranis* (1777). Among several volumes constituting Falconcini's original manuscript about the biographies of many of the personalities of Volterra, only the one, containing, among others, Damiani's life, has come down to us. Cristiano Balducci (2005) recently translated the manuscript on behalf of Istituto Nazionale di Geofisica e Vulcanologia (INGV). The incipit of the manuscript section dedicated to Damiani says:

> Mathias Damianius ex honestissimâ Populari Familiâ Volaterris ortus est postridie Kalendas Maias anno MDCCV. Hic etsi haud parem Virtuti fortunam obtinuerit, adeptus tamen est tantam nominis celebritatem, divino penè ingenio, scriptisque commendatam suis, ut Iure inter illustriores, atque honoratiores Nostrates recenseatur. Magnos enim homines, ut in Eumene aiebat olim Cornelius Nepos, Virtute, non fortunâ metimur.

Mattia Damiani was born in Volterra in May 2, 1705, from a very respectable family of the popular classes. Although the destiny he received was not peer to his virtue, all the same he reached so high reputation—a celebrated fame—because of his almost divine talent and because his writings—to be rightfully counted among our more illustrious and honored fellow citizens. Indeed, as Cornelius Nepos once said in *The life of Eumene*, we esteem the greatness of men on the basis of their virtue, and not on the basis of their fortune.

The Damianis were an educated family of Volterra (Tuscany), whose father, Girolamo—married to Francesca Cetti—was a country doctor. Mattia's two brothers, the elder, Pietro, and the younger, Nicola, become a chemist and a surgeon, respectively. Having completed the study of Latin and rhetoric at the age of fifteen, the precocious Damiani began attending the University of Pisa, from which he obtained a degree in literature and jurisprudence in 1726. The eminent Newtonian scientist and mathematician, Guido Grandi (Cremona 1671–Pisa 1742) gave him lessons in mathematics and the sciences. His graduation thesis was, indeed, devoted to meteorological subjects, with the title *De Hygrometris et eorum defectibus disputatio* (*Disputation about hygrometers and their defects*) and was recalled in Falconcini's manuscript as a very careful work, full of erudition. After his graduation and three years spent in Volterra, Damiani decided to turn to the legal profession. He moved to Rome as a guest of his brother, Pietro, the chemist, at which time he met Pietro Metastasio, who became his friend and then confidant for life.

In 1733, after five years in Rome, Damiani consented to the request of the Bishop of Volterrra, Ludovico Maria Pandolfini, to accept an appointment as headmaster of the Bishop's Seminary high school of the town. He was ordained in 1735. In his biography, Falconcini writes that Damiani's teaching was of great quality and effectiveness, but that he never abandoned literature, adopting the style of Metastasio as a model and working on poetry at night. At this point, Falconcini writes that Bishop Pandolfini, lacking in moderation, tended to revive old disagreements with the local nobility. Some excessively severe punishments and recriminations by Pandolfini against the young noble scholars led to Damiani's decision not to become involved, and to accept the proposal of Francesco Gaetano Incontri, then appointed Bishop of Pescia, to act as his vicar in 1738. After three years, Incontri was transferred to Florence, and Damiani become plenipotentiary vicar of Pescia. After the appointment of a new Bishop, however, some disagreements with this superior led Damiani to leave Pescia and accept the proposal of Vincenzo Riccardi to become tutor to his two sons, Carlo and Cosimo. In this happy period of his life, Damiani helped Giovanni Lami, Riccardi's librarian, to publish the journal *Novelle Letterarie* (*Literary Newsletter*). Other friends of these years were Anton Francesco Gori (1754), an expert in Etruscan antiquities, and Andrea Pietro Giulianelli, the deputy director of the Laurentian Medicean Library. All these friendships allowed Damiani to be co-opted and accepted as a member of the *Florence Academy* and the *Apatisti Academy*, both interested in an amalgamation of literature, science, and art.

After performing the task of tutor to the complete satisfaction of the Riccardi family, Damiani—thanks to the good offices of Riccardi—obtained an ecclesiastical allowance in Volterra. This was a moderate life annuity in return for the small job of performing religious services as a bestower of the sacraments at communions, weddings, etc., for the pupils of the "Maria Maddalena" Hospital–Boarding School for Women in Volterra.

In 1745, he returned to the town of his birth. At the same time, the Great Duke of Tuscany, Francesco di Lorena, became the German Emperor, and culture flourished in the Grand Duchy. Then, the *Accademia dei Sepolti* (*Academy of the Buried*) was revived in Volterra, to which Damiani was appointed the permanent secretary. The quarrelsome Bishop Pandolfini died in 1746, and a series of tumultuous events allowed a dear friend of Damiani, Iacopo Gaetano Inghirami, to become the regent of the Volterra diocese. Damiani's life then became easier and

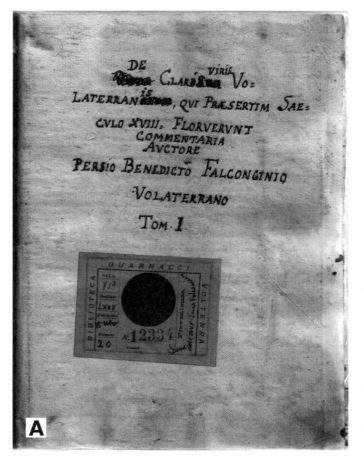
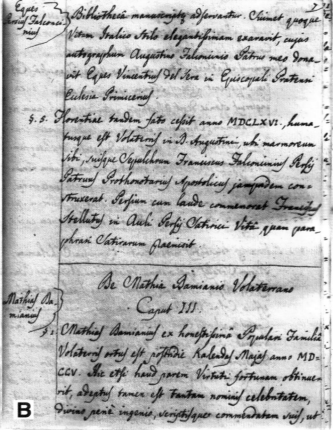

Figure 2. (A) The frontispiece of the manuscript of Persio Benedetto Falconcini. The manuscript is kept at the Guarnacci Library in Volterra. (B) The incipit of the section of Falconcini's manuscript that is dedicated to Damiani's biography.

was divided among a new appointment to teach in the seminary, the practice of law, especially as a highly-regarded justice of the peace, poetry, and natural philosophy. In this period, Damiani became a cultural point of reference in his town, having friendships and correspondence with many important men and women of his time (correspondence with Ludovico Antonio Muratori, 1672–1750, for instance). Falconcini writes:

> … any more or less famous man of letters, Italian or English or from elsewhere, coming to Volterra stimulated by an erudite curiosity to visit the vestiges of this glorious antiquity, sought a meeting with Damiani, or a meeting was sought by him and welcomed with high regard, courtesy, and respect.

After the appointment of Iacopo Inghirami as Bishop of Arezzo in 1755 and the nomination of the fellow-citizen, Filippo Nicola Cecina, as regent of the Volterra diocese, Damiani's delicate health started to decline. A progressive contraction and stiffening of his legs afflicted him for the next twenty years of his life. It is difficult today to understand the real nature of his illness. Well-grounded hypotheses can indicate a slipped disk, fibrosis of the ligaments of rheumatic origin, or autoimmune progressive fibrosis. The associated shooting pains referred to by Falconcini make the first hypothesis most probable. When his physical condition became too severe for him to continue to work, in 1768, Damiani asked to be replaced at the seminary by his nephew, Lorenzo Cetti. The Grand Duke, Pietro Leopoldo, visited Volterra in 1773, but Damiani could not meet him. However some of his odes—written on this occasion in honor and celebration of the Grand Duke—were sung by excellent choristers. Mattia lost his battle with death on 27 July 1776, when he was seventy-one years old.

THE LITERARY AND SCIENTIFIC EIGHTEENTH CENTURY

Science was not completely separated from philosophy and literature in the eighteenth century, and the Enlightenment produced an increased interest in the sciences and in the spread of science (Consoli, 1972; Borsellino and Pedullà, 2004). As a consequence, a number of examples of literary works about astronomy, physics, and the life sciences can be quoted from different parts of Europe. In Italy, the best known of these compositions included the *Dialogues* (1737) of Francesco Algarotti (1712–1764) (*Il neutonianismo per le dame* [*The Newtonian theories explained to the ladies*]), enlarged in the subsequent editions that came out titled *Dialoghi sopra l'ottica neutoniana* (*Dialogues about Newtonian Optics*). A series of seven dialogues (six in the first edition) discusses the different topics of the scientific revolution. Algarotti was a member of the Institute of the Science and Arts of Bologna, founded by Luigi Ferdinando Marsili (1658–1730) in a cultural climate in which an integration of art, literature, and science started to be more efficaciously favored in Bologna by Cardinal Gabriele Paleotti (1522–1597) and Ulisse Aldovrandi (1522–1605) (Vai and Cavazza, 2003).

In the eighteenth century, most literature was influenced by reaction to the immediately preceding Baroque literature, which was "experimentalist," redundant, transformist, and an expression of a luxurious life style (Battistini, 2005). The reaction sought a simpler stylistic form and arrived at the end of the seventeenth century, embracing what looked the opposite style; namely, a bucolic life away from the affectation and intellectualism of the cities. At least three examples are known of cases of noneducated people (shepherds or sons of very poor people, such as Giandomenico Peri [1564–1639]; Benedetto Di Virgilio [1602–1666]; and Jacopo Martino [1639–1656]), who were able to compose fine pieces of prose and poetry (Tiraboschi, 1829). Giandomenico Peri especially, born near Siena, was the composer of pastoral dramas. He never abandoned his shepherd's garb. These cases roused admiration and imitation and were one of the manifold causes of the success of the Arcadian movement. The Catholic church monopolized the Arcadian Academy because of the need to drive the intellectual energies of the time and to attenuate the emerging rationalism.

In Tuscany, the need for a simpler and clearer way of writing went toward a link with the tradition begun with the *Vocabolario della Crusca* (Dictionary of the Crusca Academy) and by Galileo Galilei's scientific-literary works (Galilei, 1632), and in this Italian region, a greater number of people pursued the aim of diffusing scientific results by adopting a plain and more understandable style. In the seventeenth century, Galilei, Cesi, Redi, Magalotti, and the reports of the Academies of Lincei (Rome) and of Cimento (Florence), became models for subsequent writers of the eighteenth century, in which Arcadia was combined with the Enlightenment.

While common opinion (but biased by a philosophic preconception) is that these didactic compositions never reached great heights, in a few cases the pedagogic purpose linked masterfully to Arcadian loveliness and neoclassic forms. The best known examples are works of the already mentioned Francesco Algarotti (1712–1764; *Dialogues*, 1737), and of Carlo della Torre di Rezzonico (1742–1796; *Il Sistema de' Cieli*, 1775) and Lorenzo Mascheroni (1750–1800; *L'invito a Lesbia Cidonia*, 1793). But Mattia Damiani's writing is different from all these, assuming a form that was completely original because all the arts—literary, visual, and musical—are combined together and to the science.

THE KNOWN WORKS OF DAMIANI

The Physical Muses

The literary productivity of Damiani (1747, 1754, 1758, 1761, 1765a, 1765b, 1770–1771, 1772, 1773) began ca. 1747 with a *Componimento pastorale-filosofico detto in Firenze nell'Accademia degli Apatisti* (*Pastoral-philosophic composition declaimed in Florence at Apatisti's Academy*); the second main work was the *Le Muse Fisiche* (*The Physical Muses*), a collection of scientific poems that, during the years preceding their printing in a single book in 1754, probably were presented on different occasions. The work is a set of nine poems on different scientific arguments:

1. *De' Satelliti di Giove* (*About the satellites of Jupiter*);
2. *Della vicendevole gravità de' Corpi, o sia delle Forze Attrattrici* (*About the gravity of the bodies, namely the forces of attraction*);
3. *Del Suono* (*About Sound*);
4. *Della Luce, e sue proprietà* (*About light and its properties*);
5. *Della Vita, e della Fecondazione delle Piante* (*On the life and on the fecundation of plants*);
6. *Dell'Azione de' Corpi Celesti* (*On the action of the celestial bodies*);
7. *Della Pluralità de' Mondi* (*About the plurality of the worlds*);
8. *Dello scioglimento de' Corpi in Fiamma* (*About the dissolution of the bodies into flame*);
9. *Della Natura dell'Acqua* (*About the nature of water*).

The text style is not that of a normal poetic composition but follows the form of libretti designed to be set in music. The dialogues among the personages—deities, shepherds, nymphs—alternate groups of verses of different length, hendecasyllabic and seven-syllabic, followed by a *cantabile* section in which the verses are shorter, regular, and all seven-syllabic. Damiani's project is superior to many of the other contemporary attempts to make public the new concepts of science. Indeed, although the poems can be merely read and enjoyed as they are, in them works a strict amalgamation of new science, poetry, music, and theatrical spectacle designed for a particular social target—an educated audience.

The nine poems of *Le Muse Fisiche* are clearly an exposition of Copernican and Newtonian theories, and contrary to expectations, the book did not incur the Inquisition's anger. Perhaps an expedient of Damiani to avoid registration in the *Index of Forbidden Books* was to omit the name of the author on the frontispiece. This was not an unusual practice. In the eighteenth century, a similar expedient was adopted in the case of a clever upholding of the Copernican and Newtonian system; namely, the (at the time) famous *Ragionamento Filosofico intorno al Moto della Terra* (*Philosophic reasoning about the motion of the Earth*) by the Apulian professor, Giuseppe Carlucci (1766). This manuscript, after long, free circulation, was published anonymously in 1766 (Marvulli, 2001; Raucci, 2001; Sisto, 2003). Damiani cleverly avoided all possible pretexts for accusing him of heresy. In fact, in a short sentence—a premise of few words contained in a single page—the declaration is made that the use of words such as "deity," "fate," and "numen," are simple poetic expressions.

All the poems bear witness to progress in the freeing of the cultural world, and in particular of the ecclesiastic reflection—Damiani was a priest—about the new sciences and the new system of the world. Formal cancellation of the Copernican books from the *Index of Forbidden Books*, as declared

Figure 3. The first page of the poem *Della Pluralità dei Mondi—Componimento Pastorale* (*On the Plurality of the Worlds—Pastoral Composition*). The first seven verses are: *The day is already languishing / and in the aged bosom / of the nearby Ocean / the Sun seems to hide. / Every twinkling star / contributes to the luminous mantle / that forms Night, / which of the Sun seems the daughter: / She [Night], with his light, has a close resemblance to the Day.*

by Pope Pio VII, happened with the publication of the *Handbook of Astronomy* by the priest Giuseppe Settele (professor at the University *La Sapienza* in Rome) in 1822 (Maffei, 1987). The poems of Damiani are a further example of the practical ineffectiveness of the Inquisition's decrees about the forbidden Galilei and Copernicus theories.

Damiani's poem, *About the Plurality of the Worlds,* is a particular example of the exposition of Copernican ideas giving them the form of complete generalization—an idea proposed by the philosopher Giordano Bruno (1548–1600) and followed by Bernard le Bovier de Fontenelle (1686) and many others, raising a never-ending collective cultural discussion that is today focused upon the solution of the so-called Enrico Fermi (1901–1954) paradox (Webb, 2002). This is also an anticipation of modern trends in planetary geology. These concepts on the existence of many worlds—possibly an infinity of them—were considered nearer to heresy than the concepts of Copernicus or Newton. Very cautiously, Damiani adopted a dissimulation strategy. In a footnote to the last cues of the poetic dialogue among Uranio, Tirsi, and Elpino, Damiani wrote—insincerely—that the words of Elpino deride the plurality of the worlds and that he himself considered this idea a mere hypothesis. In reality, the Elpino personage is persuaded of the truthfulness of the plurality idea throughout the course of the seventh poem, and his last cue is a confirmation—in a form of joke—of his conviction. This kind of strategy is analogous to that adopted by Galileo Galilei during his trials to defend his *Dialogue concerning the two chief world systems, Ptolemaic and Copernican* (Galilei, 1632) from the accusation that it was in complete favor of the Copernican system.

The Miscellaneous Poetry

Damiani was author of a large number of poetic works—whether philosophic or dedicated to famous people of Tuscany and Volterra—canzoni, short lyrics, cantatas, and some melodramas in music. The poetic works were reprinted several times with additions (Damiani, 1758, 1765a, 1770–1771). In the last edition of 1770–1771, the editor uses four letters of Metastasio as a foreword. The philosophic lyrics (*e.g. Philosophy*, *Liberty*, etc.) are little gems and among them, the poem, *Time*, is particularly exquisite. Galilei's argument of the relativity of motion is applied to time and to its apparent shifting from the past to the present and the future. The conclusion is drawn that time-motion is not real but, instead, time is the progressive transformation of reality and of ourselves. This transformation produces the sensation of time passing. Damiani's concept, eliminating the paradoxical motion of time, is of particularly high value and anticipates ideas that are much more modern. The Appendix to this paper offers selected parts of the lyric, *Time*, with its translation. Damiani's concern for time in the eighteenth century is not only important for physics but also in the particular contexts of the geological time scale and the age of Earth, which were going to change dramatically with Buffon (1707–1788) (*Epochs of Nature*, 1774, the Earth's age increased to 75,000 years) and James Hutton (1726–1797) (*Theory of the Earth*, 1788, the Earth as being indefinitely old).

Besides the manuscripts of Mattia Damiani's published works, a number of unpublished manuscripts of lyrics—donated by Pietro Damiani (the great-grandson of Nicola, Mattia's younger brother) in 1870—are conserved in the files of the Guarnacci Library in Volterra. Examples of them are the nostalgic sonnet, *Dedicated to Volterra*, which was written on an occasion when Damiani was going away, far from his birth place for a long time, and the sonnet, *The Trinity*, based on the theological paradox of God as one and trine.

The Eulogy of Genovesi

The true unification of Italy was already complete in the cultural field a century before political unity was achieved. All the personages of the Italian Enlightenment were linked to each other by mutual esteem and cultural exchanges. Indeed the "renowned Doctor," Mattia Damiani (1772), felt himself honored to write the funeral eulogy in verse for the Neapolitan economist and philosopher, Abbot Antonio Genovesi (1712–1769). The Apulian Abbot, Domenico Forges-Davanzati (1742–1810), wrote the foreword and footnotes to the eulogy, which was published in 1772. Forges-Davanzati was the nephew of Bishop Giuseppe Davanzati (Bari 1665; Trani 1755), and in 1774 and 1789, he was the publisher of his uncle's manuscript, *Dissertazione sopra i Vampiri* (*Dissertation on Vampires*) (Davanzati, 1774)—a long text addressing the stupid superstitions that afflicted the church and the people in Europe at that time. G. Davanzati devoted some pages of his *Dissertation* to statements in support of the Copernican system and was the author of a lost *Treatise on Comets*.

The works of Antonio Genovesi on economic science and philosophy were listed in the *Index of Forbidden Books* (Anonymous, 1824), and the eulogy of Genovesi, written in 1772 by a scholar of his work, Giuseppe Maria Galanti (1743–1806), suffered a similar fate (decree of November 1773; Anonymous, 1824). This indicates the risk to which Damiani was exposed in publishing his eulogy of Genovesi and in participating in the more liberal practices of Italian cultural life, but on the other hand, it also bears witness to an epoch when the cultural climate in Tuscany—one favored by the Grand-Dukes of Tuscany—was liberal and benevolent toward novelties coming from the great streams of French and American ideals (Montanelli and Gervaso, 1970).

THE JANSENISM OF DAMIANI

Religious arguments are often touched on by Damiani in his works, and some deductions about his adherence to some current ideas of eighteenth-century theology can be made. A first clue lies in a *cantabile* part of the fourth poem of *The Physical Muses* (1754), *About Light and its Properties*. In this cantata, some statements in favor of a rigid determinism are made: the first cause of all being the Deity. A second stronger clue is found in the poem *On the Existence of God*, contained in the *Collection of Different Poems* (1770–1771). Many statements are written about the driving power of divinity on human behavior. Evil happens in the human reality to give greater prominence to the good. The creative act of God is not limited to a moment in the past—the beginning of the world—but is continuous and is realized also in the continuance of human behavior and thinking.

This collective evidence suggests that Damiani held a theological position similar to Jansenism, which was very common among the Italian adherents of the Enlightenment, and Tuscany was the Italian center of this theological movement, under the leadership of Bishop Scipione de Ricci. In many cases, those following the idea of grace, justification, predestination, and determinism had the aim of limiting the decisional power of the papacy. Jansenism was strongly opposed by the Jesuit order, coupled with more or less degrees of Vatican decisiveness. The doctrine of Jansen, however, was nearly coincident with Augustinian theology, and it was not substantially different from the Christian doctrine traditionally followed for sixteen centuries. It is common opinion that, under the pressure of the Jesuits to renew Christian thinking in response to post-Renaissance Humanism, the church developed the political solution of censuring Jansenism because it could not condemn the traditional Augustinian doctrine (Iemolo, 1928).

It is impossible to know with certainty if Damiani was simply following the old traditional Augustinian doctrine or adhering with awareness to Jansenism. An historian of the church (Iemolo, 1928) writes that, in the first half of the eighteenth century, the Jansenists-Augustinians were tolerated without being marginalized and that many of them reached prominent positions in the church hierarchy. The famous synod of Pistoia—which gathered the Italian Jansenists—was held in 1786, ten years after Damiani's death. Only after this synod was the Jansenist movement practically defeated in Italy (Stella, 1972).

THE FORMERLY LOST *DISPUTATION ABOUT HYGROMETERS*

In 2002, thanks to the kindness of an Antiquarian of Florence, Dr. Paolo Pampaloni, a copy of the public dissertation of Mattia Damiani at Pisa University, published in the year 1726, was made available to INGV. A complete Italian translation of the Latin text, comprising thirteen theses and a foreword in the form of a letter addressed to the Bishop of Volterra, the Prince Ludovico Maria Pandolfini, has been made.

Damiani's rediscovered dissertation, *De Hygrometris et eorum defectibus disputatio* (*Disputation about hygrometers and their defects*), is a critique of the experimental methods used in meteorology to measure the amount of humidity in the air. The history of these criticisms has never ended, and a more modern example can be read, for example, in a paper by Aliverti (1929), in relation to the hygrometers of the beginning of the twentieth century. The *Disputation* is also a discussion of the two opposing ways of explaining the meteorological phenomena, by Newton laws or Descartian whirls (see De Fontenelle, 1732, Section IX). Damiani took the side of Newtonian physics.

A number of different ideas on the nature of clouds existed at the beginning of the eighteenth century (Hamblyn, 2001). The most credited theory claimed that water particles, under the action of sunlight, could transform themselves into little bubbles filled by low-density air, the so called "aura." These bubbles, like ballonets, tended to rise and accumulate at high-altitude, forming clouds. The rupture of these little spheres produced rain and closed the cycle by returning the moisture to Earths' surface.

Others held the view that acids circulating in the atmosphere corroded the dissolved water in the air and that this corrosion became visible as clouds. An analogy was between the irregular shapes and colors of clouds and the corroded zones of metals exposed to acids.

Another idea on the production of water vapor and clouds attributed the role of elevating the vapor to hypothesized igneous particles transported by light rays. These lighter-than-air particles adhered to the vapor drops like a life-belt, transporting them skyward. When detachment of the igneous particles occurred, rain was generated.

The concepts discussed by Damiani belong to yet another school of thought. He hypothesized that air was composed of filaments wrapped in spherical spirals. Particles of water vapor were considered to be endowed with gluten and thus able to adhere to the air particles. In this way, winds could transport a

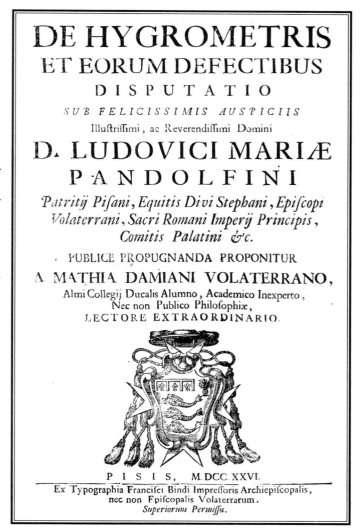

Figure 4. The frontispiece of the Damiani's dissertation *De Hygrometris et eorum defectibus disputatio* (*Disputation about hygrometers and their defects*) published in 1726.

large quantity of vapor to great heights, thus forming clouds. Similar transportation would be more difficult for dust particles, he reasoned, because they are not glutinous. After a long discussion of these ideas, Damiani, in the small thirteenth section of the third thesis, inserted a short quotation about the above-mentioned hypothetical igneous-particle—"life-belt" mechanism—which he judged a plausible idea.

Although in sixteenth century a better way of reasoning was inaugurated by Nicolaus Copernicus (1473–1543) and Giordano Bruno (1548–1600), who thought correctly that the air was simply a participant in Earth's rotation, in the century preceding the birth of Damiani, nascent Galilean physics was mixed with ancient Aristotelian concepts. Using these still-uncertain concepts, thoughtful individuals tried to place atmospheric phenomena in a context of the new awareness of the movement of Earth. An example of these resulting misinterpretations is the explanation that a follower of Copernicus, Father Paolo Antonio Foscarini (Foscarini, 1615; Romeo, 1992), gave about the phenomenon of the equatorial winds being constantly directed toward the west. The ghost of the *Primum Mobile* is still present in his ideas. Foscarini wrote to Galilei (Caroti, 1987; Scalera, 1999):

> I have judged that, under the equinoctial line, the phenomenon of the perpetual west-directed wind is due to nothing but a little resistance of the air against the motion of the Earth, which, in it [the air] contained and moving in accord with it from west to east, produces the night and the day.
>
> [...]
>
> But if the air that contains the clouds is moving with the same motion of the Earth, by which cause do the clouds not move by the same motion? And if somebody will propose the wind as cause, I will reply that, in comparison to the impetus of the wind, the impetus that transports the natural motion of the air together with the Earth is greater. Since the circle of the air is greater than the circle of the Earth, if the Earth moves at eight hundred miles/hour or more, the air will move at 1500 miles/hour. But no wind—however impetuous it can be—can reach this velocity.

In the case of Foscarini and many others (Scalera, 1999), Earth's motion was considered a primary phenomenon, but discussion about atmospheric motions and clouds was soon influenced by the Descartian system of the world in which the primary cause of planetary motion was the whirling of very thin matter pervading the universe. The French philosopher's ideas were also fonts for erroneous explanations of meteorological processes.

In the continuation of his third thesis, Damiani confuted the mechanical process of cloud formation—like a centrifuge effect—as proposed by some followers of Descartes. Descartes' space filled by material whirls was judged by Damiani as a complicated way to describe nature, which actually could divert thinking from a simple and realistic explanation of the phenomena. In the case of the rising of the water vapor that gives rise to clouds, Damiani strongly criticized the action of the whirling thin matter proposed by the followers of Descartes:

> [...]
>
> Indeed the existence of the abovementioned very thin matter—that is present in all the space between the Earth and the Moon—is uncertain. And also if any evidence was adduced of the existence of this matter, the circular motion of this matter around the Earth would nevertheless be uncertain, and the cause that produces the movement itself and its conservation without decrement would be obscure. Moreover, it is not known with certainty if the velocity of the thin matter was constant or would become greater thanks to the velocity of the bodies that are pushed and driven in its whirl. [...]
>
> Moreover, if a circular movement of this kind is admitted, it is difficult to explain the mechanism by which the Moon—dragged by the whirl of matter moving around it—can complete both the rotation on its axis and the revolution along its orbit in 27 days, 7 hours, 43 minutes. Finally, it cannot be understood how the Earth complete the movement of rotation around its axis in 24 hours. And given that [the uprising of the vapours] is founded on all these hypotheses, all the same the opinion that the uprising of the vapours was produced by the centrifugal force of the material whirls must be judged uncertain and doubtful.
>
> *... like in a building, if the ruler itself is wrong,*
> *if the set-square is not rectilinear,*
> *and if in some part of the level the wire is curled,*
> *all the house will be built incorrectly,*
> *bent, twisted, bowed and deformed.*
> *(Lucretius, De Rerum Natura, IV)*
> *[...]*

Damiani (like Algarotti and many others) emerges from all the misunderstandings of the seventeenth century with a clear preference for the physical concepts of Newton (1642–1727).

Throughout the *Disputation*, a delightful amalgamation of scientific and humanistic literature is pursued. Alongside the discussion of the properties and limits of the contemporary hygrometers, the text is interspersed with quotations of verses on natural phenomena, mostly from poems of the classic age—a prelude to the author's future personal engagement in writing scientific verses. These quotations are inserted for a purely aesthetic purpose, because they do not constitute the basis on which to found the reasoning. The quotations have then different general sense from the *ipse dixit* of the Aristotelian tradition. An example of the old way of quoting old sentences from writings of classic authors can be easily found in the 43rd chapter of *Il Saggiatore* (Galilei, 1623), in which a long section of a work of Lotario Sarsi (the nickname of Orazio Grassi) is transcribed. The work of Sarsi—containing innovative concepts about the comets—is example of how the old literary and methodological customs are mixed with new science. Only Galilei and other members of the Accademia del Cimento and Accademia dei Lincei inaugurated a more modern way to write scientific works (e.g., Magalotti, 1667).

The seventeenth century gave birth to modern climatology with the Torricelli barometer. In the attempt to measure the quantity of water dispersed in the air, a series of different scientific instruments were developed and tested. Damiani discusses a number of these apparatuses, highlighting their limits and flaws. The nonlinear behavior of the instruments in response to a linear variation in the moistness is especially emphasized.

After he had established the inadequacy of the human senses to evaluate precisely the water-vapor content of the air, Damiani, in the fifth thesis, started his analysis of the hygrometers. First, he classified hygrometers into two types:

1. Hygrometers that absorb vapor from the air and indicate the water quantity by an increase in weight or other effects; and
2. Hygrometers that are able to collect the water vapor, condensing it in a graduated glass.

The First Type of Hygrometer

In the sixth thesis, the use of salt, sponge, wool, cotton, and other hygroscopic materials is scrutinized. The author criticizes the trials using an increased weight of these substances because of lack of linear response, and because, in his opinion, when the pores of the materials become filled with water, their sensitivity to further increase in moisture becomes zero. In this thesis, the experiments of Cardinal Nicola da Cusa (1401–1464), Father Francesco Lana (1631–1687), and learned members of the Academy of Nuremberg are quoted and confuted.

In the seventh thesis, strings of hemp and catgut are considered. The nonlinearity of their response to a linear increase in humidity is shown. The savants involved were Christian Wolf (1679–1754), Leonhard Christoph Sturm (1669–1719), Jean-Baptiste Duhamel (1624–1706), Ogier Ghislain De Busbecq (1522–1592), and Ludovico da Ripa. This last author is quoted by Damiani as author of *De vi vaporum in Hygrometris. Disertat. Geometric. Mechan.*, but is not inserted in Italian Encyclopaedias and not mentioned in classic compilations on the history of the Hygrometry (Agamennone and Cancani, 1885).

In the eighth thesis, the fibres of the ears of oat are discussed. The confutative arguments are similar to the ones he used for ropes of hemp and catgut in the seventh thesis. The personages referred to are Anastasio Kircher (1601–1680), George Sinclair (?–1696), M. Gottfried, and Ferdinand Helfrich Lichtscheid (1661–1707).

In the ninth thesis, the change of the sound of a catgut string in response to moisture variations is examined. Some investigators tried to measure the variation in humidity by counting the number of semitone changes with respect to the original tone compared to a reference flute note. Confutation of this method, besides the arguments raised in the seventh thesis, cites the possibility that, in turn, the flutes also change their notes in response

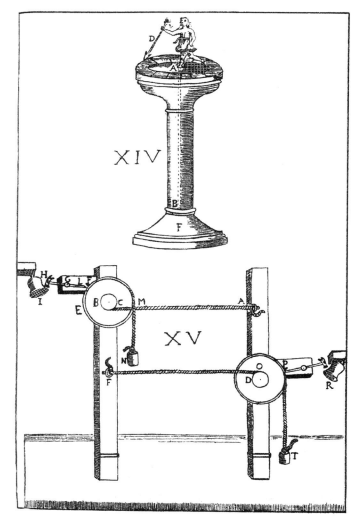

Figure 5. A first type of hygrometer discussed by Damiani in his *De Hygrometris et eorum defectibus disputatio* (*Disputation about hygrometers and their defects*). The engraving represents an hygrometer built by Francesco Lana using a long vegetal fiber contained in the cylinder. The torsions of the fiber in response to the variation of the humidity is indicated on a graduate scale by the arrow in the hand of the little paper-figure (Fig. 14, Table 5, in Francesco Lana, *Prodromo all'Arte Maestra*, chap. 8th, 1670). The other instrument on the bottom of the table is able to alert with bell strokes for an arbitrary high value and an arbitrary low value of humidity. It is based on the property of the ropes to lengthen or shorten in response to humidity variations. Little teeth on the gears arm the hammers.

to the contraction or dilation of wood and metal due to variations in moisture and temperature. In addition, the limits of the human auditory apparatus to perceive variations in sound are noted. The quoted scientists are Sturm, Lana, Claude Perrault (1613–1688), and groups at the Academies of Florence and Paris.

In the tenth thesis, hygrometers are examined in relation to contraction and dilation of little wood bars that they contain. The simplest one—consisting only of two wooden bars—was the instrument constructed by Christian Wolf, and the most elaborate one—having many little wood bars and gears—was that developed by Gottfried Tauber. Confutation of this thesis relies on the impossibility of finding two equal wooden bars, because of the inhomogeneous nature of wood. This counterargument is accompanied by a list of the unfavorable characteristics of wood that make it unsuitable for use in precision instruments. An experiment of Perrault and the unsuccessful attempts to reproduce it is quoted as supporting evidence.

The eleventh thesis—using reservations similar to those in the seventh thesis—deals with the behavior of sheets or ribbons of paper, parchment, and leather.

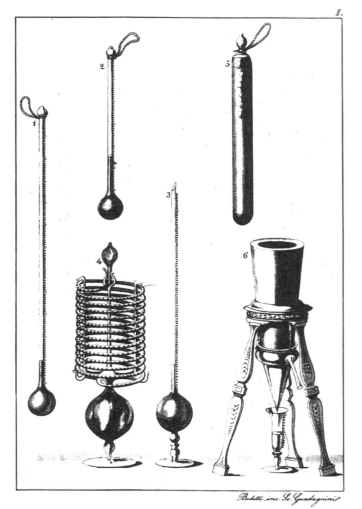

Figure 6. A second type of hygrometer discussed by Damiani in his *De Hygrometris et eorum defectibus disputatio* (*Disputation about hygrometers and their defects*). This hygrometer—Figure number 6 in table II of the volume on the experiments of the *Accademia del Cimento* of Florence, 1841, third edition (Magalotti, 1667)—is the instrument described in the Damiani's *Disputation* as drawn by the engravers Bedetti and Guadagnini.

Finally, in the twelfth thesis, a hygrometer made by Guillaume Amontons (1663–1705) is described. The instrument was based on the expansion or contraction, under the effect of the amount of water vapor contained in the air, of a leather sphere filled with mercury. The mercury emitted by the contracting sphere was canalized into a vertical narrow pipe, and the level of the mercury in the pipe was read on a graduated scale. The shortcoming of this instrument was again the nonlinear response and the effect of other problems similar to those described in the preceding theses.

The Second Type of Hygrometer

The second type of hygrometer is the subject of the last, or thirteenth, thesis. Only the hygrometer constructed by scholars of the *Accademia del Cimento* (Academy of Florence) is described. The apparatus is based on the condensation of vapor on cold surfaces. In this hygrometer, an upside-down cone filled with ice collects the condensed drops of water in a graduated tube (Magalotti, 1667). The measure of humidity is determined as the rate of increase of the water in the tube in a given unit of time. The critical argument of Damiani is that, although correctly based in principle, the results of this experiment are temperature-dependent and if the air temperature is less than the ice temperature, the apparatus does not condense vapors at all. The criticism is based on the experience of drinkers in a tavern in winter, when no water vapor condenses on glasses.

Damiani uses this thirteenth thesis as the conclusion of his *Disputation*. He was perfectly aware that he was only making a critique of the status of the art of hygrometric science. Moreover, he was aware that he had nothing new to propose and had only been able to identify problems and urge scientists to find new solutions and design new instruments. Following a traditional literary custom, this quotation from Horace's (Quintus Horatius Flaccus, 68–8 B.C.) *Ars Poetica* concludes the book:

My role will be that of the whetstone, which sharpens the iron that was previously unable to cut.
I will teach the task and the duty without writing anything.

Damiani's approach through Galilean and Newtonian physics and his views on the origin of meteorology (a field to become geologically important in understanding, for example, ancient climates and climate change, chemical and physical weathering and erosion, sediment generation and deposition, and soils) would seem as if he anticipated Hutton, Lyell, and Darwin, who in setting forth their respective principles, spread the net of their evidence so very widely, reasoned in similar fashion to Damiani, and would not subscribe to divine intervention or metaphysical forces.

DISCUSSION AND CONCLUSIONS

The individuality of the personage of Mattia Damiani has probably not facilitated proper acknowledgment from either his contemporaries or posterity, but some additional external conditions could also have contributed to his neglect.

The literary productivity of Damiani was not vast, but an important part of what he studied and wrote about follows a path between science and literature (Battistini, 1977). Important scientists and philosophers have overcome the boundary between science and literature many times in the past. Galileo Galilei (1632) is remembered for the titanic literary enterprise of his *Dialogo sopra i due massimi sistemi del mondo, tolemaico e copernicano* (*Dialogue concerning the two chief world systems, Ptolemaic and Copernican*), but there are older literary compositions of Galilei, such as the commentaries about Torquato Tasso and lessons on the topography of Dante Alighieri's hell (Banfi, 1940; Spongano, 1956; Sansone, 1965), are also deserving of close attention. Tommaso Campanella (1568–1639) is still studied for his poems (Ruschioni, 1980). Even the Russian chemist, Michail Lomonosov (1711–1765) is remembered as a composer of poems. In general, all post-Renaissance European science and philosophy is marked by sporadic links between science and art.

In Damiani's work *Le Muse Fisiche*, however, overcoming the barrier between the "two cultures" (Snow, 1970) is more complete. This is evident not only in the direction of poetry, but also theater, with its links with the visual arts, and music. Indeed, Damiani lived at the time of the birth of melodrama (Mila, 1956), and he was certainly fascinated by the fusion among the arts that it realized. His object was to add science to an amalgam of images, costumes, scenery, poetry, and music. While I cannot be certain, there is a distinct probability that some "pastoral compositions" from *The Physical Muses* were actually performed in a Tuscan theater. Today, the music of most of the melodramas of the eighteenth century is lost, and only conjectures about the style, musicality, and instrumental ensembles used in the performances is possible (Mila, 1956; Giuntini, 1994).

Some logic can now be seen for the start of the creative process of the nine compositions in *The Physical Muses* of 1754. Their interior elaboration can be carried back to the time of collecting the poetic material about natural phenomena that was quoted in the *Disputation about hygrometers and their defects* of 1726. This material in verse—from Lucrece, Virgilio, Horace, etc.—constitutes the ideal basis on which to found Damiani's inspiration and his achieving of the artistic aims of all his predecessors toward an enlarged multimedia—or better, blended-arts—characterization.

During his life, Damiani always showed great modesty, and he attributed his decision to become a poet to his friendship with Pietro Metastasio. The sincere amity of this relationships has been mistaken by a prejudicial critic (Losavio, 1925) as intellectual dependence upon Metastasio, and the same critic, basing his judgment on idealistic prejudices, disdained the mere presence of scientific topics in the poetry. Only a later critic, Silvano Bertini (1965), reached a more balanced appraisal of the value of the Damiani's poetic work. Losavio's harsh criticisms did not distinguish the different levels of philosophic elaboration of the two friends. The recently rediscovered *Disputation about the hygrometers* makes it clear that Damiani did not regard himself entirely a poet; rather, his cultural background made him more a philosopher of nature *sensu lato*. The spirit of his activity—Damiani's target—was not purely aesthetic, as were Metastasio's aims, but it had, beside the Enlightenment component, a strong component of the Renaissance spirit. Damiani can be defined as a "Renaissance man" in that he starts from the observation of nature in defining and limiting his own thoughts. It is not coincidental that an ode of eulogy by him was dedicated to Antonio Genovesi, who in his philosophical works attempted a dissection of all the "systems of the world," demonstrating that most of them were based on preconceptions and not on observations (Galanti, 1772). This attitude of Damiani—coming from his long lapses of time spent in Renaissance centers of Pisa and Florence—is harmoniously blended to the more philosophical attitude of the Enlightenment *Weltanschaung*.

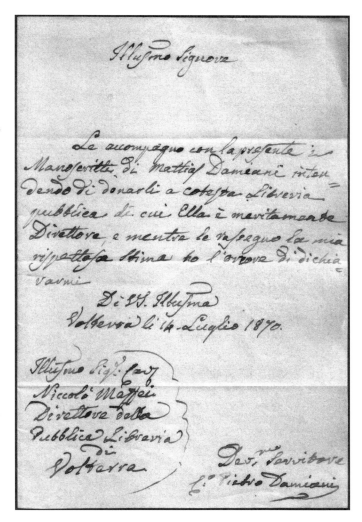

Figure 7. The collection of manuscripts bequeathed by Mattia Damiani. The donation was made by Pietro Damiani on 14 July 1870 to the Public Library of Volterra. The letter was addressed to the librarian, Niccolò Maffei. The text says: *Enclosed with this letter are the manuscripts of Mattia Damiani, which I wish to donate to the Public Library of which you have the well-deserved position of Head Librarian …*"

The manifold causes of the oblivion of Damiani's work can be attributed to:

1. Some vicissitudes in the Italian cultural scene. Influential people, like the Jesuit Giambattista Roberti (1719–1796) (Borsellino and Pedullà, 2004), recommended only a parsimonious introduction of physics into poems. The Arcadian movement in the eighteenth century was under the cultural control of the church, trying to curb Libertinism and the nascent cult of the "Deity Reason," so rationalistic works like *The Physical Muses* were probably not regarded benevolently. The idealistic conservative critics of the first half of twentieth century considered the introduction of physics into the pure aesthetic poetry as not legitimate (e.g., Losavio, 1925).

2. Damiani's ideas were close to the theological positions of the Jansenists, and the possibility exists that a hidden suppression of dissident voices operated in literature and theology, especially after the Pistoia Synod of 1786. Even today, in some recent compilations about the history of the church, the Jansenist movement is almost completely ignored (Fernandez, 2000), including the bloody struggle between the Jansenists and the Jesuits in the seventeenth century, mostly in France, of which the Port Royal des Champs vicissitudes, a Jansenist stronghold until the nuns were expelled in 1709, provides an example. The time was one of transition from the old to the new Christian attitude—from predestination to free will.

3. The founder of the Italian history of literature, Girolamo Tiraboschi (1731–1794), was a Jesuit, and it is possible that some cultural filter was applied by him and transmitted to those historians who followed him. Indeed, clues supporting this possibility can be found in the introduction of Antonio Lombardi's (1768–1847) *History of the Italian Literature of the XVIII Century* (Lombardi, 1827–1830).

As for the implication of Damiani's work in demanding logical explanations of natural phenomena, a tendency to dismiss theological explanations (mostly Diluvianism), which was then dominant in European earth science, in favor of a pure scientific approach, already quite common in Italy, is reconfirmed.

Damiani was follower of a new conception of time, adoption of which by the western scientific community was of primary importance in the expansion of the geological time and the age of the earth from few thousands of years (biblical computations stated only 6000 yr) to millions or billions of years.

I am personally convinced that, at present, cultural life is being irresistibly spread under the influence of multimedia, and so a revaluation of the work of Mattia Damiani—late Renaissance man, follower of the Enlightenment, accomplisher of the most complete banishment of the "two culture" separation—is both necessary and urgent. The present paper is a small contribution to that end, and further revaluation of Damiani's contribution will be continued by research on the original manuscripts deposited in the Guarnacci Library of Volterra. Marrucci (1997) has already given a short biographical sketch of Damiani, and now the recent translation (Balducci, 2005) of the fortuitously preserved Falconcini manuscript (in Latin), which contains Damiani's biography, has allowed the biographical sketch in this paper to be presented.

Finally, it should be stressed that the rediscovered *Disputation about hygrometers and their defects* opens a new window on the personages involved and on the evolution of the meteorological concepts in Europe in a context of Galilean and Newtonian physics. Moreover, the text is a precise source of information on the probably seventeenth-century published work (quoted by Damiani as *De vi vaporum in Hygrometr. dissertat. geometric-mechan.*) of Ludovico da Ripa on hygrometers—which today is lost, but profitably might be sought by historians of science.

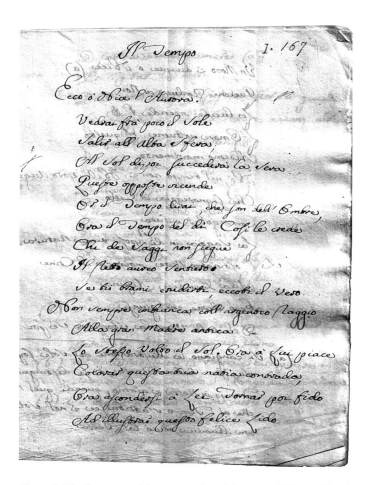

Figure 8. The first page of the manuscript of the poem *Il Tempo* (*Time*). This poem deals with the mystery of physical time. Damiani confutes the concept of the passing of time, and he proposes that the sensation of passing time is an illusion produced by the continued evolution and transformation of ourselves and of our surrounding environment. Because of the progressive diffusion of this new conception of time, the geological time and the age of the Earth expanded in the western scientific thought from few thousands of years (biblical computations stated only 6000 yr) to millions or billions of years.

APPENDIX. Selected parts of the poem *Time*, by Mattia Damiani.

IL TEMPO	TIME
Ecco, o Nice, l'Aurora:	Lo, o Nike, Aurora:
Vedrai fra poco il Sole	Shortly you shall see the Sun
Salir all'alta sfera;	Climb to the noble sphere;
Al Sol dipoi succederà la sera.	Then shall the eve follow the Sun.
Queste opposte vicende	These opposing events
Ora il tempo dirai, che son dell'ombre,	Now you will say time is of darkness,
Ora il tempo del dì. Così le crede	Now of the day. So it is believed
Chi de' saggi non siegue	By those who do not follow
Il retto aureo sentiero;	The straight golden path of the wise;
Se tu brami erudirti, eccoti il vero.	If you desire to be enlightened, here is truth.
Non sempre imbianca coll'argenteo raggio	Not always does the Sun cast his rays
Alla gran Madre antica	On the same face of the great Ancient Mother.
Lo stesso volto il Sol. Or a lui piace	At times he likes
Colorar questa tua natìa contrada,	To paint this your native village,
Or ascondersi a lei; tornar poi fido	Or hide from her; then to return confident
Ad illustrar questo felice lido.	To illuminate this happy shore.
Un moto è dunque, o Nice,	Motion, o Nike, is then
Quel che produce a noi	What produce for us
Le lucide vicende,	The events of light
L'ombre notturne. E questo	And darkness
Alterno movimento	And this alternating movement
No, nel Tempo non è. [Esso è] Nell'Astro augusto,	Is not in Time but in the august Star,
Che signoreggia in Cielo,	That reigns in the Sky.
[... ...]	[... ...]
E se miri la vite	If you behold the vine
Spiegar su gli orni i già fecondi tralci,	Spreading its already fecund shoots on the ash
Dirai, che giunse il Tempo	You will say that the Time has arrived
Sacro al Nume Tebano,	That is sacred to the Theban God,
Nume, che anima i vili,	The God who animates the faint-hearted
Rasciuga al mesto i pianti,	Who dries the tears,
Fuga il rossor dai timorosi amanti.	Flees the blushes of timorous lovers.
Ma queste chi produce	But who creates
Mutabili vicende	Changing events
Nel regno di Natura? Agita il Sole	In Nature's realm? The Sun is what arouses
Allor che più si volge	As he casts his beams
Sull'Arcadi Campagne,	On the Arcadian Countryside,
Il pigro umor racchiuso	The idle humor enclosed
Nel sen del verde suolo, e ne' riposti	Within the bosom of the green soil,
Del Mirto, e della Rosa	Of Myrtle, and the Rose
Infiniti ricetti; ed ecco il ramo	Infinite secret haunts; and lo the branch
Cinto di nuova fronda; ecco sen riede	Girded in fresh foliage; now here is
L'aurea Stagion de' fiori, ecco ritorna	The golden Season of the flowers, now returns
La tormentosa Estate,	The tormenting Summer,
L'Autunno a Bacco sacro, il Verno rio,	The Autumn sacred to Bacchus, the wicked Winter,
Se cangiando sentier l'Astro del giorno,	If the Star of day, changing its path,
In questa vasta sullunar Sostanza	In this vast sublunar substance
Novelli riconduce	Brings back new
Opposti cangiamenti,	Opposing changes,
Ch'or tu chiamasti, o Nice,	Which, o Nike, you called
Primavera felice,	Now happy spring,
Ora feconda Estate,	Now fecund Summer,
Autunno a Bacco amico,	Now autumn the friend of Bacchus,
Or crudo Verno ai bei piacer nemico.	Now crude Winter, the enemy of sweet pleasures.
[... ...]	[... ...]

(continued)

Quei dì, ch'a te promette a mille a mille La Parca non avara, altro non sono Che un mutabil successo ora di lieti, Or d'infelici avvenimenti. I giorni, I mesi, e gli anni, e i lustri Non succedon fra lor, moto non hanno, Il corso non gli affretta .... Basta, o Nice, t'intesi. Ma per vie più fiorite Guidar ti vo', per cui tu giunga al Vero: Se udirmi ancor t'aggrada, ecco il sentiero. Se di vago naviglio Sedesti mai sulla dipinta prora Dolce cantando al mormorio dell'onda, Sembrato ti sarà, che l'alta riva Da te sen voli, e fugga; E che la nave tua dell'acque in seno Immobile riposi. E sai perché? Non t'accorgesti allora, Che dividea quell'onda Fuggitivo il naviglio, e il moto impresso Nel condottiero legno Ascrivesti delusa al fermo lido, Mercé del ciglio tuo, duce malfido. Fingi or, che sia del tempo L'immobil ripa la verace imago; La nave, che sen vola, Sia dell'umane cose Il mutabil successo; e t'avvedrai Del conceputo inganno, E non sarai fra quei, che il ver non sanno. [... ...] Così se delle cose Al variabil progresso Volgi il pensier, conoscerai, che il moto Il cangiamento, il corso Non van del tempo in compagnia; ma l'orme Sieguono delle cose, Poiché il moto Natura in lor ripose. [... ...]	Those, days, promised to you by the thousands By Fate who is not mean, are none other Than a changing succession now of happy, Now of unhappy events. The days, The months, and the years, and the lustrums Happen not by themselves, motion they have not, Their course hastens them not … Enough, o Nike, I have understood. But along more flourishing trails I want to lead you, so that you may reach Truth: If you are still pleased to hear me, here is the path. If you ever sat on the painted prow Of the roaming vessel Softly singing to the murmuring waves, It must have seemed to you, that the high bank Flew from you, and fled; And that your ship remained motionless, In the bosom of the water. And do you know why? Then you did not perceive, That the craft cleaved that river And the motion given To the beam of the prow. Crestfallen, you ascribed to the motionless shore, At the mercy of your brow—untrustworthy leader. Now you feign, that the motionless shore Is the true image of time; That the ship, that flies forth, Is the changeable result Of things human; and you will perceive The deception conceived, And will not be among those who know not Truth. [... ...] So direct your thought To the variable progress Of things, and you will discover, that motion, the change, the elapsing Do not go hand in hand with time; but they follow The traces of things, Since Nature has placed motion in them. [... ...]

ACKNOWLEDGMENTS

Many people have helped me develop this work by assisting me in a highly professional manner. Luca Pini, Giancarlo Baroncini, and the late Angelo Marrucci facilitated my research in the Guarnacci Library of Volterra. Paolo Pampaloni has kindly provided a photocopy of the *Disputation about hygrometers*. Tiziana Bison and Gianni Biazzo translated the *Disputation* and then assisted me in the scientific revision. Cristiano Balducci provided a high quality paleographic interpretation and translation of the Falconcini manuscript. David Branagan and the editors provided invaluable linguistic suggestions and corrections. Phil Rand revised the section "Biographical Sketch"—which was added in course of editing—and translated the poem *Time*. Andrea Battistini has suggested invaluable additions and improvements. Michele Marvulli kindly provided information and a photocopy of the rarest Carlucci book. To all these, and to many others, I extend my thanks.

REFERENCES CITED

Aliverti, G., 1929, Sui metodi di misura dell'umidità (On the measurement methods of humidity): Il Nuovo Cimento, v. VI, no. 6, p. CXVII–CXXIII.

Agamennone, G., and Cancani, A., 1885, Contributo alla storia ed allo studio dell'igrometria (Contribute to the history and to the study of the Hygrometry): Annali dell'Ufficio Centrale di Meteorologia Italiana, Ser. II, vol. 7 (parte I), p. 1–35.

Anonymous, 1824, Index Librorum Prohibitorum: Neapoli, Ex Typographia Xaverii Jordani, 351 p.

Astaldi, M.L., 2001, Metastasio (a biography): Milano, Fabbri Editori, 359 p.

Balducci, C., 2005, Transcription and translation of the manuscript of B.P. Falconcini: De claris Viris Volaterranis [Unpublished manuscript]: Roma, Istituto Nazionale Geofisica Vulcanologia, 21 p.

Banfi, A., 1940, Galileo Galilei, Quaderni di Analisi Letteraria: Milano, Vallardi Editore, 185 p.

Battistini, A., editor, 1977, Letteratura e scienza: Bologna, Zanichelli, 234 p.

Battistini, A., 2005, La cultura del Barocco, in Malato, E., ed., Storia della Letteratura Italiana, Vol. V (I): Roma-Milano, Salerno Editrice e Il Sole 24 Ore, p. 463–560.

Bertini, L., 1965, Mattia Damiani poeta volterrano amico del Metastasio: Volterra, anno IV (7/8) luglio-agosto, p. 4–7.

Borsellino, N., and Pedullà, W., 2004, Storia generale della Letteratura Italiana vol. VII—Il secolo riformatore: Poesia e ragione nel settecento: Milano, Motta Editore, 744 p.

Carlucci, G. (but anonymous), 1766, Ragionamento Filosofico intorno al Moto della Terra (Philosophic reasoning about the motion of the Earth): Napoli, Per Vincenzo Flauto, 103 p.

Caroti, S., 1987, Un sostenitore napoletano della mobilità della Terra: il padre Paolo Antonio Foscarini, in Lo Monaco, F., and M., Torrini M., eds., Galileo e Napoli: Napoli, Guida Editori, p. 81–121.

Consoli, D., 1972, Dall'Arcadia all'Illuminismo: Bologna, Cappelli Editore, 220 p.

Damiani, M., 1726, De Hygrometris et eorum defectibus disputatio—Sub felicissimis auspiciis Illustrissimi, ac Reverendissimi Domini D. Ludovici Mariae Pandolfini: Pisa, Ex Typographia Francisci Bindi Impressoris Archiepiscopalis, 46 p.

Damiani, M., 1747, Componimento pastorale-filosofico detto in Firenze nell'Accademia degli Apatisti.

Damiani, M., 1754, Le Muse fisiche—al chiarissimo sig: Pietro Metastasio romano poeta di sua maestà imperiale: Firenze, Appresso Giovan Paolo Giovannelli, XVI + 231 p.

Damiani, M., 1758, Poesie liriche (con lett. ded. "A sua eccellenza la signora duchessa di Choiseuil"): Parma, XI + 180 p.

Damiani, M., 1761, Componimento drammatico pastorale sulla morte di S. Ottaviano, Volterra.

Damiani, M., 1765a, Cantata a due voci in morte del chiarissimo signore cavaliere Lorenzo Guazzasi: Arezzo, Appresso Michele Bellotti, 8 p.

Damiani, M., 1765b, Delle poesie del celebre sig. dottore Mattia Damiani: Firenze, Appresso Andrea Bonducci, 338 p.

Damiani, M., 1770–1771, Raccolta di poesie diverse del signor dottore Mattia Damiani di Volterra: Terza edizione notabilmente accresciuta: Livorno, Stamperia di Gio. Vincenzio Falorni, 3 vls., v. 1, XV + 280 p.; v. 2, IV + 228 p.; v. 3, 217 p.

Damiani, M., 1772, Componimento in morte del signor abate Antonio Genovesi pubblico professore nella Regia Universita di Napoli: Del celebre signor dottore Mattia Damiani di Volterra (con lettera prefazione di Forges Davanzati, Domenico [1742–1810]): Napoli, Nella Stamperia Raimondiana, 31 p.

Damiani, M., 1773, Per la venuta in Volterra di s.a.r. Pietro Leopoldo arciduca d'Austria granduca di Toscana—cantata dell'abate Mattia Damiani: Firenze, Appresso Stecchi-Pagani, 16 p.

Davanzati, G., 1774, Dissertazione sopra i Vampiri (Dissertation about Vampires): Reprinted in 2000 by Giacomo Annibaldi, G., ed.: Nardò, Besa, 159 p.

De Fontenelle B. Le B., 1686, Entretiens sur la pluralité des mondes (Conversations on the plurality of worlds): Italian translation by E. Concanari, Roma-Napoli, Edizioni Theoria, 130 p.

De Fontenelle, B. Le B., 1732, Théorie des tourbillons Cartésiens, avec des réflections sur l'Attraction (Section IX—Sur les Atmosphères des Corps célestes, p. 282–302), in Œuvres de Monsieur De Fontenelle, des Académies, Françoise, des Sciences, des Belles-Lettres, de Londres, de Nancy, de Berlin & de Rome, Nouvelle Édition (1766). Tome IX : Paris, Chez les Libraries Associés, p. 143–326.

Falconcini, P.B., 1777, De claris Viris Volaterranis: Manuscript conserved in the Library Guarnacci, Volterra, p. 7–21.

Fernandez, F.G., 2000, I movimenti dalla Chiesa degli apostoli a oggi: Milano, Rizzoli, 344 p.

Foscarini, P.A., 1615, Lettera sopra l'opinione de' Pitagorici, e del Copernico: Della mobilità della Terra e stabilità del Sole e del nuovo Pitagorico Sistema del Mondo: Napoli, Lazaro Scoriggio, Reprinted by Romeo, L., ed., in 1992, Montalto Uffugo (Cosenza), Grafiche Aloise, IX + 64 p.

Galanti, G.M., 1772, Elogio storico del Signore Abate Antonio Genovesi, in Actis-Perinetti, L., ed., 1973, Gli Illuministi Italiani: Torino, Loescher Editore, 159 p.

Galilei, G., 1623, Il Saggiatore: Translation from Latin by Libero Sosio, 1979: Milano, Feltrinelli Editore, 330 p.

Galilei, G., 1632, Dialogo sopra i due massimi sistemi del mondo, tolemaico e copernicano (Dialogue concerning the two chief world systems: Ptolemaic and Copernican), in Sosio, L., ed.: Torino, Einaudi, 1970, 593 p.

Giuntini, F., 1994, I drammi per musica di Antonio Salvi: Bologna, Società Editrice il Mulino, 269 p.

Gori, A.F., 1754, Sig. Abate Damiani amico onorabilissimo: Foreword letter to "Le Muse Fisiche," in Mattia Damiani: Le Muse fisiche—al chiarissimo sig. Pietro Metastasio romano poeta di sua maestà imperiale: Firenze, Appresso Giovan Paolo Giovannelli, p. IX–XV.

Hamblyn, R., 2001, L'invenzione delle nuvole (The invention of clouds): Milano, Rizzoli, 366 p.

Iemolo, A.C., 1928, Il Giansenismo in Italia prima della rivoluzione: Bari, Laterza, 439 p.

Lana, F., 1670, Prodromo all'Arte Maestra, new annotated edition by Battistini, A.: Milano, Longanesi, 380 p.

Lombardi, A., 1827–1830, Storia della Letteratura Italiana nel secolo XVIII: Modena, Presso la Tipografia Camerale, v. I, XXI + 521 p., v. II, VI + 352 p., v. III, VIII + 456 p., v. IV, VII + 501 p.

Losavio, F., 1925, Un poeta volterrano del settecento: Mattia Damiani, Rassegna Volterrana, anno II (2) settembre, p. 92–97.

Maffei, P., 1987, Giuseppe Settele, il suo diario e la questione galileiana: Foligno, Edizioni dell'Arquata, 583 p.

Magalotti, L., 1667, Saggi di naturali esperienze, fatte nell'Accademia del Serenissimo Principe Leopoldo di Toscana e descritte dal Segretario di essa Accademia: Lorenzo Magalotti, Terza edizione con aggiunte 1841: Firenze, dai torchi della Tipografia Galileiana, p. 133–183-XC.

Marrucci, A., 1997, Damiani Mattia, in I personaggi e gli scritti: Dizionario biografico e bibliografico di Volterra: Fondazione Cassa di Risparmio di Volterra, p. 9659–66.

Marvulli, M., 2001, Il declino dell'Università di Altamura in un inedito di Luca de Samuele Cagnazzi: Altamura, Rivista Storica/Bollettino dell'A.B.M.C, no. 42, p. 175–217.

Metastasio, P., (Pietro Trapassi), 1847, Lettere inedite dell'abate P. Metastasio a Mattia Damiani poeta volterrano (53 letters), *in* Metastasio, P., Opere: Volterra, Tipografia all'Insegna di S. Lino, v. XXVII, 102 p.

Mila, M., 1956, La nascita del melodramma, *in* Il Sei-Settecento, a cura della Libera Cattedra di Storia della Civiltà Fiorentina: Firenze, Sansoni, p. 167–194.

Montanelli, I., and Gervaso, R., 1970, L'Italia del Settecento: Milano, Rizzoli, 702 p.

Natali, G., 1964, Il Settecento: Milano, Vallardi, v. I, 585 p.

Raucci, B., 2001, Uno scienziato nel Regio Studio di Altamura: Luca de Samuele Cagnazzi: Altamura, Rivista Storica/Bollettino, dell' A.B.M.C., no. 42, p. 151–172.

Romeo, L., ed., 1992, Reprint of: Foscarini, P.A., 1615, Lettera sopra l'opinione de' Pitagorici, e del Copernico: Della mobilità della Terra e stabilità del Sole e del nuovo Pitagorico Sistema del Mondo: Napoli, Lazaro Scoriggio, IX + 64 p.

Ruschioni, A., 1980, Tommaso Campanella filosofo-poeta: Brunello (VA), Edizioni otto/novecento, 256 p.

Sansone, M., 1965, Considerazioni su Galileo letterato: Bari, Annali della Facoltà di Lettere e Filosofia, Pubblicazioni dell'Università di Bari, v. X, p. 395–423.

Scalera, G., 1999, I moti e la forma della Terra (Motions and shape of the Earth): Roma, Tangram-Istituto Nazionale di Geofisica, 195 p.

Sisto, P., 2003, I nostri illuministi tra scienza, ideologia e letteratura: Fasano (Brindisi), Schena Editore, 118 p.

Snow, C., 1970, Le due culture: Milano, Feltrinelli, 105 p.

Spongano, R., 1956 Galileo scrittore, *in* Il Sei-Settecento, a cura della Libera Cattedra di Storia della Civiltà Fiorentina: Firenze, Sansoni, p. 107–122.

Stella, P., 1972, Il Giansenismo in Italia: Bari, Adriatica Editrice, 168 p.

Tiraboschi, G., 1829, Storia della Letteratura Italiana: Milano, Per Antonio Fontana, v. XXIX, XVIII + 232 p.

Vai, G.B., and Cavazza, W., editors, 2003, Four centuries of the word "geology": Ulisse Aldovrandi 1603 in Bologna: Argelato (Bologna), Minerva Edizioni, 327 p.

Webb, S., 2002, If the universe is teeming with aliens…where is everybody?: Fifty solutions to the Fermi paradox and the problem of the extraterrestrial life: New York, Copernicus Books & Praxis Publishing Ltd., 288 p.

Manuscript Accepted by the Society 17 January 2006

The "classification" of mountains in eighteenth century Italy and the lithostratigraphic theory of Giovanni Arduino (1714–1795)

Ezio Vaccari*

Dipartimento di Informatica e Comunicazione, Università dell'Insubria, Via Mazzini 5, I-21100 Varese, Italy

ABSTRACT

During the eighteenth century, scientific literature devoted to the earth sciences documented a significant increase in the study of the composition and formation of mountains and above all their stratigraphical sequence. The diverse and widely ranging philosophical theories of the late seventeenth century on the origin of Earth were gradually replaced by new concepts based on field research on both a local and regional scale. This new approach analyzed the lithology and the fossil content of the rocks, the geomorphology of the area, and in some cases helped to determine the chronological sequence of mountain formation. Nicolaus Steno's idea of superimposition of strata (1667–1669) was followed by most of the late eighteenth-century scholars in earth sciences, who developed subdivisions of mountains from the point of view of their formation and also included a classification of the rocks. These subdivisions supported the idea of relative chronology of the formation sequence of the studied strata: the most recent or the most ancient formation could be deduced from its position in the sequence as well as from its external lithological features. In this context, the role of scientific terminology, which was gradually established in eighteenth-century geological science, became very important: the terms "primary" (or "primitive"), "secondary," and "tertiary" were used for indicating the categories of mountains as well as for stratigraphic units. In the second half of the eighteenth century, the work of Giovanni Arduino contributed decisively to the development of basic lithostratigraphic "classification" of rocks and mountain building. His lithological studies, a result of twenty years of fieldwork in the mountains and hills of the Venetian and Tuscan regions, were also supported by a specialized knowledge of mining. The new "classification" into four basic units called "ordini" (1760) was based only on lithology (without using paleontological indicators) and included different rock types, which formed three kinds of mountains and one kind of plain, in a regular chronological order: "primary" (underlain by "primeval" schist considered by Arduino to be the oldest rock type), "secondary," and "tertiary"; the "fourth" and younger chronolithological unit included only alluvial deposits. Arduino's system is still regarded by the geological world as being one of the starting points for modern stratigraphy.

Keywords: stratigraphy, mountains, G. Arduino, eighteenth century, Italy.

---

*E-mail: ezio.vaccari@uninsubria.it.

Vaccari, E., 2006, The "classification" of mountains in eighteenth century Italy and the lithostratigraphic theory of Giovanni Arduino (1714–1795), in Vai, G.B., and Caldwell, W.G.E., The origins of geology in Italy: Geological Society of America Special Paper 411, p. 157–177, doi: 10.1130/2006.2411(10). For permission to copy, contact editing@geosociety.org. ©2006 Geological Society of America. All rights reserved.

INTRODUCTION: THE HISTORIOGRAPHICAL CONTEXT

Among the contributions made in the eighteenth century by the Italian scientific community to the origin of historical geology or stratigraphy, Giovanni Arduino's system of subdividing rocks into four basic units called "ordini" attained a unique level of precision. This lithostratigraphic subdivision, which represents the most famous aspect of Arduino's geological work, has been discussed by many historians of the earth sciences since the nineteenth century (overview in Vaccari, 1993, p. 1–22). The reason for this may be twofold: the remarkable geological intuition of Arduino, which was much appreciated by the scientists of the following centuries, and the fact that his theory was placed at the center of the question of the "classification" of mountains, that is to say, the trend to subdivide the mountainous rocks into different units (usually two or three) that marked subsequent stages of formation and were often defined with the terms "primary" or "primitive," "secondary," and "tertiary" (see also Vai, Chapter 7).

The "classification" or subdivision of mountains can be considered a central topic in the history of eighteenth-century geology because it gradually contributed to the introduction of the idea of geological time connected to a complex history of Earth based on the study of its structural elements (Guntau, 1996, p. 221–223; Rudwick, 1996, p. 272–275). This history was made up of a series of important successive changes in Earth's surface that could be observed in its most prominent geomorphological features: the mountains. Consequently, the different kinds of rocks, the strata or massive formations which form hills and mountains, became the indispensable keys for recognizing the path of such a long history. This happened well before the establishment of the systematic use of fossils as chronological indicators, which was an early nineteenth-century achievement and eventually led to what is known today as stratigraphy.

The question of the "classification" of mountains also expanded several important aspects of the eighteenth-century earth sciences, such as the investigations of extinct volcanoes and igneous rocks, the problem of the origin of springs, the interpretation of strata and their lithological components, the morphology of mountain ranges, and the debate on the possible chronology for the history of Earth. All these questions were treated and discussed in the eighteenth century Italian geological literature, as the result of the research activities of an extremely lively and productive scientific community, especially in the second half of the century, when a great quantity of information of an empirical nature was accumulated, analyzed, and published. Moreover, the extraordinary geological variety of the Italian territory (from the active volcanoes in the south to the Alps in the north) certainly inspired and facilitated much research in the field, which started as observations of the strata and later produced subdivisions of mountains and classifications of their rock types in different units, from the point of view of their formation.

During the last four decades, after the early studies by Frank Adams (1938, p. 329–398), the historical development of the theories on the origin of mountains and on their subdivisions had been investigated within several works of history of geology and geography, particularly by some French scholars (Broc, 1969, p. 99–145; Ellenberger, 1978; Gohau, 1983, 2003). However, in spite of a great abundance of primary sources, the question of the "classification" of mountains connected to early lithostratigraphical studies is a subject that has not yet attracted in Italy the same interest as shown, for example, in France or in Germany. In order to contribute to the filling of such an evident historiographical gap, this paper aims to outline some of the main topics of a monographic work, presently in preparation, on the development of historical geology in eighteenth-century Italy.

THE ORIGIN OF MOUNTAINS AND THE "THEORIES OF THE EARTH"

Ancient authors such as Pythagoras, Ovid, and Seneca asserted that the mountains were formed and continued to form for diverse reasons, such as subterranean winds or volcanic eruptions, but later, between the Middle Ages and the Renaissance, other scholars emphasized the role of sedimentary and erosive phenomena, as in the case of Avicenna, Ristoro d'Arezzo, and Leonardo da Vinci. Using some of these hypotheses, the French philosopher René Descartes (1596–1650) described the origin of mountains following the breakup (due to the force of subterranean air) of a crustal layer of the lightest superficial material, which was then tilted to form the flanks of the mountains, underneath of which enormous caverns filled with air and water were opened (Descartes, 1644, IV, § 37–44). This well known Cartesian interpretation presented a "theory of the Earth in a secular version" (Ellenberger, 1994, p. 13), in which the formation of the terrestrial globe was explained within a natural mechanistic philosophy that postulated a constant movement of the component elements of the terrestrial bodies in order to continually refill the void left by the displacement of this material (Roger, 1982).

In the second half of the seventeenth century, this "secular" definition would, however, be opposed by the "theories of the Earth" published in books often entirely dedicated to the formation of the terrestrial globe and its actual surface (Roger, 1973; Taylor, 1992; Vaccari, 2001). In these works, based on scholarly erudition and only occasionally on fieldwork, frequent reference to or comparison with the pages of the Bible did not leave much possibility open for an interpretation of the origin of mountains, especially in the case of those considered to be more ancient: the mountains were thought to have either been made by God at the time of the Creation or they were formed due to the Deluge. In any case, the birth of terrestrial relief was always directly referred to a divine intervention. Thus, according to some scholars, such as the German Jesuit Athanasius Kircher (1601–1680), the mountains were modeled by God in an inscrutable way at the Creation of the world and therefore they remained substantially intact up to the present, despite the perturbations caused by the Deluge (Kircher, 1664; Morello, 2001). According to others, such as English scholar Thomas Burnet (1635–1715), God made the present mountains by means of the great catastrophe of the Deluge; the originally flat terrestrial surface was fractured by the diluvial waters rushing forth from their subterranean abyss (Burnet, 1681; Pasini, 1981). A third hypothesis, put forward by John Woodward (1665–1728), integrated the previous two in theorizing that the terrestrial orography created by God was completely dissolved by the diluvial waters, which subsequently deposited new rock strata and modeled the present mountains (Woodward, 1695; Ellenberger, 1994, p. 114–124). A rather different theory less dependent on the Bible was outlined by the Bolognese scientist Luigi Ferdinando Marsili (1658–1730) in a late manuscript (Vai, Chapter 7, this volume).

It is important to keep in mind this double interpretation of the origin of mountains (created directly by God or formed by the Deluge) because this would not be confined to the seventeenth century but would be particularly important in the following century, in Italy as in Europe, when the problem of the "primitive" mountains in the absolute sense was shown to be without a simple solution outside of a Biblical position.

In Italy, since the early eighteenth century, the gradual enlargement of field research, which included mountains, hills, and other land bodies, was determined by a "Stenonian heritage" (that is to say, the necessity of fieldwork in a regional context), as well as an influence by the "theories of the Earth." The Italian case shows, in fact, that the development of historical geology through the "classifications" of mountains was not a simple linear process, which sharply moved from a stage dominated by theoretical models to a new period exclusively based on field research. The "Stenonian heritage" and the "theories of the Earth" were instead equally influential especially during the first decades of the eighteenth century (Vaccari, 2001, p. 26–33).

In this period, the Italian scholars who were involved in studying Earth's surface seemed gradually to adopt Nicolaus Steno's (1638–1686) methodological view, which necessarily was based on numerous data collected in the field, as well as his theory of superimposition of strata, which stated that all the strata were originally deposited horizontally and every stratum is always younger than the underlying stratum, but older than the overlying stratum (Steno, 1669). Although there is still much historical work to do on the impact of Steno's ideas, it is already known that his writings (published in Latin) circulated widely in several scientific milieus (Vaccari, 2001, p. 29–30).

Besides Steno, the *Mundus Subterraneus* of Kircher, who was based in Rome, and the "theories of the Earth" of Burnet and Woodward had a particularly dramatic reception within the Italian scientific community. They were debated, rejected (Burnet), or partially or fully accepted (Kircher, Woodward), but in some significant cases they were also able to stimulate new research in the field. For example, the naturalist Giuseppe Monti (1682–1760), a distinguished member of the Academy of Sciences of Bologna, clearly supported Woodward's idea of the universal dissolution of Earth's crust due to the water of the Deluge (Monti, 1719); consequently, all the present orography was considered by Monti of "secondary formation," because the "primitive" original mountains, created by God, had been dissolved by the Deluge, which deposited regular strata. According to Monti, these strata later cracked to form new series of hills and afterward all the water flowed back into the central abyss of Earth.

Monti was stimulated to do fieldwork by his theoretical "Woodwardian" approach: during his numerous naturalistic excursions into the Bolognese hills and Apennines, he interpreted rock strata and fossils as clear evidence of the Deluge. This fieldwork also provided the specimens for the constitution of the first Italian paleontological collection, the *Musaeum Diluvianum*, established between 1710 and 1730, which still exists in the University of Bologna (Sarti, 1988, 1992).

The rejection of the "theories of the Earth" produced fieldwork and more interest in mountain relief and in rock strata, as in the case of Antonio Vallisneri (1661–1730), a distinguished professor of medicine in the University of Padua (Rappaport, 1991; Arpiani, 1999). He corresponded with the Swiss naturalist Johann Jakob Scheuchzer (1672–1733), who had widely explored the Alps from 1702 to 1711 and had contributed to the diffusion of Woodward's "theory of the Earth" in northern Italy (Vaccari, 2001, p. 31–32). However, Vallisneri firmly rejected Woodward's theory; the content of his book, *De' Corpi Marini* (Vallisneri, 1721), shows that he adopted a Stenonian attitude in traveling and doing fieldwork through the northern Apennines. On the other hand, he seemed also to accept the idea of "primitive" and permanent mountains created by God (eventually just slightly eroded or increased by several floods and earthquakes) as described in the *Mundus Subterraneus* by Athanasius Kircher. These "primitive" mountains, according to Kircher, were the real bearing-structures of Earth, a "geocosm" which had remained unchanged since the date of the Creation. Therefore, it is not by accident that Vallisneri called the mountains and their strata the "bones" of Earth (Vallisneri, 1721, p. 49, 73–74) and consequently rejected the orogenetic action of a unique universal flood as theorized in different ways by Burnet, Woodward, and Monti.

Vallisneri's emphasis on the importance of the study of the "anatomy of mountains" (*la notomia de' monti*: Vallisneri, 1715, p. 66) and consequently of the "use" of the underground for geological purposes was also supported by Luigi Ferdinando Marsili, who had studied several mining exploitations in the Danube region during the years 1693–1694 and later made some significant observations on the Swiss Alps in 1705 (Vaccari, 2000, p. 166–172; Vaccari, 2003a; Vallisneri, 1991, p. 296: letter to Marsili, dated 20 February 1705). After this fieldwork, Marsili attempted to subdivide mountains into three basic morphological units: "ruined" mountains (made of debris), massive rock mountains, and stratified mountains. This subdivision—never published by Marsili—was purely structural and still did not place its three units in an historical sequence of formation (Gortani, 1930, p. 259–262; Vaccari, 2003a, p. 182–183). In later manuscripts, however, Marsili introduced a clear lithostratigraphic triple classification of (1) "primary mountains," (2) "alluvial, earthquake, bituminous and volcanic hills," and (3) "earthy plains" (Vai, Chapter 7, this volume). It must be recalled that Marsili generally agreed with his contemporaries about the Biblical concept of "primitive" mountains, although he also supported Vallisneri's attempt to reduce the geological role of the Deluge within the history of Earth.

At this stage we can clearly recognize that, between the late seventeenth and the early eighteenth century, the concept of "primitive" had been suggested by the common acceptance of the Biblical chronology. Within this framework, the "primitive" mountains were exclusively the relief created at the beginning of the world. Consequently, interpretations could not diverge about the origin of these mountains, but only about their later status. According to Monti and other Woodwardian diluvianists, they had been dissolved by the Deluge, while according to Marsili or Vallisneri, they were still in existence. In spite of these different theoretical positions, during the first thirty years of the eighteenth century, Italian scholars gradually established the basis of a new method of "classification" of terrestrial landform bodies. They extended observations in the field on the complex structure of mountains (*la struttura de'monti*: Vallisneri, 1715, p. 29) and this trend was determined by an association of different causes, which contributed to the focusing of many studies on the morphological and lithological distinction of strata. Within this context, the debate on the origin of springs, which took place in Italy particularly between 1715 and 1730, was very important for the development of the detailed study of the rock-strata (Vaccari, 2003b).

EARLY "CLASSIFICATIONS" OF MOUNTAINS

Around the middle of the eighteenth century, Anton Lazzaro Moro (1687–1764) and Giovanni Targioni Tozzetti (1712–1783) attempted in different ways to classify the mountain forms within two units. In the book *De'Crostacei* (Moro, 1740), Anton Lazzaro Moro, a clergyman from Friuli in the northeast of Italy strongly attacked the theories of Burnet and Woodward, as had Vallisneri (Baldini et al., 1988). He developed an original "classification" of mountains into "primary" and "secondary" divisions. The "primary" mountains (*monti primarj*), pushed up from the bottom of an ancient sea by underground heat, like submarine volcanoes, were composed of massive stone, not stratified; they were considered the highest mountains in the Alps and their shape was usually pictured as very sharp (Fig. 1). The "secondary" mountains were formed of strata deposited on the terrestrial surface by volcanic eruptions of the former "primary" mountains during different ages: this new type of relief could include petrified shells and other marine remains from the top of the "primary" mountains that had been mainly formed by parts of the bottom of the ancient sea (Moro, 1740, p. 271–272, 321–326).

Moro's classification of mountains has been regarded as the first to establish the use of the terms "primary" and "secondary" (Adams, 1938, p. 368). However, this is only partially true, because this binary terminology had already been used by Thomas Burnet, one of the targets of Moro's criticism, who had identified a category of *montes primarii*, formed by fractures

Figure 1. Primary mountains according to A.L. Moro (1740).

that shaped Earth's surface at the time of the Deluge and made of massive stone, but also a most recent group of *monticulos secundarios*, formed by debris of the *primarii* after earthquakes or floods (Burnet, 1681, p. 94–95). The original element, instead, was to link Moro's "classification" to the role of a central fire, considered as the main orogenetic agent that had produced all the mountains at different stages of the history of Earth and could still be the cause of the present formation of new relief. Moreover, the "primary" mountains had appeared on the already formed surface of Earth and had not been created by God or built by the catastrophic event of the Deluge. Moro had only explored the mountains and the hills of his region; nevertheless, his theory was particularly inspired by the reports on the birth of a new small volcanic island in 1707, near Santorini in the Aegean Sea (Moro, 1740, p. 214–230).

The content of the *De' Crostacei* by Moro, much more than Vallisneri's *De Corpi Marini*, caused a strong reaction from the Italian diluvianists, such as Giuseppe Costantini (1692–1772), who published a 500-page-long treatise in order to reaffirm "the truth of the Universal Deluge" (Costantini, 1747). On the other hand, Moro's habit of looking at mountains and classifying them in an historical sequence was well accepted by probably the top follower of Steno in Tuscany, the Florentine naturalist Giovanni Targioni Tozzetti (Arrigoni, 1987).

Between 1751 and 1754, Targioni Tozzetti published six volumes of *Relazioni* (reports) on his travels through Tuscany, combining historical, economical, and scientific information, but paying particular attention to intensive regional geological research (Targioni Tozzetti, 1751–1754). In this work and especially in the outline of a planned (but then never published) major work on the "physical topography" of Tuscany (Targioni Tozzetti, 1754, p. 11–42), he attempted to classify the Tuscan mountains into two units: the *monti primari* or *monti primitivi*, "primitive" mountains formed by the most ancient thick strata called *filoni*, which were irregular, winding and composed of schistous rocks with mineral veins; and the *colline*, hills of more recent formation, made of regular horizontal strata of tuff, clays, and sandstones and containing fossils.

However, he realized that this general subdivision could only be related to the present surface of Earth, and that perhaps what were called "primary" mountains may instead have been "secondary" or "tertiary," having been formed from the debris of former ancient mountains. This careful statement on the relative age of the "primary" mountains was not due to uncertain observations or to the adoption of a new theoretical framework. According to Targioni Tozzetti, the data collected in the field permitted him to reconstruct only some stages of the long history of Earth's surface, and those data were mainly lithological and geomorphological (rocks and strata), while the use of fossils was rather restricted.

About the origin of mountains, Targioni Tozzetti, who was a diligent observer of volcanic phenomena (Targioni Tozzetti, 1779), did not accept the "ultra-volcanic" theory of Moro and gave more importance to a supposed action of currents within the primeval sea and to erosion due to meteoric water. Targioni Tozzetti was also well aware that only a very small part of all the geological changes that had occurred on Earth's surface could be traced in the field, even through long and accurate observations. This rigorously empirical approach cannot be explained with the influence of the Stenonian heritage alone, as Targioni Tozzetti was not only an eclectic naturalist and a polymath, but also a mining expert (Vaccari, 1996; Vaccari, 2000, p. 173–175).

THE MAKING OF ARDUINO'S LITHOSTRATIGRAPHICAL "CLASSIFICATION"

Mining experience usually provided an invaluable knowledge of strata, mineral veins, and rock formations, directly observed below Earth's surface: the eighteenth-century Italian figure who best represented this interaction between science and mining technology was Giovanni Arduino (1714–1795). Born in Caprino, near Verona, he came from a strictly technical mining background and had explored—in a sort of ideal link with Marsili, Vallisneri, and Targioni Tozzetti—almost the same geographical area of the Apennines and Tuscany, which then led to the fundamental geological excursions in the Venetian region between 1757 and 1769, particularly in the provinces of Padova, Vicenza, and the eastern part of Verona (Vaccari, 1993).

In the first half of the 1750s, while Targioni Tozzetti was elaborating his "classification" of mountains and was finishing the last volumes of his reports of travels in Tuscany, the then unknown Veronese mining expert Arduino visited for the first time the area of Montieri in the hills nearby Siena, in order to evaluate for a private mining company from Livorno the possible presence of mineral veins for exploitation (Vaccari, 1996). Having acquired a good technical education in mining and metallurgy, but otherwise being self-taught in mineralogy and chemistry, Arduino eventually held the position of land surveyor at Vicenza before being nominated agricultural superintendent of the Republic of Venice in 1769, where he served until the end of his life (Vaccari, 1992). Contrary to Targioni Tozzetti and Moro, and also to Vallisneri, Arduino soon abandoned the theoretic studies initiated at Verona to begin at the age of 18 a profitable apprenticeship as a technician in the mines of Tyrol. His training was completed in Tuscany and the Modenese Apennines after gaining eight years experience in mining activity in Vicenza. This constituted an indispensable basis for his scientific development.

Arduino acquired a remarkable amount of practical and technical knowledge from his mining experience, including a specialized lexicon rich in borrowed German terms, but above all he gained the capacity to observe and study the structure of the mountains in order to obtain a precise idea of their possible mineral richness. Thus, in the depths of the Vicentine, Tuscan, and Modenese mines, the Venetian mining expert began to collect those indispensable data that would later allow him to evaluate the different origins of rocks and minerals. The mining activity and technical apprenticeship clearly formed the basis of

Arduino's geological studies in the 1760s and above all of the lithological "classification" of four units (*ordini*), outlined for the first time in two letters sent to the Paduan naturalist Antonio Vallisneri Jr. (1708–1777) and published in Venice (Arduino, 1760).

In a little less than one hundred pages, based on information gathered during his excursions in the Vicentine area between 1758 and 1759, Arduino treated many arguments, such as the quality of the mineral water of Recoaro, the formation of stalactites in the caves of Costozza, and the origin of fossil coal (*carboni fossili*), but he presented especially a new general stratigraphic subdivision—not only of mountains—based on essentially lithologic characteristics (Arduino, 1760, p. clviii-clxix.). It dealt with a general "classification" of four orders (*ordini*) of rocks, which included mountains defined as "primary" (*primari*), "secondary" (*secondari*), and "tertiary" (*terziari*), as well as the terrain of the alluvial plane, which was considered as belonging to a "fourth" (*quarto*) unit. Clearly better articulated in lithological terms than the subdivisions proposed by Moro and by Targioni Tozzetti, the classification by Arduino was based on about twenty years of observations of the Venetian Prealps, the Tyrolian and Lombard relief, on part of the Modenese Apennines, as well as the metalliferous hills of Tuscany. It was the results of a very refined lithologic analysis that attempted to describe as precisely as possible the characteristic rocks of each individual order. In addition, a "primeval" rock type (*roccia primigenia*) was identified at the base of all the relief examined and therefore was considered by Arduino as being of older origin with respect to the others.

The *Due Lettere* of 1760 also contained some significant references to Arduino's Tuscan experience, citing on several occasions the *Relazioni* by Targioni Tozzetti. In particular, among the references to the exploration of the Tuscan territory he emphasized the Apuane Alps observed above Pietrasanta and Serravezza, studied also by Antonio Vallisneri in 1704 (Perrucchini, 1722–1726), which

> *hanno la loro base di quella pietra fissile talco quarzosa [...] da me creduta primigenia, e ne sono in molti luoghi composte sino a circa la metà della loro altezza.* (Arduino, 1760, p. cxlv)

have at their base that fissile, talc quartzose rock [...] which I believe to be primitive, and it is to be found in many places even up to half of their height.

Arduino had therefore already identified in the Apuane Alps a basal element of his lithostratigraphic theory, that is to say, the rock presently called crystalline schists. Additionally, he inserted as a principal element of his "tertiary order" (*terzo ordine*) "the hills of tuff and clay of Tuscany" (*le colline di tuffo e d'argilla della Toscana*), which had been defined as "secondary" by Targioni Tozzetti, "which he had judiciously observed and described" (Arduino, 1760, p. clxv). With regard to this point, it is interesting to note the explicit use on the part of Arduino of the term *secondarie* for indicating the *colline* (hills) described by Targioni Tozzetti in his *Relazioni* of some travels in Tuscany: this could represent a forced interpretation with regard to the caution expressed by the Florentine scientist (who had not explicitly used the term "secondary" for the hills), but on the other hand, Arduino's attitude highlighted an attempt to make a comparison with a theory—such as that of Targioni Tozzetti—and not at opposing it.

The lithostratigraphic classification elaborated by Arduino was much more complex and better articulated than all other analogous attempts that had previously been expressed within the Italian and European scientific milieu. First of all, the four "orders," also considered as four large strata (*quattro grandissimi strati*), uniformly superimposed and internally composed of many minor strata (*strati minori*), were regarded as successively formed in different times and circumstances. It is appropriate to quote the crucial passage of the Second Letter, which effectively synthesized the results of Arduino's fieldwork:

> *Per quanto ho potuto osservare, [...] la serie di questi strati, che compongono la corteccia visibile della terra, mi pare distinta in quattro ordini generali, e successivi, senza considerarvi il mare. Essi quattro ordini si possono concepire essere quattro grandissimi strati (come sono effettivamente) che in qualunque luogo, dove sono scoperti, si veggono l'uno sopra l'altro collocati, in modo costantemente uniforme. Quantunque ogn'uno d'essi grandi strati risulti dall'unione d'innumerabili altri strati minori, composti di materiali di molti generi, specie, e varietà, contuttociò considerati in complesso tutti i componimenti d'ogni uno di detti ordini, e strati principali, e confrontato un'ordine coll'altro, vi si vede tale diversità di natura, e d'accidenti, che da' chiaramente a conoscere d'essere stati formati, non solo in tempi, ma anche in circostanze assai diverse.* (Arduino, 1760, p. clviii)

From what I have been able to observe, the series of these strata which make up the visible crust of the Earth, seem to me to be distinguished in four general orders, and to be successive, without considering the sea. These four orders may be imagined as four large strata (as in effect they are) which in all places where they are exposed, may be seen positioned one above the other, in a constantly uniform way. Although each of these large strata are a union of numerous other minor strata, composed of many types, species and varieties of material, nevertheless considering all the components as a whole of each of the said orders, and principal strata, and comparing one order with another one can see the diversity of nature, and of events which clearly make known that they were formed, not only in different times but also in quite diverse circumstances.

With this assertion, the Venetian scientist illustrated that he understood the complexity of the geological agents that had modified Earth's surface in different periods. Later, he confirmed that the times of formation of the rocks and mountains were not rigidly four, as to the number of his lithological units, but many more, recognizable within each "order" according to the various effects on Earth's crust (Arduino, 1774, p. 256).

Arduino's "classification" scheme considered as "primeval" (*primigenie*), that is the oldest, the already mentioned "talc quartzose rocks" (*pietre talco-quarzose*), known by the eighteenth century German miners as *schieffer-stein* (slates) or *hornstein* (horn-rock) and in Tuscany were called *lardaro*, *lavagna*, or *sasso morto* ("dead stone"). According to Arduino, these rocks were found at the base of all the mountains examined by him, for example Mount Spitz at the northern end of the Agno Valley, and may be described as follows:

La sua base visibile [del monte Spitz] costa di quella specie di pietra scissile, o fissile, sfogliata come il Talco, di colore lustro metallico, o di squame di Pesci; tutta tramezzata ed intrecciata di vene di quella selce bianca conosciuta oggigiornio col nome di Quarzo, e talvolta tutta di Quarzo come imbevuta ed impastata. Questa pietra, di cui se ne veggono moltissime varietà, è di natura vetrificabile; partecipe di tenuissime sostanza di ferro, della cui ruggine trovasi spesso infetta: e talvolta contiene, o dentro la sua sostanza, o tra i suoi letti, e fenditure, dei Cristalli Granati, ed altre simili pietre, e vene di metalli, e di minerali d'ogni specie. (Arduino, 1760, p. ciii-civ)

The visible base [of Mount Spitz] is formed of that type of scissile or fissile rock, layered like talc, of a lustrous metallic color or of fish scales; all traversed and crossed by veins of that white flint known today as quartz and sometimes entirely drenched and mixed with quartz. This rock, of which there are many varieties, is by nature vitrifiable; composed of thin iron substances which are often affected by rust: and sometime contains, or within its substance, or between its beds, and fractures, some crystals of garnets, and other similar stones, and metal veins, and every type of minerals.

Later, in his writings published in the 1770s and in particular in the *Saggio Fisico-Mineralogico di Lythogonia e Orognosia* (Physical-Mineralogical Essay of Lithogony and Orognosy: Arduino, 1774, p. 243, 260–281), Arduino assigned the name *schisto* (schist) to this "primitive" rock, now known as a metamorphic crystalline rock composed essentially of quartz and mica. Arduino also observed it under the Apuane Alps (crystalline schist), but above all at the base of the Alps and Venetian Prealps, where today this rock is called micaceous-quartziferous phyllites (*fillade micaceo-quarzifera* in Italian), which normally form the superior horizon of the "Azoic, immediately below the oldest fossiliferous terranes" (Artini, 1969, p. 686–691).

The "first order" of the lithostratigraphic classification comprised therefore the "primitive or primary mountains" (*monti primitivi o primari*), also called the "mineral mountains" (*monti minerali*) as they contained various metalliferous deposits. According to Arduino, these mountains occurred in layers above the "primary" rocks and were formed of sandstones and conglomerates ("a mixture of pebbles, sand, and dust of the mentioned primeval rocks" [*un impasto di ciottoli, arene, e polvere di dette pietre primigenie*]), as well as intrusive rocks such as granite. The complete lack of fossils within these rocks distinguished the stratified rocks of the first unit from those of the second, which formed the secondary mountains, composed of

strati sopra strati di marmi, e di pietre di natura calcaria, la maggior parte partecipi di reliquie di marini viventi (Arduino, 1760, p. clxii)

layer upon layer of marble and of stones of a calcareous nature, the greater part including relics of living marine beings,

but absolutely without minerals. The rocks of the "second order" can therefore be retained as being prevalently fossiliferous limestone formed during the great marine sedimentation of the Mesozoic (Arduino, 1760, p. clxi-clxiii).

In the "third order" Arduino placed the "tertiary" mountains and hills in as far as

la maggior parte de' loro materiali sono guscj, frammenti, ed arene di marini testacei: e frantumi, ciottoli, arene, e polvere nate dal disfacimento di grandi porzioni de' monti Primarj, e Secondarj (Arduino, 1760, p. clxv)

the greater part of their material is shells, fragments, and sands of testaceous marine animals: and fragments, pebbles, sands and dust originated from the decay of large portions of the primary and secondary mountains,

that is to say limestones, sandstones, clays and conglomerates with recent fossils. The remarkable eruptive events that occurred in the Venetian Prealps during the Oligocene were also perceived by the Veronese scientist, although they had not yet been fully recognized and described:

colli interi di dette glebe e tufi vetrificabili, tutti buccherati e spugnosi di vari colori, che sembrano pomici, e lave, d'antichi Vulcani (Arduino, 1760, p. clxvii)

the entire hills of those vitrifiable clumps and tuffs, all perforated and sponge-like of various colors, which seem like pumice and lava of ancient volcanoes

(and which also contained basalt, examined in the Euganean Hills); Arduino included them in the third "order," together with the above cited "hills of tuff and clay of Tuscany," described by Targioni Tozzetti, as well as with the layers of lignite called *carbon fossile* (Arduino, 1760, p. clxii–clxiii, clxv–clxix). The volcanic phenomena were considered by Arduino as both orogenetic and lithogenetic agents and consequently inserted, although with a certain caution, within the category of "tertiary" mountains.

This was a different approach than that adopted by Targioni Tozzetti, who had instead excluded these phenomena from his "classification" of terrestrial relief after having studied them in detail.

The "fourth order," outlined by Arduino, as

comprensivo di tutte le pianure, che sono anch'esse formate a strati sopra strati, per alluvioni, e deposizioni di materiali, condotti giù da' monti dall'acque de' fiumi (Arduino, 1760, p. clxix)

inclusive of all the plains, which are also formed by layer upon layer, by floods, and deposition of material brought down from the mountains by the waters of the rivers

can be finally identified in the alluvial deposits and in the dismantlement due to erosion during the Neozoic or Quaternary era, although he never used the word *quaternario*.

Although the *Due Lettere* presented the results of extensive research in the field in different geographical areas, the decisive turning point toward a broader reflection on the classification of rocks and mountains can clearly be linked to a trip that Arduino undertook at the end of October 1758 in the Agno Valley in the upper Vicentine area. There, the geological and geomorphological characteristics observed in the ~20-km journey up the valley (explored in different diversions to the mountains on both sides) definitively convinced Arduino to elaborate on a general theory that also capitalized on what he had observed up to that point in various Alpine, pre-Alpine, and Apennine regions. Among the manuscripts relative to his geological explorations in the Venetian area (Vaccari, 1994, p. 304–307, 331–334), a beautiful geological sketch of the Agno Valley stands out (Fig. 2). This was drawn after the trip in 1758 and is presently housed in the Civic Library of Verona in Italy (Fondo "G. Arduino," busta 760, IV.c.11). Why this sketch was never published is not known; perhaps printing it was too expensive or Arduino intended to prepare a better version for a second edition of the *Due Lettere*. This second edition was never realized.

The profile represented in this sketch is very important for the history of geology as it clearly represents the visual support of Arduino's lithostratigraphic classification, even if the captions did not refer directly to the four "orders" described two years later, but only to the "primitive" rock (which may be seen on the lower left corner of the sketch, labeled with a capital "A," including three mineral veins marked by lower case "a"). Arduino represented all the rock types and mountains he had observed on the two flanks of the valley in this cross section. The criteria for differentiating the rocks forming the various strata identified in the section, distinguished into 15 units with letters ranging from A to R, was based on their external characteristics (form, color, etc.), on their eventual reaction with fire—which confirmed that Arduino subjected his mineral and rock samples to some chemical "experiments" (*esperienze*) in the laboratory, and on the presence or absence of "shells" (*conchiglie*) or "marine bodies" (*corpi marini*).

Concerning the latter aspect, it should be noted that Arduino did not include in the *Due Lettere* of 1760 a true and proper study of paleontological characteristics to accompany the eventual chronological reconstruction of the history of Earth. His theory utilized the presence of "remains of marine inhabitants" (*spoglie de' marini abitatori*) solely for marking the difference between the "primary" and "secondary" rocks. Apart from this, not even the typological differences were specified, as, for example, between the fossils contained in the strata of the second or third "orders."

The refinement of the lithological subdivision was not repeated in paleontological terms, as the words used for defining the fossils often remained more or less quite general (*produzioni del mare*—marine products). Certainly, in the case of Arduino, one cannot speak of what is today referred to as biostratigraphy even if, in comparison to Moro and Targioni Tozzetti, the Venetian geologist had highlighted an increased interest in the relationship between fossils and the strata or the rocks that contained them, noting in particular a progressive similarity of the fossils to the still-living species, as one gradually passed from older to more recent strata. This important intuition was clearly expressed in an unpublished letter sent to Antonio Vallisneri Jr. on the 7th of July 1760, in which Arduino particularly emphasized

i differenti gradi di perfezione nelle dette specie d'acquatici Animali impietriti, più roze ed imperfette ne' più bassi strati delle montagne da me distinte per secondarie [...] e più perfette negli strati superiori di mano in mano, secondo l'ordine della loro successiva formazione, tanto che negli ultimi strati, cioè in quelli che formano i monti e colli terziarj, si veggono esse specie perfettissime, ed in tutto simili a quelle che nel moderno mare si ravvisano. (unpublished letter kept in the Civic Library of Verona, Fondo "G. Arduino," busta 757, I.b.1)

the various grades of perfection of the said species of petrified aquatic animals, rougher and more imperfect in the lowermost strata of the mountains that I have distinguished as secondary [...] and gradually more perfect in the upper strata, according to the order of their subsequent formation, so that in the last strata, those which form the tertiary mountains or hills, there are perfect species, and all are similar to those recognized in the modern seas.

In fact, in the same letter Arduino specifies that he had noted in the deeper strata (therefore, among the oldest ones containing fossils) the presence of "innumerable immense rough products, as many embryos or attempts" (*innumerabili immense abbozzate Produzioni, come tanti embrioni o tentativi*). This concept was continued in another unpublished work, the incomplete *Risposta Allegorico Romanzesca* (Allegorical Fictional Reply) on the genesis of Earth's surface, started at the end of 1771 and directed to the Swedish scientist Johann Jakob Ferber (1743–1790). Arduino stated in this *Risposta* that the strata of the "secondary" mountains:

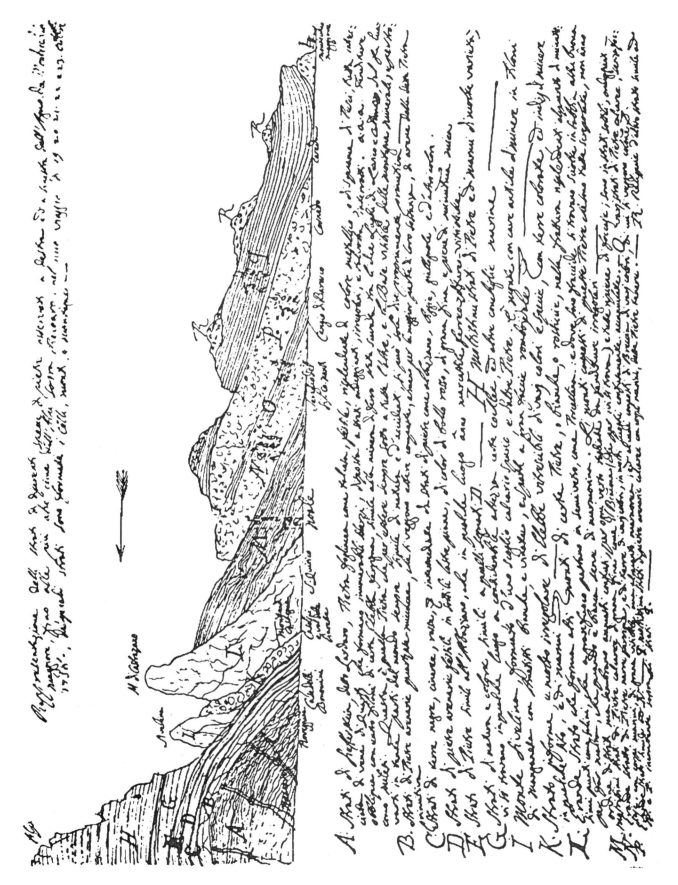

Figure 2. Cross section of the Agno Valley, near Vicenza, drawn by G. Arduino in 1758 (Civic Library of Verona, "Fondo G. Arduino," bs. 760, IV.c.11).

sono per la maggior parte ripieni di Marine Conchiglie, le quali variano di specie quasi di strato in strato; contando dalle più basse visibili radici, fino alle più alte sommità delle Montagne, che de' medesimi sono composte. Le specie più basse sogliono essere quelle ignote, dette perciò da' Litologi Conche anomie; ma da certe altezze in sù vannosi sempre maggiormente alle forme di quelle moderne approssimando. (Vaccari, 1991, p. 202)

are mostly full of marine shells, which vary in species almost from stratum to stratum; counting from the lowermost visible roots up to the highest summits of the mountains, which are composed of the same. The lowermost species used to be the unknown ones, called for that reason by the lithologists anonymous shells; but from a certain point upwards they show much more similarity to the modern forms.

The extent and the quality of the fieldwork carried out by Arduino in the Agno Valley emerges also through a comparison (Fig. 3) of Arduino's sketch with a modern geological profile of the section taken between the mountains of Civillina and Castrazzano (Barbieri et al., 1980), which corresponds to the contents of the box drawn on the left of Arduino's sketch. It should be noted that by *Monte di Castrazano*, Arduino almost certainly indicated the present-day Mount Scandolara, higher than the mountain today known as Castrazzano and slightly lower than Civillina. In Figure 3A, the enclosing frame and the demarcation lines in bold of the four "orders" as defined by Arduino (numbered 1–4) have been added to the 1758 section by the present author for clarity, recognized on the basis of an analysis of the manuscript captions compared with the text of the *Due Lettere*. On the modern geological profile (Fig. 3B), the boundaries in bold between the rocks classified today as Paleozoic (1), Mesozoic or Secondary (2), and Tertiary (3) have also been drawn in by the present author.

Both profiles represent in the lower left-hand corner of the figures the crystalline basement of Civillina and its mineral veins and its diverse and complex sandstone strata. The upper part is made up of compact limestone; next are the almost vertical strata of vitrifiable clumps (*glebbe vitrescibili*), which today are considered to be eruptive rocks (such as rhyolites, dacites, and *latititi*) that separate Civillina from Scandolara Mountain (or Castrazano according to Arduino); next, the large calcareous stratum L, which curves toward the top to form the peak of Castrazano/Scandolara; then the series (M) of many thin, more inclined calcareous strata (called *Biancone* by Arduino and today indicated by the same name), which are found on the right flank of Scandolara; and finally, the beginning—on the right-hand side of the box—of the N-O-P strata of "black iron rocks" (*pietre nere ferrigne*), or rather, basalts and other volcaniclastic rocks. The subdivisions between the first three lithostratigraphic "orders" by Arduino and the first three modern geological periods correspond substantially.

In transferring all that he had seen and interpreted in the field onto paper, Arduino demonstrated without doubt a great capacity for geological observation, sharpened by a mining eye and carried out with extreme rigor. Nevertheless, the sketch of 1758 does not yet allow us to understand if, for example, Arduino had even then recognized the igneous origin of the rocks called *vetrescibili* (including the "primitive" schist) and *ferrigne* (iron-like), such as those basalts indicated in the top (R; Figs. 2 and 3) of minor relief toward the end of the valley.

With regard to the classification of the rocks, Arduino subdivided the rocks into three large classes, already outlined by Carl von Linné (Linnaeus: 1707–1778) in the *Systema Naturae* (1735), called *vetrescibili, calcarie*, and *apyre*, a division based on the different reactions of rocks to fire in as far as the first rock type become vitrified, the second calcined, and the third neither burns nor melts. Certainly, Arduino kept closely to the taxonomic terminology of Linnaeus, although the "classification" outlined by the Veronese scientist did not develop toward a systematic definition (above all in species and varieties) of all the rocks observed. However, he quite soon applied that "classification" to the original observations on the position, the age, and the origin of the rocks. In later works, the main subdivision between vitrifiable and calcareous rocks would in fact be based on their different origin: the first as a result of fire due to their igneous character; the second of sedimentary type due to the action of water (Arduino, 1774, p. 242–260).

FIRE AND WATER: ARDUINO'S THEORY ON THE "EPOCHS" OF EARTH AND THE BUILDING OF THE MOUNTAINS

The question of the origin of the rocks and mountains inserted in the lithostratigraphic "classification" had not yet been fully confronted by Arduino in the 1760s, but would be in the following decade when it would intertwine with the problem of geological time or, to use an Arduino expression, with "the great turn of the centuries" (*grande giro de' secoli*), during which the geological deformations of Earth's crust occurred.

At the start of the 1770s, Arduino tried to establish, without ever having completed or published it, a general subdivision of the history of Earth into four "epochs" (*epoche*). He intended to correlate these epochs with the orogenetic and lithological processes first outlined in his "classification" of 1760, later described in more detail in the *Saggio Fisico-Mineralogico* in 1774. It should, however, be pointed out that this quite broad chronological scheme was never really quantified by Arduino in terms of years or centuries and remains confined to the manuscript pages of the *Risposta Allegorico Romanzesca* to Ferber, which was never completed nor published by its author (Vaccari, 1991, p. 189–211).

In the first "epoch" of this lithogenetic theory, Arduino had identified as the first phenomenon the formation of the "primitive" schistose rocks, which consolidate after the cooling of Earth's surface from its original state of fusion (*antica sua vetrosa viscida fluidità*). Subsequently, on Earth's crust, already fractured and devastated by sudden cooling due to the large quantity of

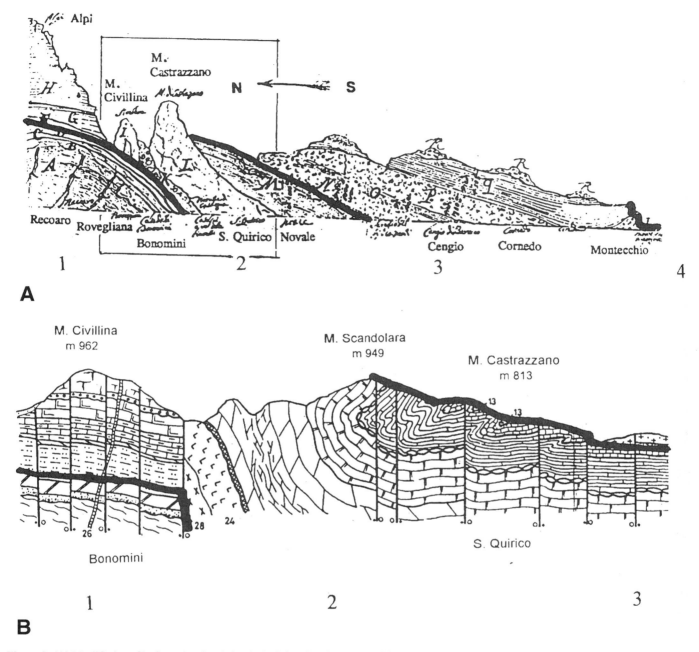

Figure 3. (A) Modified profile from the sketch by G. Arduino (1758). (B) Modified profile from G. Barbieri et al. (1980).

rainfall, volcanic activity determined the emersion and expulsion of new rock material, which produced the oldest group of "primary" mountains:

> que' tanti sterminati Massi, e Monti di ogni sorta di Porfidi, di Serpentini, di Graniti, di Diaspri, di Gabbri e di consimili pietre di natura vitrea, che or sopra, or a lati, or nell'interno delle primigenie montagne, senza ordine simetrico, collocati si osservano (Vaccari, 1991, p. 195).

Many immense blocks and mountains of every type of prophyry, serpentine, granite, jasper, gabbro and similar rocks of a vitreous nature could be seen positioned without any symmetrical order above, laterally or within the primitive mountains.

The superficial part of these mountains, being less compact, was detached and broken up following the combined erosive action of water and wind, which formed a second group of "primary" mountains composed of quartz, spar (*spati*), and other types of rocks.

In the second "epoch," the slow sedimentary process, which occurred in the water that had gradually covered the planet, caused the formation of the "secondary" relief composed of "marbles" (*marmi*) and "calcinable rocks" (*pietre calcinabili*), which were then partially modified in their morphological and lithological structure by the resumption of the volcanic activity. The "secondary" mountains were therefore defined as "secondary works of Neptune" (*opere secondarie di Nettuno*) in relation to that part of the "primitive" mountains formed also by the action of water: in general, with respect to the "primitive" mountains, the "secondary" were considered to be

> *d'ordinario molto meno di quelle colorate, e per lo più biancheggiano; i loro strati sono molto più regolari, più paralleli, e senza confronto, in lungo ed in largo, più estesi, e tra di loro omogenei.* (Vaccari, 1991, p. 202)

ordinarily much less than the colored ones and mainly whitened; their strata are much more regular, more parallel and without comparison, in length and width of greater extent and more homogeneous.

Unfortunately, the interruption of the manuscript right at the start of the third "epoch" does not allow complete verification of the correspondence between the great chronological scenario outlined by Arduino in the *Risposta Allegorico Romanzesca* and the lithostratigraphic "classification" of 1760. The detailed work carried out by the Veronese scientist from the end of the 1760s was without doubt confined to the first two units: "primary" mountains produced by fire and modified by wind and water; and "secondary" mountains produced by water and modified by fire. It is not clear if this restriction was due to a personal choice by Arduino or if he was forced by external factors, such as his enormous workload as agricultural superintendent of the Venetian Republic beginning in the early 1770s.

However, the contents of the incomplete *Risposta Allegorico Romanzesca* clearly found a non-casual parallel in the structure of the *Saggio Fisico-Mineralogico di Lythogonia e Orognosia* (Arduino, 1774). In this work, Arduino confirmed and reinforced the main points of his lithostratigraphic "classification" proposed 14 years earlier, limiting himself to concentrating on a lithological subdivision within the "primary" mountains, which was already implicit in the first "epoch" outlined in the *Risposta Allegorico Romanzesca*. At the beginning of the *Saggio*, the Veronese scientist had rather efficiently summarized the true tripartite distinction of the reliefs, with the addition of the fourth "order" reserved for the plain:

> *Furono da me compresi nell'ordine de' primitivi quei monti, che per essere formati di quelle fossili materie, ch'essere sogliono gli ordinarj originali ricettacoli delle miniere metalliche, vengono da' pratici col nome di Monti Minerali distinti da quelli di altre qualità: ed ho considerate secondarie le montagne di marmi, e pietre calcarie a strati, nelle quali tali miniere sono sommamente rare, e forse sempre avveniticcie, e accidentali; ma che di pietrificate organiche produzioni del mare comunemente sono ripiene. Al terzo ordine io ho riferiti quei bassi monti, e colli, che si ravvisano essere composti di Ghiaje, di arene, e di terre limose, argillacee, marnose ec.; materie quasi sempre con immensa copia di marine quisquilie mescolate.* (Arduino, 1774, p. 229) [With the word *fossili* (fossils) Arduino still indicated every stony material that may be found underground]

I included in the primitive order those mountains, which are formed by those fossil materials, which used to be the ordinary original receptacle of the metal mines, and for this reason are called the Mineral Mountains by the practical men [miners] in order to distinguish them from those of another quality; and I have considered as secondary the mountains of marble and stratified calcareous rocks, in which those mines are extremely rare, and probably always occasional and accidental; but they are commonly full of petrified organic marine products. I referred those low mountains and hills to the third order which are seen to be composed of gravel, sandstone, muddy, clayey and marly soils etc.; material with always a large abundance of mixed marine minutia.

Therefore, Arduino had updated this general scheme using the recent studies on the

> *materie vulcaniche, da me osservate nelle nostre montane situazioni, delle quali, e degli effetti di antichissimj sotterranei incendi ho poi parlato più precisamente in altre Memorie, che ho posteriormente scritte.* (Arduino, 1774, p. 230)

volcanic material that I have observed in our mountains, which together with the effects of ancient subterranean fires I have treated more precisely in other memoirs that I have subsequently written.

The studies by Arduino on ancient volcanism in the Venetian area were in fact concentrated in the second half of the 1760s, particularly in some articles that were published in the Venetian journal *Giornale d'Italia* (Vaccari, 1993, p. 203–217). But in reality, contrary to a widespread coeval interpretation of Arduino's "classification" (Ferber, 1776a, p. 28–27), which thus influenced a great majority of the historiography of the nineteenth and twentieth centuries (Vaccari, 1993, p. 5–11), the volcanic mountains (*montagne vulcaniche*) were never indicated by Arduino either in this work or elsewhere as the fifth unit (or "order") of his lithostratigraphic distinction. Instead, what emerges right from the beginning of the *Saggio* is once again the awareness of a chronological scale relative to the dating of the strata and mountain formations, which in turn originated from "effective causes" (*cause effettrici*) and "circumstances" (*circostanze*), which were also very varied among themselves:

> *Siccome però anche le pietrose, e terree materie di ciascuna di dette quattro divisioni sono tra sè molto differenti, e che considerando i modi, e luoghi delle rispettive loro situazioni, conosconsi non essere coetanee, ma effetti successivi, prodotti in tempi varj, e col concorso di circostanze diverse, dalle rispettive loro cause effettrici; così parlando dei monti primitivi, considerai primigenj quelli de' loro materiali, che pel sito ch'essi occupano, e per altri rispetti sembrati mi sono di origine più antica: non già però in riguardo alle costituenti sostanze di ciascuno; ma unicamente alle modificazioni, sotto le quali ora esistono.* (Arduino, 1774, p. 230)

As however also the rocky and terrigenous material of each of the four said divisions are very different from each other, and considering the ways and respective places where they occur, knowing that they are not contemporaneous, but successive results, produced in different times and with the contribution of diverse circumstances by their respective effective causes; thus speaking of primitive mountains I considered as primeval those of their material, for the position that they occupy, and for the others aspects, they seemed to me to be of an older origin: not however with respect to the constituent materials of each one; but only with regard to the modifications under which they now exist.

It may be remembered, in fact, that in the manuscript of the *Risposta*, Arduino had outlined various lithogenetic and orogenetic scenarios (linked to the isolated or combined action of fire, water, or wind) and he had distinguished them into two different levels of intensity: catastrophic during volcanic eruptions and also in the fast process of consolidation with fractures of Earth's crust, following the violent downpour of rain on the fluid incandescent surface; and slower and uniform during the sedimentary deposition that occurred within the waters of the ancient sea. Besides, Arduino was convinced that

> *che le parti visibili di questo Globo non sieno primordiali, rispetto alla presente loro forma, e struttura; ma che dipendano da varie rivoluzioni e metamorfosi dallo stesso sofferte.* (Arduino, 1774, p. 230)

the visible parts of this Globe were not primordial with respect to their present form and structure, but that they depended on the various revolutions and metamorphosis that the same [Earth] had undergone.

Therefore, the terrestrial orography could no longer have been exactly the original formed on the surface of Earth after the cooling of the vitreous, fluid, and incandescent material. In support of this interpretation Arduino quoted some passages from Linneaus' *Systema Naturae* (from the 12th edition: Linné, 1766–1768, vol. 3):

> *anche il celeberrimo Linneo, che la stratificata struttura della Terra, giudica essere lavoro dell'Oceano, cioè prodotta con successivi suoi sedimenti, dopo averne indicati i modi, conclude: "ideo et Rupes saxeas altissimas aevi veros filios esse dum omnia obmutuere, ipsi loquantur lapides." Poscia egli replica: "saxa non primaeva, sed temporis filios esse, abunde evincunt strata montium."* (Arduino, 1774, p. 233–234).

also the most celebrated Linnaeus, who judges the stratified structure of the Earth to be the work of the Ocean, that is its sediments were produced successively, after having indicated the ways, concludes: "*ideo et Rupes saxeas altissimas aevi veros filios esse dum omnia obmutuere, ipsi loquantur lapides.*" Later he replicates: "*saxa non primaeva, sed temporis filios esse, abunde evincunt strata montium.*"

Moreover, Arduino also recalled other authors, whose "ingenious" (*ingegnose*) or "absurd" (*assurde*) hypotheses on the modifications that had shaped Earth's crust had in common "the belief that the modern form of the fossil Kingdom [in the sense of mineral kingdom] was not "primitive" (*il credere non primitiva l'odierna forma del Regno fossile*: Arduino, 1774, p. 233–234).

Thus, among those who had theorized a "reformation of Earth on the ruins of the primeval" (*riformazione della Terra sopra le ruine della primeva*), Arduino mentioned Thomas Burnet and Francesco Patrizi. Other scholars, such as Gottfried Wilhelm Leibniz, Buffon, Nicolaus Steno, William Whiston, Benoît de Maillet, Henri Gautier, Johann Gottlob Lehmann, Mikhail Lomonosov, Agostino Scilla, Fabio Colonna, Antonio Vallisneri, Anton Lazzaro Moro, Giovanni Targioni Tozzetti—and other less well-known Italians, such as Giuseppe Baldassarri, Giovan Battista Passeri, Jacopo Odoardi, Giovan Giacomo Spada—were instead included among the authors who had conjectured that the changes that had affected Earth's surface were due to several different causes: "the long presence of water over the entire face of the Globe" (*lungo soggiorno dell'Acqua sopra l'intera faccia del Globo*); the gradual lowering of Earth's crust; the "so-called universal flood" (*detto universale inondamento*); the "igneous fusion" (*all'ignea fusione*); and the action of volcanoes (Arduino, 1774, p. 231).

According to Arduino, the writings of all the scholars included in this second group showed that

> *le idee di primitiva fusione, di sotterranei posteriori incendj, di vulcaniche eruttazioni, di sollevamenti, di*

avallazioni, rovine, e subissamenti sono state da molti Geologi combinate in diversi modi con quelle di universale, e di parziali innondazioni. (Arduino, 1774, p. 232)

the ideas of primitive fusion, of later subterranean fires, of volcanic eruptions, of uplifts, of depressions, ruins and collapses, were combined by many Geologists in different ways with the ideas of universal and partial floods.

This attempt to combine the action of water and fire in periods of diverse duration and by events of diverse intensity, within a reconstruction of the history of Earth based on the observations in the field, was a fundamental outline of the research method adopted by Arduino. In light of this persuasion, one easily understands the coldness with which Arduino presented the third and last group of authors, as they strictly adhered to the Woodwardian theory of the terrestrial dissolution caused by the waters of the Deluge: Johann Jakob Scheuchzer, Giuseppe Monti, Louis Bourguet, and Tommaso Gabrini. Nevertheless, even if Arduino remained distanced from and indifferent to the universal Deluge, he considered the Biblical catastrophe an event totally unrelated to geological investigations. Arduino still considered Woodwardian theories of the early eighteenth century only because the great universal flood and dissolution had been followed by a reconstruction and, in some cases, a gradual successive modification of the orographic structure of Earth, after the action of erosive agents.

On the other hand, Arduino simply ignored the Kircherian "theory of the Earth," which was centered on the absolute, unchangeable structure of Earth in a chronological scheme rigidly anchored to the Holy Scriptures, neither citing nor criticizing it in the *Saggio Fisico-Mineralogico*. Arduino did not seek confrontations or polemics with conceptions that he considered to be on a completely different level from that of a true research methodology. A rigorous attitude free from compromises appears evident here, undoubtedly supported by an impressive amount of data collected in the field and by the practical mentality of Arduino. Nevertheless, the most interesting and original element of the *Saggio* lay not in the brief examination of the coeval geological literature, but in the internal subdivision of the first "order" of the lithostratigraphic classification.

According to Arduino, it was fire that molded the first group of "primary" mountains composed of older rocks (*pietre e terre vitrescenti*) considered as *primitive*: among these he numbered various rocks that were not "of a calcareous nature" (*di indole calcaria*), such as

gli Schisti cornei, o sia talcoso-quarzosi, e li semplicemente micacei, e gli argillosi; li Graniti, e Granitelli, e Porfidi, e Serpentini, Diaspri rupestri, Selci cornee, e quarzose, e certi Basalti, e Trapp, e Pietre argillose, steatitiche, e talcose. (Arduino, 1774, p. 243)

the horny or rather talcose-quartzose schists, the simple micaceous and the argillaceous ones; the granites and granitelli, porphyrys, serpentines, rocky jasper, horny and quartzose flintstones, and certain basalts, and trapp, and argillaceous, steatite and talcose rocks.

According to Arduino, these early "primary" mountains, without any regular symmetry of the strata and totally lacking in fossils, displayed in fact signs of

di sofferta fusione, di tumultuarj bollimenti, di accidentali miscugli, e di sconcerti accaduti in tempi varj, ed a quelli de' moderni Vulcani non poco somiglianti. (Arduino, 1774, p. 244)

having undergone fusion, of turbulent boiling, of accidental mixtures and of confused occurrences over various times, quite similar to those of modern volcanoes.

These mountains were composed of irregularly inclined "veins" of massive rock (*filoni*), sometimes almost perpendicular to the soil, and usually containing mineral veins.

The force of the fire had not modeled only the oldest "primary" mountains, but had subsequently interacted with the water in the formation of the "primary" mountains of the "second subdivision" (*seconda suddivisione*), considered therefore by Arduino to be

di mista formazione, risultata da Vulcanici, e pelagici effetti; dove disgiunti; dove in varie guise alternati; dove senza ordine bizzarramente confusi. Alcune parti di essi monti sono composte di pietre, e terre vitrescenti, in modi analoghi a quelli delle sopraddette prime montagne, e senza forma stratosa. Altre sono fabbricate a strati sopra strati tra se paralleli; ora di un solo genere di pietre o di terre variamente indurite, ora di più generi. Le pietre arenareo-micacee, la calcarie, le argillacee, le marnose, gli Schisti argillosi, limosi, e di altre simili qualità, e le terre variamente colorate, e di diversa natura, e consistenza, sono, complessivamente parlando, le sostanze, delle quali essi strati sono formati. (Arduino, 1774, p. 245).

of mixed formation resulting from volcanic and pelagic effects; somewhere disjointed, somewhere alternated in various ways; somewhere strangely confused and without order. Some parts of these mountains are formed of stones and vitrescent soil, in a similar way to those of the above-mentioned first mountains, and they are of unstratified form. Others are made of strata upon strata, parallel to one another; sometimes made of only one type of rock or of variously hardened soil, some other times made of many types. The arenaceous-micaceous rocks, the limestones, the shales, the marls, the clayey schists, silty rocks and others of similar qualities, the soils of various colors and of diverse nature and consistency are, speaking overall, the substances of which these strata are formed.

It is interesting to note how, with respect to the remarks contained in the *Risposta Allegorico Romanzesca*, which seemed to postulate a rigid subdivision of diverse orogenetic causes for the two categories of "primary" mountains, here instead the role of volcanic activity is better clarified within the processes of the formation of the second subgroup of primitive mountains. These mountains, placed at the boundary between the "primary" and "secondary" "orders," were considered by Arduino to be

diverse da quelle calcarie a strati, quasi sempre regolari, e molto uniformi, e spessissimo ripieni di pietrificati marini Testacei ecc., da me nell'ordine secondario collocate; e più ancora da quelle del terzo ordine. (Arduino, 1774, p. 247)

different from those calcareous strata, nearly always regular and very uniform and often full of petrified marine testaceans etc., which I placed in the secondary order; and these [primary mountains] are even more different from those of the third order.

But the common element to the two internal "subdivisions" of the "primary" mountains was undoubtedly the presence, in both, of mineral or "metalliferous" veins: this was why, according to Arduino, the "primary" mountains could indifferently be called "metallic or mineral mounts or mountains" (*monti, o montagne minerali, o metalliche*).

The mountains of the second "order," formed mostly of "marbles and calcareous rocks in strata one upon the other," rested therefore against the flanks of the primary mountains, or rather, they formed a complete ring around them, covering their "roots." Although these "secondary" mountains "could be recognized as true children of the ancient ocean, born after the above-mentioned ones [primary mountains]" (*s'abbiano a riconoscere per vere figlie dell'antico Oceano, nate posteriormente alle anzidette* [*montagne primarie*]), the regular order of their horizontal and parallel strata was sometimes found to be interrupted by

certe Lave, ed altre bruciate materie, eruttate da antichissimi spenti vulcani, tra le squarciature, e crateri, che tratto tratto vi si scuoprono, e le Breccie, le rotture, e sconvolgimenti dalli medesimi causati. (Arduino, 1774, p. 237)

certain lava and other burned material, erupted by ancient extinguished volcanoes, between the cracks and craters which step by step were uncovered, and the openings, the breakages and cataclysms caused by the same [volcanoes].

The volcanic rocks therefore represented a clear element of interruption (lithological and morphological) already present in the "secondary" mountains, but also in the "tertiary," considered as stratified low mounts or hills, which also were

opere del Mare molto meno antiche, costrutte con stratificate deposizioni d'innumerabili eterogenee materie [*contenenti*] *materie Vulcaniche* [...] *in certi luoghi abbondantissime a segno di costituire dei tratti non piccioli di bruciati Poggi.* (Arduino, 1774, p. 238)

the products of a much less older sea, constructed of stratified deposits of innumerable heterogeneous material, [containing] volcanic material […] extremely abundant in certain places where they form a series of burnt hillocks.

The lithological composition of the "tertiary" mountains included, as had already been outlined in the *Due Lettere* of 1760, strata of "calcined sand, vitrescent sandstone, gravel, shale, clay, marl, pebbles rounded by long transport in water" (*di sabbie calcinabilj, di arene vitrescenti, di ghiaje, di argille, di crete, di marne, di ciottoli da lunghi movimenti delle acque ritondati*) together with a large quantity of fossils (Arduino, 1774, p. 237).

Finally the plains, classified within the fourth "order," were

formate dalle alluvioni delle acque, discendenti da tanti secoli giù dalle parti montuose, colle materie sassose e terree dalle medesime condotte, triturate, e sparse; ed in parte anche dal Mare colle sabbie, e materie limose, ecc.; da esso di continuo respinte ai lidi, e dentro le salse paludi, e sopra le terre basse, che frequentemente inonda, depositate. (Arduino, 1774, p. 238)

formed by the floods of the water, descending for many centuries from the mountain flanks, with stony and terrigenous material brought down, ground and spread by the same waters; and in part also by the sea, with sandy and muddy material etc.; deposited by its continuous crashing on the shore and within the saline lagoons and above the low land, which it frequently invaded.

In summary, Arduino's position with regard to the different "physical causes" (*cause fisiche*) of the formation of his four lithostratigraphic "orders" (and therefore of the three classes of mountains) emerged quite clearly in the *Saggio Fisico-Mineralogico* (1774), a text that certainly represented the more mature expression of the geological thinking of the Veronese scientist and that developed, although without directly using the chronological subdivision in "epochs," some significant intuitions included in the *Risposta Allegorico Romanzesca* (Table 1).

Fire alone was responsible for the formation of the oldest "primary" mountains of the "first subdivision," as well as for the "primeval" rocks placed at the base of all the visible rock formations, while water was the only geological agent that determined the deposition of the alluvial and coastal terrains in the planes and marine coasts, respectively, composing the fourth "order," which was of more recent formation with respect to the others. Between these two lithogenetic stages (respectively, oldest and latest) of Arduino's "classification" were placed the "second subdivision" of "primary" mountains (produced by fire and modified by wind and water), the "secondary" mountains (produced

TABLE 1. ARDUINO'S LITHOSTRATIGRAPHICAL THEORY (1760–1775)

Units (*Ordini*)	Mountain type	Rock type	Causes
1	Basement/primeval rock (*roccia primigenia*)	Crystalline schist *schisto*	Fire Cooling of the original Earth surface
1	Primary or mineral mountains (*monti primari o minerali*) a. First subdivision b. Second subdivision	Granite, porphyry, and mineral-bearing crystalline rocks (*rocce vetrescibili*); sandstone and conglomerates without fossils	Fire, wind, and water a. Volcanism b. Volcanism and erosion due to wind and water
2	Secondary mountains (*monti secondari*)	Marbles and stratified limestones with fossils; stratified rocks like *vetrescibili* but without mineral veins	Water and fire Marine sedimentation and modifications due to the reprise of volcanism
3	Tertiary mountains (*monti terziari, colline*)	Gravel, clay, fossiliferous sand, volcanic material	Fire and water Volcanism and sedimentation within sea waters
4	Plains (*pianure*)	Alluvial deposits, sometimes stratified	Water Erosion caused by rain and rivers

by water and partially modified by fire), and the "tertiary mountains" (of analogous origin to the "secondary," but sometimes also only volcanic). Therefore, according to Arduino, water and fire had acted alternately on all these orographic units, which had formed successively within a complex chronological and orogenetic scheme. This had been only partially reconstructed in the *Risposta Allegorico Romanzesca*.

Arduino was quite aware of the difficulties and the obstacles to completing a detailed and comprehensive interpretation of the history of Earth. Consequently, his caution is not surprising, as he intended to avoid the risk of a possible rigid interpretation that could link too strictly the number of his general lithostratigraphic subdivisions to the same number of times in which mountain-building occurred.

Non vorrei però si credesse, che avendo distinti li monti in soli tre ordini, possa essere persuaso che siensi anche in tre soli tempi formati. Egli è troppo chiaro a vedersi, che tante debbon essere le Etadi, quanti sono gli effetti diversi, che nelle montagne gli uni agli altri soprapposti si succedono. Nelle lunghissime, ed altissime nostre Alpi calcarie, per esempio, composte a strati sopra strati, dalle loro più basse visibili radici, fino alle più eccelse cime, è forza concepire tanti periodi successivi di tempo della ripetitiva loro formazione, quanti sono in numero essi strati, tutti patentemente composti di calcinosi sedimenti delle Acque marine le quali in contrassegni autentici delle sterminate loro fabbriche, vi hanno dentro impastate Conchiglie in copia immensa, e di moltissime specie, che per lo più sono diverse da strato a strato. (Arduino, 1774, p. 256)

I do not wish that it is believed that having distinguished the mountains in only three orders that one be convinced that they were also formed in only three times. It is too easy to see that the periods had to be many, as the effects are diverse, which occur in the mountains one after the other. In our very long and high calcareous Alps, for example, composed of strata upon strata, from their lowest visible roots, up to the highest peaks, it is inevitable to conceive of many periods successive in time due to the repetition of their formation, correspondent to how many strata there are in number, all obviously composed of calcined marine sediments that mark the end of their formation, full of shells in immense quantities, and of many species that for the most part are diverse from one strata to another.

The "primary" or mineral mountains, as already mentioned, were composed, according to Arduino, of large "veins" of massive rocks (*filoni*), irregularly inclined or sometimes almost perpendicular to the ground and containing mineral veins. It dealt with an interpretation that was quite widespread in the second half of the eighteenth century, especially among Italian, German, and Swedish scientists who were familiar with the mines. Among these scholars it was generally accepted that the "primary" mountains could be distinguished from the "secondary" in as far as the first were made of massive rocks with mineral veins and the second of strata with fossils.

FROM MINING EXPERTISE TO GEOLOGICAL CHRONOLOGY: ARDUINO AND LEHMANN

With regard to the general subdivision outlined above, it is interesting to recall the analogy, proposed by Arduino in the *Saggio Fisico-Mineralogico*, between his oldest "primary" mountains of completely igneous origin and the *Gang-Gebürgen* (or mountains of veins) considered by the German mineralogist and chemist Johann Gottlob Lehmann (1719–1767) to be the oldest reliefs at the base of his "classification" of mountains (Arduino, 1774, p. 244). In the *Versuch einer Geschichte von Flötz-Gebürgen* (Essay on an history of the stratified mountains: Lehmann,

1756), published four years earlier than the *Due Lettere* by Arduino, Lehmann had presented a subdivision of the mountains in three "classes," based on the different times and causes of their building, as well as on their different lithological features, emphasizing a strong "diluvianist" position that determined the strict adhesion to an absolute time scale (Adams, 1938, p. 374–378).

The aim of this book was to discuss in detail the terrestrial mountains made of strata (*Flötz-Gebürgen*) in order to achieve a better knowledge of the internal structure of Earth and its mineral resources. On the one hand, Lehmann intended to provide a new working tool for better future exploitations of the Prussian mines, which were often found within a stratified terrain; on the other hand, he was prompted to outline a general "theory of the mountains" by the contemporary debate on the history of Earth. For this reason, Lehmann's work was based on a combination of two main elements: first, the results of extended observations in the field, due to a wide practical knowledge of Earth's surface and its underground, which he had gained during several years of intense travels for supervising mining activities in Saxony, Thuringia, and Silesia (in central-eastern Germany) in order to establish a *Geographia Subterranea* (subterranean geography) on a regional basis; and second, the analysis of some theoretical models (*systemen*) on the changes undergone by Earth's surface (*veränderungen*), with particular attention to the building of mountains and strata, such as the theories exposed by Woodward, Burnet, Whiston, Moro, and Élie Bertrand (1752).

Lehmann's "classification" of mountains was introduced by a discussion of some contemporary theories of Earth and orogenetic models: he firmly rejected the idea that all the mountains had been formed by the Deluge (as put forth by Burnet and Whiston) or entirely dissolved and redeposited with a new shape by the same catastrophic event (as put forth by Woodward). Likewise, Lehmann also refused the "ultra-vulcanistic" orogenetic theory of Anton Lazzaro Moro (which he probably read in the German edition: Moro, 1751), although Lehmann accepted the activity of the so-called "subterraneous fires" (Lehmann, 1753).

Like Arduino, Lehmann had a solid, practical mining background (von Freyberg, 1955), but unlike him, he had fully adopted the Biblical chronology. According to the German scientist, the "primary" mountains had been created directly by God and were overlain by the "secondary" mountains, formed by the Deluge, which were eventually overlain by the post-Deluge "tertiary" mountains. The first class (*erste classe*) included the primitive veined mountains (*Gang-Gebürgen*), which had existed since Creation: they were the most elevated and were composed of inclined or vertical irregular masses of granite and other crystalline rocks, lacking in fossils, but rich in mineral veins. Then, the Deluge washed away the fertile ground from the top of the primitive mountains and gradually deposited on their flanks several strata of the secondary mountains during the downflow of the waters over Earth's surface. This second class (*zweyte classe*) of relief included stratified mountains composed of regular horizontal fossiliferous strata (*Flötz-Gebürgen*). Finally, the minor reliefs of the third class (*dritte classe*), mainly stratified, were formed in different times after the Deluge and following the accumulation of the debris from the older mountains by local phenomena, such as erosions due to rain and wind, landslides, earthquakes, floods, and volcanic eruptions.

On the contrary, Arduino's classification positioned its four orders chronologically, according to their relative construction, not with reference to fixed stages such as the Creation and the Deluge, as put forth by Lehmann. Arduino considered the "primary" mountains to be a result of a formation, not of a creation, even if the present terrestrial orography could not have been exactly the original that formed on the surface of Earth after the cooling of the fluid and incandescent vitreous material. If Lehmann considered the action of the Deluge as fundamental for the formation of the "secondary" mountains, Arduino instead simply ignored the Biblical catastrophe, which he considered a supernatural event separate from geological investigations. Nevertheless, despite a strongly "diluvianist" view, the observations carried out on the strata in the eastern sector of the Harz mountains in Thuringia (Ilfeld and Mansfeld) allowed Lehmann to draw a lithostratigraphic cross section that identified thirty types of stratified secondary and tertiary deposits (*Flötz-Gebürge*) overlying each other on the flank of a primitive mountain (*Gang-Gebürge*), pictured at the extreme left in Figure 4. This profile also showed a great graphic analogy with the unpublished representation of the strata observed in the Agno Valley, drawn by Giovanni Arduino in 1758, which describes sixteen different lithostratigraphical units from the base of "primary" rock (on the left in Fig. 2) to the plain. It is also worth noting that, in spite of Lehmann's detailed lithostratigraphic description that identified a greater number of secondary strata, the interpretation of their genesis remained exclusively based on the action of the waters during the Deluge.

Arduino was not aware of Lehmann's *Versuch* while drawing this sketch (Fig. 2) and preparing his *Due Lettere* of 1760, as the work of the German scientist is not quoted at all. Instead, a detailed synopsis of the French edition of the *Versuch*, translated into French by D'Holbach (Lehmann, 1759), is found among Arduino's manuscripts and reveals a later careful reading of the work, presumably during the preparation of the *Saggio Fisico Mineralogico*, before 1774 (manuscript kept in the Civic Library of Verona, Fondo Arduino, bs. 758, II.f.16, II.f.16bis).

The two profiles by Arduino and Lehmann can also be regarded as a real visual synthesis of their classifications into four and three lithological units, respectively. And if it is proved that Lehmann did not influence Arduino, then it may be suggested that this kind of visual language was already in common use among German and Italian mining experts, who had been in contact at least since the sixteenth century, in particular in the Venetian region and in Tuscany (Vergani, 1979; Cipriani and Tanelli, 1983, p. 260). In his section, Arduino had recognized a series of rock formations that are now identified from the upper Palaeozoic to the Tertiary, while Lehmann worked on Carboniferous and Permian rocks, probably up to the Lower Triassic (Ellenberger, 1994, p. 252–253). They both adopted a chemical approach to the study of rocks and minerals, but also used mining

Figure 4. Cross section of the eastern sector of the Harz mountains in Thuringia, published by J.G. Lehmann (1756).

terms (the only terms then available) for defining the different rock units; for example, *zechstein*, was used by Lehmann and is now adopted for a regional European chronostratigraphic unit corresponding to the Upper Permian. It must be also considered that the region studied by Lehmann generally has very little inclined or dislocated geological formations, thus suggesting research and interpretations on regular depositions of long series of superimposed strata; Arduino, on the other hand, worked on a series of very different terrains within a short distance, including volcanic intrusions through the sedimentary strata.

In his *Saggio Fisico-Mineralogico*, Arduino proposed a structural analogy between his oldest "primary" mountains and the primitive *Gang-Gebürgen* of Lehmann, both at the base of all the other formations and with steep-lying mineral or metalliferous veins. But for Lehmann, as for other eighteenth century German scholars, water was the principal and almost absolute orogenetic agent, while for Arduino it was the force of fire that modeled the oldest primary mountains and interacted with water in the formation of the second subdivision of the primary mountains, as well as in the case of the secondary and tertiary mountains.

TERMINOLOGICAL QUESTIONS AND THE RECEPTION OF ARDUINO'S THEORY

Arduino did not find in the contemporary literature a "classification" that could be considered similar to his own, apart from the above-mentioned point of contact with Lehmann regarding the primitive "mineral mountains." The lithostratigraphic succession proposed by Linnaeus in the *Systema Naturae*, for example, was criticized because it regarded water as responsible for the formation of all the mineral kingdom; moreover, Linnaeus had placed sandstone (*cote*) in the deepest position, overlain by the schist, by marble with fossils, again by schist, and by other rocks. According to Arduino, that scheme was unacceptable because

> *le Coti, o Pietre arenarie (Lapis Molaris Agricolae), da me non si videro mai occupare, nei monti, il sito più profondo; ma le ho sempre osservate soprastratificate ad altri diversi generi di pietre. Nel Trentino, e nei monti minerali Vicentini, ed in alcuni altri Luoghi, gli strati delle medesime sopraggiacciono immediatamente allo Schisto quarzoso-micaceo.* (Arduino, 1774, p. 257, note 2)

the *Coti* or arenaceous rocks (*Lapis Molaris Agricolae*), for me would never occupy the deepest position in the mountains; I always observed them overlying other diverse types of rocks. In Trentino and in the Vicentine mineral mountains and in some other places the strata of those [arenaceous rocks] immediately overlay the micaceous-quartzose schist.

In reality, the eighteenth-century question of a geological and in particular lithological nomenclature still in the making, which was far from being standardized, made the comparison between the "classifications" of Arduino and Linnaeus rather difficult. In fact, Arduino recognized that

> *lo Schisto, di cui parla il signor Linneo, numerandone tredici specie, è da quel genere di pietre Schistose, che ho considerate tra le primigenie, molto diverso.* (Arduino, 1774, 257, note 2)

the schist mentioned by Linnaeus, of which he numbers thirteen species, is very different from that type of schistose rock which I have considered primitive.

The Veronese scientist was well aware of the problem that with the same term a completely different rock could be indicated. For this reason, he dedicated the second part of his *Saggio Fisico-Mineralogico* (Arduino, 1774, p. 260–281) to the definition of "primitive schist," the talcose-micaceous-quartziferous rock observed at the base of all the reliefs studied therefore being the oldest: it was a compact rock, absolutely lacking in fossil remains, which originated in the remote time of the "primitive fusion."

Arduino retained a cautious position regarding the presumed absolute antiquity of this "primeval" rock (*pietra primigenia*), although he did not express himself in the same way as Targioni Tozzetti, according to whom the "primitive" mountains could better be considered "secondary" or "tertiary," being possible residues of much older mountains that could no longer be found on the surface of Earth. On the contrary, the "primeval" mountains and rocks were for Arduino the first to really form on Earth's surface, destined to subsequently undergo continuous structural changes (building of new mountains and hills; geomorphological remodeling) and lithological modifications (formation of new eruptive, sedimentary, or alluvial rocks) that could in part be observed in the present, according to an attitude that would today be defined as actualistic.

The lithostratigraphic classification by Arduino and his geological works generally circulated well in Europe in the second half of the eighteenth century. The *Raccolta di Memorie Chimico-Mineralogiche* (Arduino, 1775), a miscellaneous collection of Arduino's works previously published in the Venetian journal *Giornale d'Italia*, together with writings sent to Arduino by various Italian and foreign correspondents, was translated into German (Arduino, 1778; Vaccari, 1993, p. 285–306). Among the works collected in the *Raccolta*, there were the *Saggio Fisico-Mineralogico*, the first of the *Due Lettere* to Vallisneri Jr., and a letter on extinct volcanism in the Venetian territory. Moreover, the account of Arduino's "classification" of mountains provided by Johann Jakob Ferber in his book on a scientific tour in Italy should also be recalled; it was published in German in 1773 and translated into English and French three years later (Ferber, 1773, 1776a, 1776b).

Within the varied community of Italian *orittologi* (geologists) of the second half of the eighteenth century, Arduino's classification model received diverse responses, favorable among many Venetian naturalists but criticized by other scientists working in the central and western Alps. For example, at the end of the century, some Piedmontese and Lombard scholars rejected the supposed position of the schist below the granite as a sign of an older age, supporting instead the granite as the only real "primeval" rock. But these and other discussions, which would stimulate a lively and productive debate in the course of the last decades of the century, certainly did not tarnish the recognized originality, based on results from fieldwork, of a true geologist and a mature scientist.

CONCLUSIONS

Arduino's classification of mountains is widely regarded by historians of geology as the basis for the modern chronological subdivisions of Earth's geological history. Some geologists, like Berry (1968, p. 64–65, 107–108) and Albritton (1980, p. 118), recognized the importance of Arduino's term "Tertiary," which takes its place in the modern standard stratigraphic column. Arduino did not use the formal term "Quaternary" to refer to the alluvial deposits of the "fourth order," which remains the least treated of the four lithostratigraphical-chronostratigraphical units of his system. Moreover, Arduino's reflections on the alternate action of fire and water during the long geological history of Earth's surface opened up the possibility of an actualist "third way" between the catastrophist and uniformitarian hypotheses. This went beyond the rigid antithesis between "Neptunism" and "Plutonism," which strongly influenced the study of earth science around the end of the eighteenth and the beginning of the nineteenth century.

REFERENCES CITED

Adams, F.D., 1938, The birth and development of the geological sciences: Baltimore and London, Williams & Wilkins, 506 p.

Albritton, C.C., Jr., 1980, The abyss of time: Changing conceptions of the Earth's antiquity after the sixteenth century: San Francisco, Freeman Cooper & Co., 251 p.

Arduino, G., 1760, Due lettere [...] sopra varie sue osservazioni naturali: Al Chiaris. Sig. Cavalier Antonio Vallisnieri professore di Storia Naturale nell'Università di Padova: Lettera Prima [...] Sopra varie sue Osservazioni Naturali (Vicenza, 30 gennaio 1759): Lettera Seconda [...] Sopra varie sue Osservazioni fatte in diverse parti del Territorio di Vicenza, ed altrove, appartenenti alla Teoria Terrestre, ed alla Mineralogia (Vicenza, 30 marzo 1759): Nuova Raccolta di Opuscoli Scientifici e Filologici (Venezia), v. 6, p. xcix–clxxx.

Arduino, G., 1774, Saggio Fisico-Mineralogico di Lythogonia e Orognosia: Atti dell'Accademia delle Scienze di Siena detta de' Fisiocritici (Siena), v. 5, p. 228–300 [second edition, with some corrections and additions, *in* Giornale d'Italia (Venezia), v. 11, 1775, p. 171–217].

Arduino, G., 1775, Raccolta di Memorie Chimico-Mineralogiche, Metallurgiche, e Orittografiche del Signor Giovanni Arduino, e di alcuni suoi Amici: Tratte dal Giornale d'Italia: Venezia, B. Milocco, 237 p.

Arduino, G., 1778, Sammlung einiger mineralogisch-chymisch-metallurgisch und oryktographischer Abhandlungen, des Herr Johann Arduino, und einiger Freunde desselben: Aus dem italienischen über setzt, durch A.C.v.F.C.S.B.C.R. [August Constant von Ferber]: Dresden, In der Waltherischen Hofbuchhandlung, 364 p.

Arpiani, R., 1999, Fossili, rovine del tempo: Note sulla geologia di Antonio Vallisneri senior, *in* Bernardi, W., and Manzini, P., eds., Il cerchio della vita: Firenze, Olschki, p. 165–183.

Arrigoni, T., 1987, Uno scienziato nella Toscana del Settecento: Giovanni Targioni Tozzetti: Firenze, Gonnelli, 173 p.

Artini, E., 1969, Le rocce: Concetti e nozioni di petrografia: Milano, Hoepli, 767 p.

Baldini, M., Basile, B., Conti, L., Lipparini, T., Mezzacasa, E., Piccoli, G., Piutti, R., Preto, P., and Rossi, P., 1988, Anton Lazzaro Moro (1687–1987), Atti del Convegno di Studi (San Vito al Tagliamento, 12–13 marzo 1988): San Vito al Tagliamento, Grafiche Editoriali Artistiche Pordenonesi, 156 p.

Barbieri, G., De Vecchi, G., De Zanche, V., Di Lallo, E., Frizzo, P., Mietto, P., and Sedea, R., 1980, Note illustrative della carta geologica dell'area di Recoaro alla scala 1:20.000: Memorie di Scienze Geologiche (Padova), v. 34, p. 23–52.

Bertrand, É., 1752, Mémoires sur la structure intérieure de la Terre: Zurich, Heidegger, 152 p.

Berry, W.B.N., 1968, Growth of a prehistoric time scale based on organic evolution: San Francisco, Freeman, 158 p.

Broc, N., 1969, Les montagnes vues par les géographes et les naturalistes de langue française au dix-huitième siècle: Contribution à l'histoire de la géographie: Paris, Bibliothèque Nationale, 298 p. [second edition 1991: Les montagnes au siècle des Lumières: Perception et représentation: Paris, Éditions du Comité des Travaux historiques et scientifiques, 300 p.].

Burnet, T., 1681, Telluris Theoria Sacra: orbis nostri originem et mutationes generales, quas iam subiit aut olim subiturus est, complectens: Libri duo priores de Diluvio and Paradiso: Londini, G. Kettilby, 306 p.

Cipriani, C., and Tanelli, G., 1983, Risorse minerarie ed industria estrattiva in Toscana: Note storiche ed economiche: Atti e memorie dell'Accademia Toscana di scienze e lettere La Colombaria, sér. v. 48, p. 241–283.

Costantini, G.A., 1747, La verità del Diluvio Universale vindicata da' dubbi e dimostrata nelle sue testimonianze: Venezia, P. Bassaglia, 494 p.

Descartes, R., 1644, Principia Philosophiae: Amstelodami, L. Elzevirium, 310 p.

Ellenberger, F., 1978, Le dilemme des montagnes au XVIIIe siècle: vers une réhabilitation des diluvianistes?: Revue d'Histoire des Sciences, v. 31, p. 43–52.

Ellenberger, F., 1994, Histoire de la Géologie, Tome 2: La grande éclosion et ses prémices 1660–1810: Paris, Lavoisier, 383 p.

Ferber, J.J., 1773, Briefe aus Wälschland über natürliche Merkwürdigkeiten dieses Landes, an den Herausgeber desselben Ignatz von Born: Prag, W. Gerle, 407 p.

Ferber, J.J., 1776a, Lettres sur la minéralogie et sur divers autres objets de l'histoire naturelle de l'Italie: Ouvrage traduit de l'allemand, enrichi de notes and d'observations faits sur les lieux par Mr.le Baron de Dietrich: Strasbourg, Bauer & Treuttel, 507 p.

Ferber, J.J., 1776b, Travels through Italy in the years 1771 and 1772 described in a series of letters to Baron Born, on the natural history particularly the mountains and volcanoes of that country, Translated from the German, with explanatory notes, and a preface on the present state and future improvement of mineralogy, by Rudolf Erich Raspe: London, L. Davies, 377 p.

Gohau, G., 1983, Idées anciennes sur la formation des montagnes: Cahiers d'Histoire et de Philosophie des Sciences, n. sér., v. 7, p. 1–86.

Gohau, G., 2003, Naissance de la géologie historique: La terre des "théories" à l'histoire: Paris, Vuibert, 240 p.

Gortani, M., 1930, Idee precorritrici di Luigi Ferdinando Marsili su la struttura dei monti, in Comitato Marsiliano, ed., Memorie intorno a Luigi Ferdinando Marsili: Bologna, Zanichelli, p. 257–275.

Guntau, M., 1996, The natural history of the earth, in Jardine N., Secord J.A., and Spary E.C., eds., Cultures of natural history: Cambridge, Cambridge University Press, p. 221–229.

Kircher, A., 1664–1665, Mundus subterraneus [...] Liber secundus technicus geocosmus sive de admirando globi terreno opificio, in Mundus subterraneus, XII Libros digestus: Amstelodami, J. Janssonium and E. Weyerstraten, vol. I, 346 p. [photostatic edition: Mundus Subterraneus (1678): Vai, G.B., ed., Bologna, Arnaldo Forni Editore, 2004].

Lehmann, J.G., 1753, De Aere sub terra latente causa vulcanorum.

Lehmann, J.G., 1756, Versuch einer Geschichte von Flötz-Gebürgen, betreffend deren Entstehung, Lage darinne enthaltenen Metallen, Mineralien und Fossilien: Berlin, G.A. Lange, 240 p.

Lehmann, J.G., 1759, Essai d'une histoire naturelle de Couches de la Terre dans la quelle on traite de leur formation, de leur situation, des mineraux, des metaux et des Fossiles qu'elles contiennent: avec des considérations Physiques sur les causes des Tremblemens de terre et de leur propagation, in Traités de Physique, d'Histoire Naturelle, de Mineralogie, et de Metallurgie, ouvrages traduit de l'Allemand: Paris, J.T. Hérissant, vol. 3.

Linné, C., von (Linnaeus), 1735, Systema naturae, sive regna tria naturae systematice proposita per classes, ordines, genera and species: Lugduni Batavorum, T. Haak, 7 ff.

Linné, C., von (Linnaeus), 1766–1768, Systema naturae... editio duodecima reformata: Holmiae, L. Salvii, 3 vols.

Monti, G., 1719, De Monumento Diluviano nuper in Agro Bononiensi detecto: Bononiae, Rossi, 50 p.

Morello, N., 2001, Nel corpo della Terra: Il geocosmo di Athanasius Kircher, in Lo Sardo, E., ed., Athanasius Kircher: Il museo del mondo: Roma, Edizioni De Luca, p. 179–196.

Moro, A.L., 1740, De' crostacei e degli altri marini corpi che si truovano su' monti: Venezia, S. Monti, 452 p.

Moro, A.L., 1751, Neue Untersuchung der Veränderungen des Erdbodens: nach Anleitung der Spuren von Meerthieren und Meergewächsen, die auf Bergen, und in trockener Erde gefunden werden: aus dem Italienischen übersetzet: Leipzig, Breitkopf, 464 p.

Pasini, M., 1981, Thomas Burnet: Una storia del mondo tra ragione, mito e rivelazione: Firenze, La Nuova Italia, 189 p.

Perrucchini, G.B., 1722–1726, Estratto d'alcune Notizie intorno alla Provincia di Garfagnana, cavate dal primo Viaggio Montano del sig. Antonio Vallisnieri, Pubblico Professore Primario dell'Università di Padova: Supplementi al Giornale de' Letterati d'Italia (Venezia), v. 2, 1722, p. 270–312; v. 3, 1726, p. 376–428.

Rappaport, R., 1991, Italy and Europe: The case of Antonio Vallisneri (1661–1730): History of Science, v. 29, p. 73–98.

Roger, J., 1973, La Théorie de la Terre au XVIIe siècle: Revue d'Histoire des Sciences, v. 26, p. 23–48.

Roger, J., 1982, The Cartesian model and its role in the eighteenth-century "Theory of the Earth," in Lennon, T.M, Nicholas, J.M., and Davis, J.W., eds., Problems of Cartesianism: Kingston and Montreal, McGill-Queen's University Press, p. 95–112.

Rudwick, M., 1996, Minerals, strata and fossils, in Jardine, N., Secord, J.A., and Spary, E.C., eds., Cultures of natural history: Cambridge, Cambridge University Press, p. 266–286.

Sarti, C., 1988, I fossili e il Diluvio Universale: Le collezioni settecentesche del Museo di Geologia e Paleontologia dell'Università di Bologna: Bologna, Pitagora, 189 p.

Sarti, C., 1992, Giuseppe Monti and paleontology in eighteenth century Bologna: Nuncius, v. 8, p. 443–455.

Steno, N., 1669, De solido intra solidum naturaliter contento dissertationis prodromus: Florentiae, Ex Typographia sub signo Stellae, 80 p.

Targioni Tozzetti, G., 1751–1754, Relazioni d'alcuni viaggi fatti in diverse parti della Toscana per osservare le produzioni naturali e gli antichi monumenti di essa: Firenze, Nella Stamperia Imperiale, 6 vols. [second enlarged edition in 12 volumes, 1768–1779: Firenze, Cambiagi].

Targioni Tozzetti, G., 1754, Prodromo della Corografia e della Topografia Fisica della Toscana: Firenze, Nella Stamperia Imperiale, 210 p.

Targioni Tozzetti, G., 1779, Dei Monti ignivomi della Toscana e del Vesuvio, in Dei Volcani o monti ignivomi più noti, e distintamente del Vesuvio: Osservazioni fisiche e notizie istoriche di Uomini Insigni di varj tempi, raccolte con diligenza: Livorno, Calderoni e Faina, p. VII–LIX.

Taylor, K.L., 1992, The historical rehabilitation of theories of the Earth: The Compass, v. 69, no. 4, p. 334–345.

Vaccari, E., 1991, Storia della Terra e tempi geologici in uno scritto inedito di Giovanni Arduino: la "Risposta Allegorico-Romanzesca" a Ferber: Nuncius, v. 6, p. 171–211.

Vaccari, E., 1992, L'attività agronomica di Pietro e Giovanni Arduino, in Scienze e tecniche agrarie nel Veneto dell'Ottocento: Venezia, Istituto Veneto di Scienze, Lettere ed Arti, p. 129–167.

Vaccari, E., 1993, Giovanni Arduino (1714–1795): Il contributo di uno scienziato veneto al dibattito settecentesco sulle scienze della Terra: Firenze, Olschki, 408 p.

Vaccari, E., 1994, I manoscritti di uno scienziato veneto del Settecento: notizie storiche e catalogo del fondo "Giovanni Arduino" della Biblioteca Civica di Verona: Atti dell'Istituto Veneto di Scienze, Lettere ed Arti, Classe di Scienze fisiche, matematiche e naturali, v. 151 (1992–1993), p. 271–373.

Vaccari, E., 1996, Cultura scientifico-naturalistica ed esplorazione del territorio: Giovanni Arduino e Giovanni Targioni Tozzetti, in Barsanti, G., Becagli, V., and Pasta R., eds., La politica della scienza: Toscana e Stati Italiani nel tardo Settecento: Firenze, Olschki, p. 243–263.

Vaccari, E., 2000, Mining and knowledge of the Earth in eighteenth-century Italy: Annals of Science, v. 57, no. 2, p. 163–180, doi:

10.1080/000337900296236.

Vaccari, E., 2001, European Views on Terrestrial Chronology from Descartes to the mid-18th century, in Lewis, C.L., and Knell, S.J., eds., The age of the Earth: from 4004 BC to AD 2002: London, Geological Society Special Publication 190, p. 25–37.

Vaccari, E., 2003a, Luigi Ferdinando Marsili geologist: from the Hungarian mines to the Swiss Alps, in Vai, G.B., and Cavazza, W., eds., Four centuries of the word "geology": Ulisse Aldrovandi 1603 in Bologna: Bologna, Minerva Edizioni, p. 179–185.

Vaccari, E., 2003b, Entre hydrologie et géologie: le débat italien sur l'origine des sources au début du XVIIIe siècle, in Dalby, M., ed., Colloque International, OH$_2$ "Origines et Histoire de l'Hydrologie" (Dijon, 9–11 Mai 2001): Dijon, Universitè de Bourgogne, 14 p. [CD-ROM].

Vai, G.B., 2006, this volume, Isostasy in Luigi Ferdinando Marsili's manuscripts, in Vai, G.B., and Caldwell, W.G.E., The Origins of Geology in Italy: Geological Society of America Special Paper 411, doi: 10.1130/2006.2411(07).

Vallisneri, A., 1715, Lezione Accademica intorno all'origine delle Fontane: Venezia, G.G. Ertz, 87 p.

Vallisneri, A., 1721, De' corpi marini che su' monti si trovano, della loro origine, e dello stato del mondo davanti il Diluvio, nel Diluvio e dopo il Diluvio: Lettere critiche: Venezia, D. Lovisa, 254 p.

Vallisneri, A., 1991, Epistolario, vol. I (1679–1710), a cura di D. Generali: Milano, Franco Angeli, 625 p.

Vergani, R., 1979, Lessico minerario e metallurgico dell'Italia Nord-Orientale: Quaderni Storici, v. 14, p. 55–63.

von Freyberg, B., 1955, Johann Gottlob Lehmann (1719–1767): Ein Arzt, Chemiker, Metallurg, Bergmann, Mineraloge und grundlegender Geologe: Erlangen, Verlag Universität Erlangen, 159 p.

Woodward, J., 1695, An essay toward a natural history of the Earth and terrestrial bodies especially minerals as also the seas, rivers and springs, with an account of the universal Deluge and of the effects it had upon the Earth: London, R. Wilkin, 277 p.

MANUSCRIPT ACCEPTED BY THE SOCIETY 17 JANUARY 2006

The geological work of Gregory Watt, his travels with William Maclure in Italy (1801–1802), and Watt's "proto-geological" map of Italy (1804)

Hugh S. Torrens*

Lower Mill Cottage, Furnace Lane, MADELEY, Crewe, CW 3 9 EU, UK

ABSTRACT

In mid-1801, Gregory Watt, son of James Watt, the engineer, set off on European travels in hopes of recovering his health. During the winter of 1801–1802, Watt stayed in Paris and met the Scottish-American geologist William Maclure. Together they set off early in 1802 to travel through war-torn France and Italy, but in Italy they could only venture as far south as Naples, then in a state of anarchy. Here, despite Watt's consumption, they climbed, and descended into, Vesuvius and saw other evidence of recent volcanism. Watt thought this experience would change his mind about geological, especially volcanic, processes. As a result of the trip, he and Maclure saw a lot of European geology together. Watt was then inspired on his return home to make experiments on melting basalt and to study its cooling history and to attempt a "lithological" map of Italy from Calabria to Bologna and the eastern Italian Alps. The first work led Watt to "sit on the fence" over the then much-debated question of the origin of basalt. He believed it could have originated either from the action of heat or from water. Watt's early 1804 map was a brave attempt to delineate up to 46 separate lithologies on a "proto-geological" map of Italy. The lithologies are grouped by color, but do not refer to any stratigraphical classification. The map, therefore, is still at best a proto-geological map. Watt may well have met William Smith, the English pioneer of modern geological cartography, in Bath later in 1804, just before his death, but there is no further evidence. The criteria needed for such maps to be viewed as properly "modern" and "geological" are next considered. Gregory Watt died on 16 October 1804, aged 27 yr. A year before, he became the main critic on matters geological for the *Edinburgh Review* and there published up to nine reviews, mainly on mineralogy. He wrote but a single scientific article, of which he saw only preprints before his death. This article dealt with textural variation in basalt. Watt's legacy of publication is disproportionate to his significance to the history of geology and mineralogy.

Keywords: Italy, Vesuvius, Gregory Watt, William Maclure, William Smith, Thomas Smith, William Thomson, basalt, experimental geology, geological maps, *Edinburgh Review*.

---

*E-mail: h.s.torrens@esci.keele.ac.uk.

Torrens, H.S., 2006, The geological work of Gregory Watt, his travels with William Maclure in Italy (1801–1802), and Watt's "proto-geological" map of Italy (1804), *in* Vai, G.B., and Caldwell, W.G.E., The Origins of Geology in Italy: Geological Society of America Special Paper 411, p. 179–197, doi: 10.1130/2006.2411(11). For permission to copy, contact editing@geosociety.org. ©2006 Geological Society of America. All rights reserved.

INTRODUCTION

Gregory Watt (1777–1804) is a sad figure in the history of science. Born in 1777, he was the much-loved son of one of the most eminent engineers in history, James Watt (1736–1819). His short life also sheds some fascinating new light on the parallel career of William Maclure (1763–1840), long called a "father of American geology," who was his companion on Italian explorations in 1801–1802. Watt died at the age of 27 in October 1804 from consumption (now called tuberculosis) and was only able to see hurriedly printed offprints of his single scientific publication. Despite this, he was destined to become a major figure in the history of both mineralogy and geology.

GREGORY WATT'S EARLY YEARS

Gregory was the only son of James Watt and James's second wife, Anne McGrigor (or MacGregor) (1750–1832) of Glasgow, whom he married in 1776. We get a glimpse of Gregory's early precocity from his contemporary, Mary Anne Galton, later Schimmelpenninck (1778–1856). She was a daughter of Samuel Galton Jr. (1752–1832), a member of the Lunar Society of the English Midlands. She noted how, in ~1790, Gregory

> was a youth of very precocious talents … a boy of talent … but his high estimate of himself made him at this period anything but a pleasant, though often an informing, companion. His sister Jessy [i.e., Janet 1779–1794] he held, as he did all girls, in supreme contempt (Hankin, 1860, p. 286–288).

Late in 1792, Watt's father entered Gregory at Glasgow University, which he attended until 1796 (Robinson and McKie, 1970, p. 188, 225). By October 1794, while still a student, Gregory had joined his father's firm, now under the new style of Boulton, Watt, and Sons (Dickinson and Jenkins, 1927, p. 346). At his graduation in 1796, Gregory

> obtained by far the greatest number of prizes, and degraded the prize readers most inhumanly by reading a short composition of his own, a translation [in which] the verses seemed to me better perhaps than they were in reality. He is a young man of very eminent capacity, and seems to have all the genius of his father, with a great deal of animation and ardour which is all his own (Cockburn, 1852, v. 2, p. 20).

In 1800, Gregory became a full partner in the business his father had founded, now called Boulton, Watt, and Co. (Muirhead, 1859, p. 316; Robinson and McKie, 1970, p. 274).

By 1797, Gregory was already suffering from ill-health, due first to an attack of influenza and then the scourge of the age, pulmonary consumption. So he was sent that winter to Cornwall on the recommendation of the Watt family doctor, Dr. William Withering (1741–1799), who was another member of the Lunar Society (Robinson and McKie, 1970, p. 274, 289). While there, Gregory prepared a 12-page "sketch of the mineralogy of the County of Cornwall drawn up for Mr William Withering, October 1798" (Birmingham Public Library [BPL], 6/11). This was Withering's son, William junior (1775–1832), who shared Gregory's interest in geology. Gregory also played a major part in bringing the talents of Humphry Davy (1778–1829), in Cornwall, to the attention of Thomas Beddoes (1760–1808), in Bristol (Schofield, 1963, p. 376; Davy, 1858, p. 15–32). By 1800, Gregory had become "much less affected than any [other] individuals" by Davy's nitrous oxide experiments (Davy, 1800, p. 536). In the same year, Gregory "received a remarkable set of scatological and anal erotic letters" (Hunt and Jacob, 2001, p. 491). This "naughty, irreverent exchange of letters" was with the engineer William Creighton Jr. (ca. 1779–1831; see BPL, 3/72), with whom Watt again shared a deep and abiding interest in geology and mineralogy, amid all the scatology.

Gregory's interests in mineralogy and geology seem to have initially arisen in great part as a response to his parental upbringing. For example, in 1785, his half-brother, James Watt Jr. (1769–1848), had been encouraged to continue his education in Germany. In 1787, James moved to study at the Bergakademie in Freiberg, Saxony, as student number 297 (Gottschalk, 1867, p. 232; Jones, 1999). Perhaps Gregory's early friendship with the poet Thomas Campbell (1777–1844), a fellow Glasgow University student, had played an unknown part in promoting Gregory's interest in Earth's history. In 1823, Campbell's poem "The Last Man" spoke of how the "ten thousand years hast seen the tide of human tears," a remarkably extensive view of time for this period (Robertson, 1907, p. 233) (Fig. 1).

To indulge his interest in geology, Gregory set out on 1 August 1801 to travel in Europe, at first with a carriage and servant, initially for Germany and then elsewhere on the continent. This travel was also undertaken in hopes of recovering his health. Watt's travels from 25 August to 17 October throughout Switzerland, then under Napoleonic wartime French occupation, have been recorded in detail by De Beer (1957) using Watt's own notebook.

The lawyer John Richardson (1780–1864; see *Oxford Dictionary of National Biography* [ODNB]) was among those whom Watt had met earlier in Göttingen, Germany, in August 1801. Richardson later noted how Gregory had

> proceeded to Switzerland and traversed a considerable portion of it in Company with the Savant Dolomieu. I think he permanently injured his health by over-exertion. I was in Paris November 1801 to January 1802 and while I was there Gregory arrived. He visited all the scientific persons and places accessible to him and there were few that were not so (letter from Richardson to J.P. Muirhead, 20 March 1852, BPL B and W 537/109).

Watt had met the geologist Déodat de Gratet de Dolomieu (1750–1801) at Moudon on 30 September 1801 (De Beer, 1957, p. 133). This was not long after Dolomieu's release, at the end of March

Figure 1. A miniature of Gregory Watt, 1804, by Anne Phyllis, Lady Beechey (1764–1833; see *Oxford Dictionary of National Biography*), posthumously painted according to J.P. Muirhead in 1856 "from no better helps than a black profile and Lady B.'s own acquaintance with, and recollection of, him" (see Birmingham Public Library B and W 537/115). This is reproduced by kind permission of Julian Gibson-Watt, Doldowlod, UK.

1801, from the harshest of prisons, at Messina, in which he had been placed in March 1799 by the king and court of Naples (De Beer, 1960, chapter 7). Dolomieu's letters about his release, and his thanks to Sir Joseph Banks (1749–1820) for helping to secure it, are some of the most moving one can ever read (British Library, Add. Mss. 8099, ff. 51–52 and 99–100). Naples and much of Europe, as we shall soon see from Watt's letters, were then in a state of total anarchy as a result of the Napoleonic Wars. Sadly Dolomieu's health had been permanently damaged in prison, and he died in France in November 1801, so that his meeting with Watt must have been on his last field excursion.

Watt next arrived at Lyons, France, on 19 October and duly reached Paris on 5 November 1801. There he entered into the rich life of scientific activity, which that city then offered. De Beer lists some of those he met there (1957, p. 134). On 6 November, Watt went with Count Rumford (Benjamin Thompson [1753–1814; see ODNB]) to the École Polytechnique. Humphry Davy wrote thus to Gregory's father on 29 December 1801:

I have just seen Count Rumford and I am happy to be able to inform you that he saw Mr Gregory the day before he [Rumford] left Paris and that he [Watt] had then no cough, no pain in his chest and very little debility and complained chiefly of an inordinate appetite for food (BPL B and W 537/55, with the original in James Watt Papers [JWP] 6/33, folder 2).

De Beer only very briefly discussed Watt's onward travels after Paris. Watt left Paris in mid-February 1802 for Italy, traveling via Angoulême, Limoges, Bordeaux, Montpellier, Marseilles, and Toulouse to Genoa. But De Beer nowhere realized (or mentions) that Watt, while he had been in Paris, had acquired a most interesting companion on these later French-Italian travels. This was the recently naturalized American (of Scottish birth) William Maclure.

We can follow some of Watt's adventures, not only from brief notebooks of the sort De Beer used, but also from the many surviving letters Gregory wrote on these travels to his parents and his friends. These are now preserved at BPL. On ~1 February 1802, Watt wrote to his father from Paris:

I have delayed my departure some days in expectation of receiving letters from home—None having arrived—I shall set off tomorrow for Nantes in company with Mr. Maclure who I felicitate myself in having for my permanent companion ... In letters of introduction we are most rich and with the prospect of an opening spring and delightful country, we may without being too sanguine, promise ourselves no small gratification from our journey. We shall [now] go direct to Limoges (BPL C 2/12).

Maclure has been well served by the fine edition of his European journals edited by Doskey (1988). But these surviving journals only start in 1805, so there has long been uncertainty as to how Maclure acquired his interests in geology before this. These interests led to Benjamin Silliman christening him "the father of American Geology" in 1844 (Silliman, 1844). Dennis Dean's fine review of this book (Dean, 1989) rightly questions this paternity but it could still only devote a single paragraph to Maclure's travels and activities before 1805. We can now add considerable detail (Fig. 2).

According to Watt's letter from Angoulême, dated 21 February, to his father:

Mr Maclure and myself quitted Paris on 13 [February 1802]. We proceeded to Orleans, 36 leagues [~110 miles, 180 km] in 14 hours. [They had had to avoid Limoges because the road was so bad and determined to go to Angoulême. On route] the front axle snapped and we were precipitated into a ditch. The immense machine went over very gently and the ground on which we fell being new made, the concussion was not violent and not a soul of the numerous passengers received the most trifling injury.

Figure 2. Portrait of William Maclure (1763–1840) painted in 1797 (from Doskey, 1988, frontispiece).

From here they were now aiming for Bordeaux. This letter described their travels, and the manufactures undertaken in each place seen thus far (BPL C2/12). Between Poitiers and Angoulême, Watt also noted that:

> The rock is certainly calcareous of different species varying from indurated Chalk to hard blue shistose Limestone—some strata with accompanying bands of flints—of all colours and some with cornua ammonis [ammonites] and bivalve shells,

demonstrating his awareness of the geology they had passed through. These were the sort of purely lithological observations Watt was soon to be making more extensively in Italy. Watt here noted the presence of fossils but made no attempt to use, or quantify, the information they still concealed.

On 27 February 1802, the Peace of Amiens was signed between Britain and France. Previously, Watt and Maclure had been in the unusual position of traveling through a country with which Britain, at least, was at war, but, as Déodat Dolomieu and his companion had noted when they had met Gregory a few months earlier in Switzerland, Watt was:

> le fils du célèbre Watt, inventeur de la pompe à feu, qui s'occupe beaucoup de minéralogie. Le gouvernement français, qui croit que les nations peuvent bien se faire la guerre sans que les sciences y prennent part, lui a donné un passeport pour se rendre à Paris (De Beer, 1957, p. 133; 1960, p. 79–80).

These were, if only at this level, more civilized times than today's.

Other letters show the pair passed through Toulouse on 16 March 1802, "where [Gregory] was most happy in my companions and in the perfect recovery of my health," Montpellier on 28 March, Marseilles on 2 April, and that they had reached Genoa by 28 April 1802 (BPL C2/10/12–15 and C2/12/42). Their Italian adventures now began.

ITALY

Sadly, no detailed notebook for this journey seems to have survived, unlike that of the 1801 Swiss journey reproduced by De Beer (1957). Some details are, however, given in Watt's "Memoranda in France and Italy" (BPL C/2/2). This includes a note of 12 May 1802 that he and Maclure then "left Rome for Naples." The next significant letter seems to be that dated 16 May 1802, addressed to his mother after arrival in Naples. This reads:

> My letters for some time have been all addressed to my father or [brother] James [with] information of my good health and progress. We remained only a day and half at Florence and reached Rome in two days. We only remained part of a day there and came directly to Naples where I hoped to find letters and was anxious [to] have news from you. [But] from James I have received no letter, though previous to leaving Paris I requested him to give me some introductions to his Italian friends. Perhaps a letter was lost... In one respect the season favours us for although we suffered considerably from heat at Florence and Rome and upon the road, we are here in a very moderate temperature. I am particularly happy in the change, for I was much debilitated by the warm weather and could not have conveniently made the excursions round the town.

This letter then goes on to make it very clear that their major reason for coming to Naples was to see the volcanic features for which Naples had been long famous:

> We have been this morning to the summit of Vesuvius and I descended to the bottom of the crater. This is attended with no other inconvenience than that of walking among the loose rubble in which you sink nearly to the knees, and of course the ascent which is very steep and rendered extremely fatiguing. We have also been to the Solfatara and have for tomorrow an excursion to Caserta. Italy is at this moment extremely beautiful and well merits the encomiums bestowed upon it.

But all is on the surface, for the people equally merit the universal reprobation they are under (BPL C/2/12).

Naples had long been a separate kingdom, but beginning in 1799 it briefly became a Parthenopian (Neapolitan) republic, as it was called by the occupying French revolutionary forces, until it descended into the revolutionary lawlessness to which Watt here refers (Jenkins and Sloan, 1996, p. 22, 35) and in which Dolomieu had suffered so much. Visitors to and from the kingdom or republic needed separate passports. Watt's archive contains two granted him in Naples. The first, dated 17 May 1802, was to allow Watt's return to England. It was provided by Sir William Drummond (1770?–1828; see ODNB), then the British minister plenipotentiary to Sicily and Naples. The second, dated 24 May 1802, for the same purpose, was from Sir John Francis Edward Acton (1736–1811), the Englishman who was then, remarkably, responsible for organizing the Neapolitan Navy against Napoleonic invasions and was now the kingdom's prime minister. Soon after the date of this passport, Acton himself was forced by the French to resign. He fled again, with the large English colony in Naples, which now included Dr. William Thomson (see below), to Palermo, in Sicily, where they both subsequently died (see ODNB and Acton, 1969, tavola VIII).

WILLIAM THOMSON OF NAPLES

Watt's notebook of "Memoranda in Italy, Germany and Holland," covering the period from 1 June 1802, lists, inside the front cover, relevant addresses (BPL C2/3). The first is that of "Dr William Thomson, chez Culter and Heigelin, Naples." Elsewhere in the same book is a list of letters written later, among which is one "to Dr Thomson from Verona," which helps to date Watt and Maclure's return travel north. The Watt archive also lists their letters of introduction and those for "Naples: Dr. Thomson and Breislach" had both been provided by Jean-Claude De Lamétherie (1743–1817), editor of the *Journal de Physique*, whom Watt and Maclure had clearly met in Paris (BPL C2/11B).

Dr. William Thomson (1760–1806) was the first Englishman who could be called a "vulcanologist." His expertise was clearly the reason why Watt, Maclure, and a stream of other Europeans sought him out on their Italian travels. Thomson had been born in 1760 in Worcester, England (see ODNB). He matriculated at Queen's College, Oxford, in 1776 and graduated BA in 1780. His first interests were archaeological. In 1781–1782, he studied medicine at Edinburgh University, with chemistry under Joseph Black. While there he was elected to the Royal Medical Society (1781) and was a founding member of the Natural History Society (1782). In 1784, Thomson joined the London Society for Promoting Natural History and recommended that his close friend James Macie (later Smithson [1764–1829; see ODNB], the founder of the Smithsonian Institution) tour Scotland. Thomson graduated BMed in 1785 and was elected Dr. Lee's Lecturer in Anatomy at Oxford in April 1785. He spent the winter of 1785–6 attending the London lecture course on surgery of John Hunter (1728–1793). In 1786, he graduated DM and was elected physician to the Radcliffe Infirmary, Oxford, and Fellow of the Royal Society. In 1788, Thomson became a founding member of the Linnean Society and tried to find fossils for his Scottish friend James Hutton (1726–1797), the so-called "father of modern geology." Thomson's time in Edinburgh, at such a "Huttonian" period, had aroused his interests in geology and in the origin of rocks.

In 1789, Thomson helped with those metallurgical experiments made in Shropshire iron furnaces on the agency of heat on rocks, which impressed on Thomas Beddoes the "very strong conviction of the truth of Hutton's *Theory of the Earth*" since his friend "Dr. Thomson admits the facts... and thinks my specimens justify the inferences" (letter from Beddoes to Black, 21 April 1789, Edinburgh University Library). But in September 1790, just after Thomson had delivered anatomy lectures at Oxford, he suddenly left Oxford and resigned from the Royal Society. He had now suffered "a most scandalous imputation from an *Experiment* performed on a man 4 years ago" (letter from Thomson to George Paton, 25 September 1790, National Library of Scotland). At Oxford University, some members of convocation wanted him "most publically censured on a charge of suspicion" that Thomson had been guilty of "sodomy and other unnatural and detestable practices with a servant boy" (Minutes and Register of Convocation, Oxford University archives). John Hunter and other medical experimentalists tried in vain to rally to Thomson's defense, arguing that such "indecent" physiological experiments were fair and had been conducted in a reasonable manner. But when Thomson failed to answer charges at Oxford, he was stripped first of his studentship and then of all his degrees and banished from the university in November 1790. Although H.R.V. Fox, third Lord Holland, recorded that "the memory of [Thomson's] ready eloquence and extraordinary perspicuity survived the ruin of his moral character and his consequent retirement in Italy" (Fox 1905, p. 340), Thomson now left England, never to return.

Thomson, in exile, passed via Paris, on his way to Marquis Ippolito Durazzo (1754–1816) in Genoa, then to Siena (November 1791), Florence (February 1792), and Rome (March). He had arrived in Naples by April 1792. Here he settled, having met a wide range of European savants on his travels. William now became Guglielmo, and he started again to practice medicine. Lord Holland, soon his patient there, noted remarkably how "here his medical attainments so far expiated his religious and other heresies that the Pope condescended to consult him, and, if I mistake not, to appoint him Physician-in-Ordinary" (Fox, 1905, p. 340).

Thomson now became a member of the British colony in Naples, which by the 1790s was 58 strong, and he was soon helping archaeologist Sir William Hamilton (1730–1803) to edit the publication of his second *Collection of Vases*. But Thomson's main interests were more scientific. Now living in one of the most volcanic areas of Europe, he turned his attention to this and built up fine collections of recent volcanic specimens, published classifications of volcanic products, and named several new minerals. Thomson wrote many articles for the *Giornale Letterario di Napoli* and other German, French, Swiss, and Italian journals, only a few of which were ever translated into English (in the London *Monthly Magazine*), which is why he is now so forgotten in his homeland. He was of great assistance to the Italian pioneer of volcanology, Scipione Breislak (1750–1826). Thomson's work on volcanic rocks was regarded as sufficiently important in Italy for a short summary to appear in the anonymous *Classificazione delle Rocce* of 1814 (Anonymous, 1814, p. 391–395). Breislak also referred to his discoveries and discussed them in some detail many times in his fine textbook of 1811, with a French translation in 1812 (Breislak, 1812) and in other publications.

All this time, refugee Thomson was in a city terrified by fears of Napoleonic war and invasion. In 1797, he had contemplated publishing his own *Theory of the Earth*, clearly inspired by the appearance of Hutton's in 1795, but this never appeared. In mid-1798, Thomson had fled to Palermo for three years, fearful of the appalling political situation in Naples. By September 1801, Thomson was back in Naples, where he helped Watt and Maclure, and where he published on his newly discovered Sicilian fossils, later called rudists. Thomson had also long been a student of meteorites, having been much involved in the analysis of the Siena chondrite fall of 16 June 1794 (Marvin, 1996). In 1804, Thomson published, in the *Bibliothèque Britannique*, an important paper on the Krasnojarsk pallasite. This recorded the three component structure of such meteoritic iron for the first time. It is clear that what are now called Widmanstatten structures might as well have been named after Thomson.

The fame of Thomson's fine Italian collections and the depth of his knowledge were widely known, and this was the reason he was visited in Naples by Gregory Watt and William Maclure in May 1802 and later by Alexander von Humboldt (1769–1859) in August 1805. But in 1803, Thomson, writing to Watt how a previous letter had miscarried when its carrier had been murdered in Piedmont, warned that, "if another illness is supplanted by the apprehensions of warfare, you shall hear no more of me and my studies" (letter to Gregory Watt, 6 September 1803, BPL). In January 1806, Thomson was again forced to flee to Sicily, when the French threatened to enter Naples once more. He died at Palermo at the early age of 46 in November 1806. He had been an important contact between continental and British scientists, but the circumstances of his peregrinations and publications across Europe have too long obscured his contributions. His main collections and library arrived at Edinburgh University in 1808 (having been refused by Christ Church in Oxford, "who could not do otherwise. They could not question by implication the justice of that expulsion by the University, to which their representatives in the *Caput* had been a party" [Fox, 1905, p. 340]). In Edinburgh, much survives, perhaps still supported by the endowment of half his estate. Another collection was willed by Thomson to Lady Elizabeth Anne Hippesley (1760–1843), Somerset-based wife of a former diplomat in Italy, of which a little now survives in the Natural History Museum, London.

Some details of Watt and Maclure's Neapolitan contacts with Thomson are also given in Watt's "Memoranda in France and Italy" notebook (BPL C/2/2). It includes the scandalizing note that the "Queen of Naples has been cut for a fistula and a model of the royal posterior has been sent to the principal anatomist here," which was gleefully quoted by De Beer (1957, p. 135). This note clearly emanated from, and related to, Thomson, whose medical services were then in great demand. This was in stark contrast to the medical situation from which Thomson had had to flee to Italy from England in 1790. The same notebook lists (with its original spellings) as then:

In Dr Thompson's Cabinets
1. Medals from Torre Del Greco described by Breislach [i.e., Breislak].
2. Almost all the specimens quoted by Breislach.
3. Lava from St Helena which ran into the sea and invelloped shells. It has crystalizations in its cavities.
4. Bitumen in quartz crystals.
5. Arsenate of Lead crystalized in long hexahedral Pyramids.
6. Sulfate of Strontium in quadrangular prisms truncated.
7. Capillary or acicular Calcareous Spar in the Hollow of Lava.
8. Phosphorated Lead tinged by Copper.

Another bundle in the surviving Watt papers includes a written "Memorandum for Dr Thomson," on paper with an Italian watermark, clearly made by Watt while in Naples. This lists:

1. Shropshire Bear [from the iron-hard fused base of a furnace there] with shells.
2. Mr [James] Keir [(1735–1820) of the Lunar Society; see ODNB]'s chrystalized lead.
3. Tin Pyrites and Wood tin.
4. Recomposed Basalt.
5. Crystalized aerated Barytes.
6. Sarcites [a mineral just named by Robert Townson (1762–1827) in 1799] from B[enjamin] Delesserts [(1773–1847), banker, industrialist, and naturalist who had earlier traveled in Britain].
7. Crystalized Slags.
8. Mr [Matthew] Boulton [(1728–1809) partner of James Watt senior and maker of] Coins.

This was evidently a list of the items that Watt intended to send to Thomson when he reached home. There is a further note relating to the eighth item in the first list above, that this "Blue lead from Lead Hills [Scotland] is phosphorated lead tinged by Copper" (BPL 6/16). At a later, unrecorded date after his return to England, Watt made a longer list of a hundred "specimens marked for Dr Thomson," which Watt must also later have sent to him in Naples (BPL 6/66).

VESUVIUS

Watt was still in Naples on 24 May, but by 27 May 1802 Watt and Maclure had reached Rome, from where Watt's next relevant letter, to his father, was dated. It reads:

> By the date of this you will perceive that we have terminated our journey south and begin to move northwards. At Naples I leave nothing with regret except Dr Thompson and Vesuvius. From the Dr. we have received infinite civility and the consideration of the Volcano and its products will, I believe, operate no small change in my mineralogical opinions. In the day we bestowed upon him [Vesuvius] (or rather part of two days) we visited Torre del Greco—rising like a Phenix from its ashes, followed the course of the recent lava [15 to 24 June 1794; see Nazzaro, 1997, p. 165–167] to the Bocce Nove where it burst forth at about one third of the height of the mountain. From thence to a hermitage where once dwelt a Hermit but unfortunately for travellers, the natives took one night the amusement of stripping him of all his moveables and reduced the poor devil to the necessity of absconding—we lay some hours on his floor and a little before daylight began to mount the Cone. I cannot say that I found this operation difficult. We reached the top which was all in clouds and had to wait some time before the interior of the crater became wholly visible. Our rascally guide refused to take us down the crater, led us to several impractical places and took us about half way down in one part that terminated in a precipice in hopes of fatiguing us so as to render us willing to abandon the enterprise. Intreaties failing me, we had recourse to threats and at length the villain promised to try to find a way, tho' it was going to certain death. The way found was a slope of rubbish from the lip of the crater to the bottom down which you descend in perfect ease and security sinking up to the knees each step. When down, the whole bottom of the crater is covered with scoria which seems to have been arrested in a state of ebulition for it is heaved in parts into Monticules [small conical mounds produced by volcanic action] and depressed in others into pretty profound basins. The crust presents the usual minor irregularities of a current of Lava and is divided by circular chasms that are concentric to one of the basins. There is little or no smell of sulphur and from several fumaroles escapes a vapour slightly hepatic that forms silicious stalactites.
>
> The sides of the crater are formed of alternate beds of lava and ashes. To descend was easy—but to mount was rendered prodigiously fatiguing by the looseness of this ground tho' we had descended at the lowest point where the crater is not, I think, much more than two hundred feet deep. Out however we got and made the tour of the brim and descended into the Area de Cavallo as the valley between Vesuvius and the Somma is called. Following the course of the Somma which is perpendicular on the side near Vesuvius we observed the ancient Lavas of which it is composed abounding in large Leucites. Here the same alternation of lava and ashes appears and indicate the successive overflowing erruptions of which it is the result and it also exhibits the phaenomenon of perpendicular veins of Lava intersecting the strata, this Dolomieu well explained by supposing the crater to have cracked and these cracks filled by a subsequent overflow of Lava. We have made excursions to Pompeii—Herculaneum—Portici—Solfatara—Pozz alle Baia and Caserta (BPL C2/12).

Watt ended this fascinating letter by noting the British people they had met there and concluded:

> The country people have carried off every thing from the [Boulton and Watt steam] engine [erected in 1788] at Caserta except the cylinder and pumps and they even attempted to blow up the [engine] house. All this will need renovation and when the King returns in a few weeks it will be set about and Malcolm Logan [their resident Scottish engineman, Dickinson and Jenkins 1927, p. 282] will send an order to you and demand a list of prices. Of course if anything is done it will have inexceptionable security, for the Kingdom of Naples is ripening for Hell. Now the French troops who have kept Apulia quiet are drawn out and the Russians who garrison Naples are about to be removed. They may expect a renovation of the Santa Fede instituted by Cardinal Ruffo who roasted alive the Patriots at Naples and has ever since been maintained in Calabria where all is now anarchy.

This anarchy in Calabria, and the part of the leader of the Sanfedist forces, Cardinal Fabrizio Ruffo (1744–1827), is discussed by Acton (1956, p. 359–360). For some insights into the terrible political situation here, see also Imbruglia (2000), and in particular the chapter by Antonino de Francesco (p. 167–182).

THE RETURN JOURNEY

The next news concerning the European travels of Watt and Maclure comes from a fascinating letter from Bordeaux addressed to Gregory Watt, "chez John James Esq à Naples" dated Pravial 15 An X (or 4 June 1802), with a postscript dated Pravial 23 (or 12 June). This came from another American mineralogist, Thomas Smith (1777–1802), who had met Watt and Maclure in Paris. The letter somehow reached Watt when they got to Rome. It details Smith's mineralogical tours in France, of which other details survive in diaries that deserve future study (Stapleton, 1985, p. 47):

> In my last to Mr Maclure from Moulins I informed you of my intention speedily to set off for this city in order to embark for America. I arrived here yesterday and hope in ten days to embark. I have not been able to visit Auvergne but your mineralogical interests have not much suffered by that, as I have forwarded a large case from Moulins containing all the most interesting productions of those volcanoes.

The case is directed to your brother [James in Birmingham] and contains duplicates. By this time I suppose you are satiated with the beauties of Naples. Vesuvius is not the only volcano I suppose you have visited there. I do not however ask you for specimens of the productions of any other. I hope your companion de voyage [Maclure] has not been entering craters without first prudently examining whether they were yet burning. I passed by Limoges and rested there two days but did not visit the mines as I found no means of conveyance and then it rained continuously. I saw your friend Martin but unfortunately coming after you found but little to steal—Your passage over a country is as devastating as that of the Locusts of Egypt—Emeralds, Granites, Titans, all, all are swept away by your insatiable avarice. France, as many as she has seen of them, has not beheld your equal as a delapidator. [In the postscript, Smith added that] in 6 or 8 days I embark aboard a small brig of about 120 tons without any accomodations [sic.]. I am not likely to pass my next two months as agreeably as you... Soon after my arrival in America I shall sent you enough of our minerals (rocks) to form a little kingdom for yourself (BPL C2/14).

Sadly this last promise could not be kept, as Smith was tragically killed on route home to America (as we shall see). But it is obvious that Smith thought that Watt was already far too ill to have even contemplated those volcanic expeditions which he and Maclure had made together at Naples.

On 1 June 1802, in Rome, Watt and Maclure now settled up their joint traveling expenses. Details, and all of their later financial matters up to a final settlement at Rotterdam on 14 October 1802, survive (BPL C2/11B). Soon afterwards at Florence, a note in a letter from Gregory to his brother, dated 11 June 1802, records, "Maclure is well and much astonished to find Italy approach America in temperature" (BPL C2/12/13). Details of their final days in Italy survive in a letter to Watt's father from Innsbruck [Austria], dated 13 June 1802.

At Verona I remained five days recruiting myself and enjoying the Hospitality of M. Albertini who took me to his country house on the Hills. Meanwhile Maclure made his Vincentine excursions and found even in those remote retreats Italian rascallity in greater perfection than even at Naples. They ask twenty ducats for two Poneys to go six miles and would not stir for less. He went on foot but for a very bad supper they asked two Sequins which he would not pay. He brought nothing back that would not come under the denomination of whin [basalt] but he found it strangely mixed with limestone and he found in the same bed a compleat passage from very compleat trapp or basalt, for it is often columnar, to a very porous singular stone much resembling some of the stones near Cassel [Germany] and some Scotch whins.
From Verona we went to Mantua whose depopulation and marshes made us escape out of it the morning succeeding our arrival. Proceeding to Peschiera we embarked on the magnificent lake of Garda and proceeded to Garda. Till you approach Garda the sides of the Lake are only gentle slopes, afterwards they are tremendous limestone precipices. M. Albertini surprized us agreeably by meeting us at an immense chateau he has most charmingly situated near the little village of Garda. We spent a most agreeable evening and staid all night with him. Next day, but not till evening from contrary winds, we got to Torboli at the head of the Lake under immense limestone rocks and adjoining a small but very rich valley. We got next day to Rovereto where there are many rich merchants, considerable cotton mills, and great productions of silk. We are now in the valley of Adige or, as the Germans nickname it, the Etsch and along this the road proceeds to Bolsano or Botzen. It is very romantic, but equally unhealthy for the continual heightening of the bed of the Adige renders it an undrainable morass and in autumn agues are very numerous and virulent. Till a Post [the distance traveled by horses in relays] and half beyond Mente (famous for nothing but the Council) all is Limestone, afterwards little Mounts of a brownish Porphyry sprout up in the Valley. These soon form an almost continuous range which the limestone dominates. At last the Porphyry takes the limestone on its back and lifts its head almost as high as the adjoining Mountain. Botzen is a neat town, it lies snugly at the head of the valley in the fork of the Etsch and Eisack, all the Hills round it and for a Post and half forwards are Porphyry. Then begins Gneiss and Micaceous shistus that go to half a Post beyond Brixen. Then commences mountains of a handsome black and white Granite. This lasts to near Sterzingen where we find again Gneiss etc. Near Sterzingen are mines of lead and [zinc] blende in a gang [matrix] of Micaceous shistus and Gneiss with vast quantities of ill formed Garnets. Maclure went and bought a great collection. We continued to mount to Brenner, the rocks near us all Gneiss etc, though from the immense quantity of white marble etc of a curious shistose Marble there must be rocks of them in the vicinity and from some indications I think they there alternate with Gneiss. The descent from the Brenner to the Innthal is every considerable.

Their Italian tour was over. In a footnote, Watt, whose health very clearly had been deteriorating so badly that he could no longer go on mineralogical tours, confirmed the extent of his medical problems:

Meanwhile following the advice of the Verona Doctor and of a very sensible one I have found here [Innsbruck] who has been all over Europe, I stay in perfect tranquility at Home. The Doctors both agree in giving no medicines and both saw my complaint is merely nervous and that rest and attention to Diet will soon remove it. The Innsbruch Doctor says that he has seen a great many travellers who had been in Italy reduced similarly. It was not from want of caution in Italy, for I took very little exercise for I lost many objects of curiosity rather than fatigue myself by going to them (BPL C2/12).

In a second letter from Innsbruck to his mother, dated 20 June 1802, Gregory confirmed that:

> after arriving here ... my strength were not what I could wish, I am happy in now giving you a more distinct and favourable account. [He notes his medical crisis in northern Italy and how bad the weather had been there which] brought my malady to a crisis ... All I have now to molest me are flying rheumatic pains that this Greenland of a place has produced ... Maclure, who is the most attentive of friends, is gone on a second expedition to the Zellerthal... You may judge of my mortification in passing through this most romantic and mineralogical country without being able to enjoy its scenery or make any exertion to enrich myself with its mineral wealth, except what I can plunder from the hard earned stores of Maclure (BPL C2/12).

Meanwhile Maclure was clearly equally worried about the health of his companion, and on 17 July 1802 he wrote thus from Innsbruck to James Watt Jr., Gregory's half-brother:

> It is scarce necessary to inform you that in company with your brother, I have made the tour of France and Italy for the last five months during the greatest part of which your Brother was in good health and stood the heat of the climate full better than I expected, for from his rather irritable temperament I feared the debilitating effects of heat on our return from Rome. At Florence about one month ago he complained of weakness and was fatigued by moderate exercise, at Bologna, Padua and Venice the weakness continued and rather increased, attended at some times by a small short cough. When at Verona he continued still weaker and we endeavoured to persuade him to rest sometime and try to recover strength which he would not hear of, as it would interfere with the arrangements of his time. We came slowly onto Bolsano without any change, still weak and afterwards stopped two days at Sterzingen near the summit of the Tirolean mountains where the air was cool and temperate. There he first complained of a pain in his side and the cough still continuing. I was more afraid of the cough than of the weakness and proceeded down to this place where he have been five days (BPL C2/12/51).

Late in July, the travelers carried on to Vienna, where they arrived on 1 August 1802, and then started their return, going via Prague and Dresden, to Freiberg, where in mid-September they much hoped to met the famous teacher of "proto-geology," Abraham Gottlob Werner (1749–1817). In the Werner archives at Freiberg are two 1802 letters to Werner from James Watt Jr., who had studied at Werner's Bergakademie. The first, dated 19 August 1802, describes the health problems that Gregory had had on his travels and noted that his parents had now decided to travel to Germany to meet him and bring him home. The second letter, of 20 August 1802, asks Werner to help his father and mother when they visited Freiberg (Werner Letters, v. 2, B 161–165). As it turned out, Gregory's health improved sufficiently and the Watts had only reached Frankfurt before they were able to turn back (Dickinson and Jenkins, 1927, p. 74).

But when Watt junior and Maclure reached Freiberg, they found that Werner was away and so they only stayed a short time. Werner had obtained permission, late in July 1802, to visit the warm baths of sulphurous waters at Aachen (Aix-la-Chapelle) for his own health. He then took the opportunity to continue on to Paris. There, the Dutch naturalist Martinus van Marum (1750–1837) visited him on 14 and 22 September 1802 (Schleghel, 1803, p. 162–163; Forbes, 1970, p. 370–373). The mineralogist John Pinkerton (1758–1826; see ODNB, where this activity is not noticed) also met Werner "during his short stay in that capital" (Pinkerton, 1811, v. 1, p. xlvi). Since Werner was away until the end of November 1802, at the latest, he could not have received either of the letters from James Watt Jr. until well after Gregory had left Freiberg.

The Watt-Maclure tour concluded through Leipzig to Amsterdam where Maclure and Watt separated. Watt now embarked for England, and Maclure went to Paris. Watt returned home on 15 October 1802. News of his return spread quickly. Humphry Davy wrote to the Cornish scientist Davies Giddy (later Gilbert, 1767–1839) on 26 October 1802:

> Gregory Watt is just returned from the Continent, where he has passed the last fifteen months. He has been much delighted with his excursion, but his health is at present bad. I trust, however, that English roast-beef and English customs will speedily restore it (Paris, 1831, v. 1, p. 156).

WATT'S WORK BACK IN ENGLAND

Watt soon received a letter from Maclure dated Paris, 11 December 1802 (BPL C2/14), further adjusting their finances, as he had found some coins in their joint dressing-case, and telling him how he had spent these unexpected funds (mainly on books and prints). He noted:

> Werner was here sometime and the frenchman that translated his Theory of Veins [1802, J.F. d'Aubuisson de Voisins (1769–1841)] has drawn up a long memoir on basalt which is to be published. The result I am told is that basalt is not lava or the production of fire but that lava is melted basalt by the action of the coals and other combustibles so often found near basalt in that country where Werner is best acquainted with. They deny that the great mass of Volcanic mountains have ever been in fusion [paper torn away] only that basalt [is] thrown up by action of the combustibles which were under it. If it comes out before I send off your

prints, I will put in a copy, as I consider it as the last modification of the Neptunian oppinions [sic.] (see Aubuisson de Voisons, 1803).

He then added:

Alas poor Tom, our friend, Smith is no more, the bursting of a Gun on board the ship before their arrival put a period to his existence and the well grounded hopes of his friends. Of his future utility to society I can't think of it with patience (BPL C2/14).

Maclure's next letter is dated 7 March 1803. It details the contents of the case he is about to send to Watt and discusses the difficult political situation in Paris. Maclure notes, "with more pleasure, it's probable I shall make this my home, I have offered 35 louis for a house and about 2 acres of a Garden on the Boulevard des Invalides where I shall be able to see my friends at home." This same letter noted of a nine-volume *History of Italy to 1802*, which Maclure was intending to send to Watt: "I am no admirer of history, have little faith in its truth—and still less in its utility—it's often but a cringing sycophant [who] flatters and defends the brilliant views while it consigns to oblivion the usefull [sic.] virtues." So Maclure, at least, would no doubt be shocked to discover how historical a figure he has now become and how even "useful virtues," like geology, now have historians.

WATT AND THE *EDINBURGH REVIEW*

In 1803, Watt started reviewing for the prestigious *Edinburgh Review*. On 2 July 1803, the editor Francis Jeffrey (1773–1850) wrote to his brother from Edinburgh:

I am glad you have got our Review and that you like it. I do not think you know any of my associates ... with some dozen occasional contributors, among whom, the most illustrious, I think, are young [Gregory] Watt of Birmingham and [Humphry] Davy of the Royal Institution. We sell 2,500 copies already (Cockburn, 1852, v. 2, p. 73–74).

A full list of the contributions Watt made to the *Edinburgh Review* is given here. They do not always agree with those given in the *Wellesley Index* (Houghton, 1966). As Dean has recorded, Watt became the *Review*'s resident critic on matters geological, an opportunity he used to great advantage (Dean, 1992, p. 148–149). The list of Watt's reviews follows:

1. Volume 2, July 1803, p. 391–398, review of *Nouvelle Théorie de la Formation des Filons par A.G. Werner*, Paris, 1802.
2. Volume 3, October 1803, p. 42–56, review of *Traité de Minéralogie par le Citoyen Haüy*, Paris, 1801.
3. Volume 3, January 1804, p. 295–313, review of *Lehrbuch der Mineralogie von L.A. Emmerling*, Geissen, 1799–1802.
4. Volume 3, January 1804, p. 493–497, review of *Traité élémentaire de Minéralogie suivant les principes du Professeur Werner par J.M. Brochant*, Paris, 1801–1803.
5. Volume 4, April 1804, p. 26–42, review of *Voyage Physique et Lithologique dans la Campanie par Scipion Breislac*, Paris, 1803.
6. Volume 4, April 1804, p. 117–120, review of *Specimens of British Minerals, selected from the Cabinet of Philip Rashleigh*, London, 1797–1802.
7. Volume 4, July 1804, p. 284–296, review of *Sur la Philosophie Minéralogique, et sur l'Espece Minéralogique par le Citoyen D. Dolomieu*, Paris, [1801].
8. Volume 5, October 1804, p. 64–78, review of *System of Mineralogy* by Robert Jameson, Vol. 1, Edinburgh, 1804, (Watt's authorship is clearly confirmed in several letters in BPL JWP 6/5). Francis Jeffrey "had concealed from everybody except [Francis] Horner [1778–1817] and [Henry] Brougham the share he [Gregory Watt] had in the review and ascribed his production to a London mineralogist. Promises utter secrecy" (letter, 12 August 1804), and ditto, "Jamison [sic.] has heard of his book being very severely handled and is very angry but is ignorant of the author [i.e., Watt] in the 9th number (letter, 12 August 1804; both in BPL C/2/14/126).
9. Volume 6, April 1805, p. 228–245, review of *A Mineralogical Description of the County of Dumfries*, by Robert Jameson, Edinburgh, 1805 (or 1804?). Houghton (1966, p. 437) claims that Watt could not have reviewed this last book because it was published in 1805, after Watt's death, but this is potentially negated by the copy in Werner's own library (Bergakademie, Freiberg) in which the half title and title page dates are both printed 1804.

Perhaps Jameson could have sent one of these important early variant copies in time for Gregory Watt to have reviewed it. The dedication (to Werner!) of the 1805 version is certainly dated 10 October 1804, immediately before Watt's death, but I have been unable to check whether this date is different in the much rarer 1804 edition. In his *Biographical Memoir*, Jameson's son confirmed that his father's "*Mineralogical Description of Scotland*, volume 1, was published in the same year he was appointed [Edinburgh] professor," thus in 1804 (Jameson 1854, p. 26).

Certainly Davy recorded that Gregory "devoted even his last days to promote the interests of science and he gave birth to that knowledge he had collected ... His mind seemed to become more *active* and more energetic in proportion as his bodily strength decayed" (Davy, 1858, p. 16). According to a letter from another influential *Edinburgh Review* contributor, Henry Cockburn (1779–1854), dated 6 March 1852, to J.P. Muirhead:

Gregory Watt did write the six articles mentioned [i.e., 2, 3, 5, 6, 7, and 8 above—information which had come from James Watt Jr.] and in addition to these he wrote two other articles viz [4 and 9]. Whether my note be absolutely correct or not I cannot positively say but I have no doubt that it is. It was made at the time, not by me, but by a more accurate person and one who had better means then of knowing the internal proceedings of the Review. I have always believed G.W. was the author of the whole of these 8 articles (BPL B and W 737/105).

So it seems certain that the Scottish mineralogist James Headrick (1759–1841) did not take over reviewing at the date, October 1804, when Houghton (1966, p. 435) claims. As Houghton says, the last two articles above, "seem clearly to be by the same hand," especially from their internal evidence, so perhaps Headrick only edited them for the *Edinburgh Review* after Watt's death.

William Thomson also kept in touch with Watt after his return to England. One of his letters, dated 6 September 1803, survives (BPL C2/14) with one of Gregory's faint-pressed copy-letter replies, dated 28 February 1804 (BPL 3/17). This last expressed the hope that Thomson:

> had received the letter I [Watt] wrote to you several months ago in reply to yours. It would inform you that all my volcanic collection reached England and also that I had received the box with the two Vases. I have been almost confined to the house since I wrote to you by ill health. [In arranging his collection] I have nearly followed the arrangement you pointed out in the Giornale Letterario (see Thomson, 1795a, 1795b, of which Watt probably used the 1801 edition).

Thomson's part in the "Vase business" is mentioned by Jenkins and Sloan (1996, p. 56). This letter from Watt concludes with notes of Watt's experiments on basalt and Carrara marble, to which we now turn.

WATT AND HIS EXPERIMENTAL WORK

The above December 1802 letter from Maclure, with its early news of recent French work on basalt, must have influenced Gregory Watt's next research project, which had itself been influenced by Gregory's experiences among both active and dormant Italian volcanoes. The project was to investigate experimentally the baffling problem of the origin of basalt.

Watt referred to his experiments in a letter from Soho, near Birmingham, to Humphry Davy, dated 21 March 1804:

> I have lately had a large experiment made on the regulated cooling of seven hundred weight of Rowley ragg (a basalt you know). The results have been very curious. I shall send an ample series of specimens to Mr. [Charles Francis] Greville [1749–1809; mineralogist of London, see ODNB], where you may see them, accompanied by some observations. In another experiment I made on Carrara marble confined in an iron tube, perfectly closed at one end by a screw, the tube, which was immensely strong, was burst by the expansion of gas. The gas emitted was hydrogen, and burst furiously, but the marble was partly quick, and all the outsides of the lumps coated with charcoal, as if they had been painted with lamp black. Surely this must have proceeded from a decomposition of the fixed air: the oxygen took hold of the iron of the tube, but whence came the hydrogen? Precautions were taken to expel moisture previously to closing the tube. Let me have your opinion on this subject (Davy, 1858, p. 31–32).

In another letter of 12 April 1804, still from Soho, to the professor of chemistry at Edinburgh, Dr. Thomas Charles Hope (1766–1844; see ODNB, the Huttonian mentor of the Canadian pioneer in geology, William Logan, see Torrens, 2002, XII, 2), Watt again referred to his experiments:

> It lately occurred to me that something might be observed by gradually cooling a very large mass of melted Basalt and I therefore fused seven hundred weight and cooled it very gradually—still there was inequality enough in the refrigeration to exhibit very beautifully the various stages in the transition from glass into stone and it appears to me that some of these changes have eluded former observation and that very curious and important conclusions are deducible from them.
>
> I have sent off above a fortnight ago another Box directed to you, containing three series of these gradations—each Series is accompanied by a specimen of the Rowley Ragg or amorphous Basalt which was employed—I beg you to accept one of these series yourself and to give the other two to Professors [John] Playfair [1748–1819] and Sir James Hall [1761–1832] and although I have no personal acquaintance with those Gentlemen, I trust they will not hesitate to receive this slender token of the esteem I feel for them.
>
> I have written a description of the appearances this experiment has brought to light and I have ventured to enter into a good deal of geological reasoning and to offer an hypothesis for the formation of basaltic columns etc—the whole forms a pretty long paper which I have just finished and which I propose to send to the Royal Society. I shall endeavour to send a copy of the Manuscript to you for your perusal and I shall hope to be benefited by your comments.
>
> Some experiments I have lately made upon the fusibility of Marble (under pressure) have not tended to increase my faith in the Huttonian Theory. Carrara marble is perfectly infusible in the heat of melting silver or 30 of Wedgwood, the melting point of some Basalts according to Sir James Hall, and beyond that the elastic force of the fixed air becomes so enormous that I have been unable to make Iron tubes strong enough to confine it—The marble must not be in contact with the Iron or the fixed air is decomposed and the oxygene [sic.] combines with the Iron and the charcoal is deposited on the Marble.

Hope's reply to Watt is dated 24 April 1804 (BPL C2/14/44) and comments thus on the specimens Watt had sent:

> I rejoice that you have begun to put the Geological Systems to the test of Experiment. You possess every qualification for carrying them on with Success and will probably help us to read the truth... I know you are at present no great friend to the Huttonian Theory, but I hope I shall one day see You one of Us. I have just read the review of Breislac in the Edinb. Review. I have formed the strong notion from its natural evidence that you have had a share in this article. If I am right, you must grant that Huttonianism has illustrated compleatly one Curious fact in Natural History; viz the Stony condition of Lavas. For though the hint of Beddoes (Beddoes, 1791) and the more correct views of Dr [William] Thomson [of Naples] preceded Sir J. H's [Hall's] decisive paper, they are the Emanations of Huttonism. Both B[eddoes] and T [Thomson] were in Edinb [Edinburgh] when Dr Huttons's Theory was brought into view and both knew that Dr H [Hutton] ascribed the stony state of the fossils [i.e., rocks] which he imagined to have been fluid, to their Crystallization.

Watt's paper was read to the Royal Society on 10 May 1804 (Watt, 1804a, 1804b). It also has the distinction of having been the first communication presented to the newly formed Geological Society of London late in 1807 (Woodward, 1907, p. 18) (Fig. 3).

On 30 May 1804, Henry Brougham (1778–1868), later Lord Chancellor, wrote to Watt congratulating him on his Royal Society paper, "which has since been the topic of conversation in our scientific circles." He asks if he may see the original samples now at Charles Greville's house. "Will you let me know by what conveyance I may send you a volume of Jameson's *Mineralogy* which [Francis] Jeffrey desires me to transmit." This is the review copy for number 8 in the *Edinburgh Review* list above (BPL C2/14/52).

Watt's work had built on Sir James Hall's earlier experiments, read in 1798 and first fully published in 1800, when Watt had undoubtedly read them (Hall, 1800). With the foundry facilities available to Watt at his Birmingham engineering concern, he could "expose to the action of heat, a much larger mass of basaltic matter than had ever been subject to experiment." The basalt had come from one of the famous quarries at Rowley Regis, Staffordshire, in the roadstone there called Rowley Rag (Fig. 4).

Watt had seven hundred weight of basalt (~356 kg) melted after being broken into pieces. At first it turned into a glass. Watt then experimented on the types of crystallizations achieved at differing temperatures as it cooled. He concluded that the formation of basalt by either fire or water seemed equally feasible (Dean, 1992, p. 136). Porter is wrong to have branded Watt among the "British 'Plutonists'—who saw all crystalline and Primary rocks as of igneous origins" (Porter, 1977, p. 175). Watt himself denied this in this letter to William Creighton, dated 14 February 1804 (and see below).

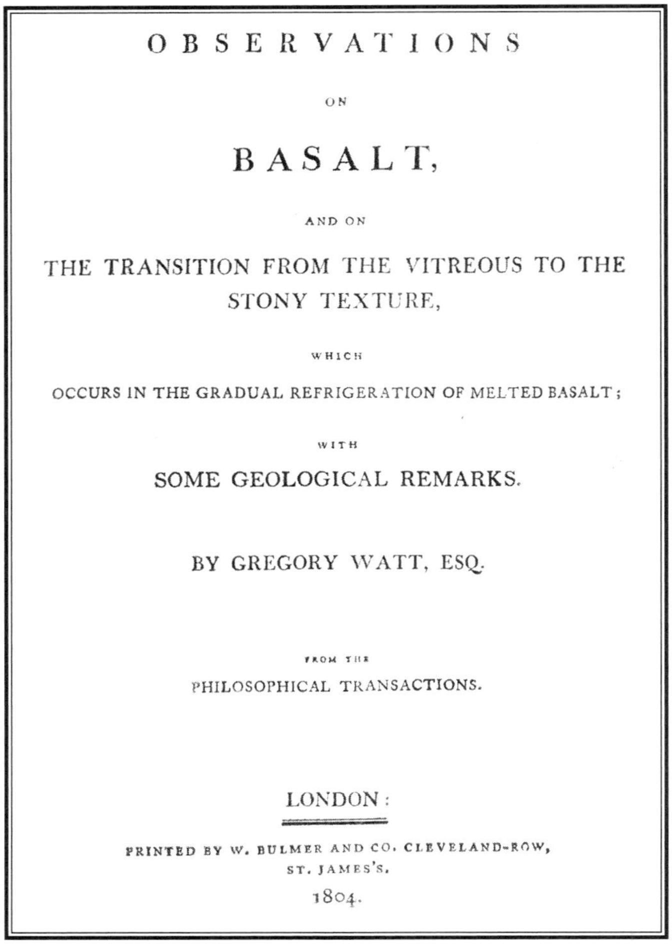

Figure 3. Title page of the rare offprint of Gregory Watt's basalt paper (Watt, 1804a, author's collection).

I have amassed an immensity of arguments against the Huttonians and shall certainly incorporate them into a Diatribe and send it to the learned professor [?John Playfair—who had recently defended Huttonism]. Among other things I have found by the [Geological Map] survey [see below], 200 specimens in my own cabinet of Fusible substances impressing or piercing refractory ones. A thing impossible on the igneous system.

Among those who have recently discussed Watt's paper in detail are Fritscher (1991, p. 57–61), who reviews German reactions to his work, and Den Tex (1996), who puts it into context in the whole debate about the "basalt controversy." The best contemporary source for a view of the depth and range of this debate is provided by Charles Daubeny (1795–1867) in 1822. He then printed his fascinating, and widely modified, *Geological Thermometer* to "shew [sic.] the opinions attributed by various Geologists with respect to the Origin of Rocks" (Fig. 5).

After it was first published in Oxford in 1822, the *Geological Thermometer* was erroneously attributed to William Buck-

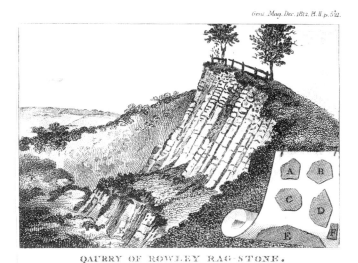

Figure 4. One of the two Rowley Rag quarries in 1812 showing the basaltic columns (from Miller, 1812, p. 513).

land (1784–1856) (Rupke, 1983, p. 16), who was Daubeny's Oxford colleague. Buckland may well have sent a later version in for publication and he must also have used it in his own lectures, but its authorship is recorded by Daubeny himself (1826, p. 451–452).

Watt's uncertainty about the origin of basalt only mirrors that of the great Swiss naturalist Horace-Bénédict de Saussure (1740–1799), whose field observations also demonstrated the complexity behind claims that all basalts were igneous in origin. He, too, vacillated about its origins (Carozzi, 2000) well before Watt started his experimental and observational work. As Carozzi and Newman conclude, "by 1794–1796 [Saussure had] reached the concept of a double origin of basalt... but by 1798 Saussure [now] considered basalt as entirely aqueous in origin" (Carozzi and Newman, 2003, p. 230). The debate, at least in Britain, about the origin of basalt was not resolved until 1816 (Wyse Jackson, 2000, p. 40–50), largely on the basis of field, not experimental, evidence.

WATT'S PROTO-GEOLOGICAL MAP OF ITALY

We first hear of Watt's next project when he writes to his "scatological," mineralogical friend, William Creighton, in the letter from Soho, Birmingham, dated 14 February 1804, quoted above. The letter opens by discussing the complex system of petrological symbols that Watt intended to use when making such maps. Next, Watt discusses the problems of making geological maps of large areas, such as Italy and Britain:

> As to [making such a map of] England I dare not think of it. Fragments of detached places must first be accumulated to a stupendous amount. I strongly recommend you to make plans on a large scale with sections of such districts as you

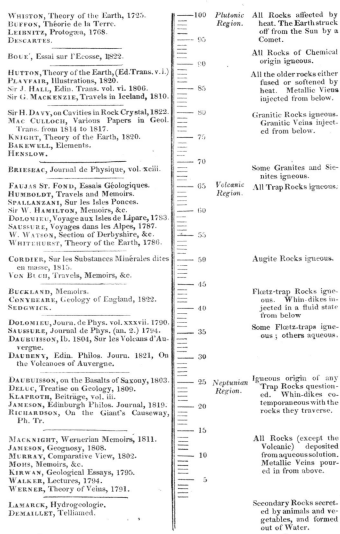

Figure 5. Charles Daubeny's original publication of the *Geological Thermometer* of 1822 (Oxford University Museum of Natural History [OUMNH], J. Phillips MSS, Box 78).

can accurately examine for example, the Paisley and Girvan Coal district then Glasgow etc [all in Scotland]. I understand there is a blind or stone coal a few miles from Glasgow on the Cumbernauld road under basalt. I should be curious to trace those ascertained and to know whether parts of the same strata not covered by basalt are of the same quality.

In South Wales parts of the same bed are coaking [sic] coal and part stone coal. By the by a most profound Welchman [George Owen (1552–1613) see ODNB], a Lord too in Q [Queen] Elizabeth's time found out that the limestone runs in two straight lines about 10 miles asunder, and nearly parallel from Milford Haven to near Bristol and mentions about twenty places on each line and observes learnedly and correctly that all the coal lies between these two lines. I have got it in print [in 1799, in the *Cambrian Register* for 1796; see North, 1931, p. 25–29; Challinor, 1953, p. 127–129] and think it the most wonderful anticipation of Geology that was ever met with. He must have been a Sir I [Isaac] Newton. However he is right, straight the two lines of lime are and the coal between them to this blessed day. Had all the learned been as clearsighted as this fellow, I should [not] have been so tormented as I have been in constructing a Geological Map of Italy (Fig. 6).

But the devil or such has [n]ever entered the country [Italy] who carried a compass except [Horace Bénédict de] Saussure and he went only to Genoa. I also was equally negligent not having found out the beauty of primitive lime; however by reading a library of books I have made out something and am got as far north as Bologna having all to the south filled. The unfathomable mud of Lombardy will cover a great deal of the remainder but the line of surrounding alps will be terrible. At the Venetian end I have abundant authority. The Paduan, the Vincentine, the Veronese and Trento are plain sailing, but the Bergamesce is totally dark and as you have read Saussure you may know it to be unprofitable to lay down on a small scale the too abundant documents he affords for the Alps. However it will give a general idea pretty well. The few strata about Genoa, Saussure mentions, are most unaccomodating [sic.] for they often run at right angles to each other making tremendous hubble bubbles [confusion] and the alternations to absolutely opposite species are so rapid as would make the map all lines. I [will] leave out the accidental ones and give [those] more that come oftenest, referring by figures to a table of omitted curiosities. The table when made will be very profound indeed (Fig. 7).

The map paper has a watermark of "J. Whatman 1794" and the index table, by the same maker, one of 1801.

THOUGHTS ON WATT'S MAP

For a map to be properly called "geological," it surely has at least (1) to use some means, whether colors or toned shading or patterning (Robertson, 1956), to discriminate properly geological, and not merely lithological, entities, and (2) more important, these geological entities must be properly stratigraphic and keyed to an ordered section of strata, showing those in use on the map in question.

Given these criteria, this map of Watt, as is clear from his own description of his methods, must be regarded as "merely" lithological, or proto-geological in nature. It measures ~80 × 80 cm. The great amount of detail on it can only be seen when it is viewed at original size (or better, enlarged). It covers the geographical areas Watt and Maclure had traveled through as far south as Naples, drawn from his and Maclure's own observations and experience. To this, Watt has added information drawn from a wide range of printed sources, of which he clearly had considerable knowledge, however unexpected for a Briton. Britain then being a country much isolated from communication with mainland Europe by war. The map shows no remotely stratigraphic key and no Wernerian features, which, as can be read in his *Edinburgh Review* articles, Watt thought were based on theoretical ideas to which Watt was wholly hostile.

The novel feature of Watt's map is the detailed discrimination of 46 separate lithologies shown on his map explanation. These are grouped together by color. Thus, all the coal strata are shown black. All the "trapp" and basaltic rocks have one color, all the serpentines and talc another; all limestones are depicted yellow; and all granites pink (see Fig. 7). Watt's Italian map is clearly still an old style proto-geological map, recording only (however dazzling they may appear) an array of lithologies.

THOUGHTS ON OTHER CONTEMPORARY MAPS

The nearest equivalent for England is the lithological map of William Maton (1774–1835) of the southwest of England (in Maton, 1797), which, however, is much more reduced, whether in scale, geography, or range of lithologies. This map, ~26 × 18 cm, shows nine lithological divisions, like chalk, clay, slate, serpentine, killas, at a scale of ~17 miles to an inch (Boud, 1975, p. 82; Butcher, 1983, p. 151–152). But in that same year another, at first apparently similar, map of Hungarian petrography was issued by another English traveler, Robert Townson (1762–1827) (Townson, 1797). This shows thirteen different "petrographic" units on a map of 60 cm × 45 cm. As Kazmer has shown (Kazmer, 1999), this is a more geological map, since it uses both color (even though only on those patches of the whole Austro-Hungarian empire that Townson had been able to visit) and is keyed to a clearly Wernerian "ordering" of strata.

Ellenberger, in two fine papers (1984, 1985) has discussed the evolution of geological maps, with special reference to those of France. He wrote, "new [cartographic] concepts can hardly arise before the time is ripe ... We would claim that the first really "geological" maps produced by French authors were inspired by the then flourishing Géognosie ... masterfully taught by Werner in Freiberg" (Ellenberger, 1984, p. 82). So we might regard the year 1797 as marking some sort of watershed in British geological cartography, when one of those first exposed to Wernerian stratigraphic ideas represented them properly on a geological map for the first time.

But Werner's stratigraphic tabulation was to prove rather inaccurate and confused (see Wagenbreth, 1967, p. 115–119). Critically, it completely failed to discriminate between similar or identical lithologies, which occurred, often repetitiously, at completely different stratigraphic horizons, so that stratigraphically important, but often recurrent, lithologies, like coal, clay,

Figure 6. The "proto-geological" map of Italy made by Gregory Watt in 1804 (Birmingham Public Library C2/7, now 3219/7/56/7).

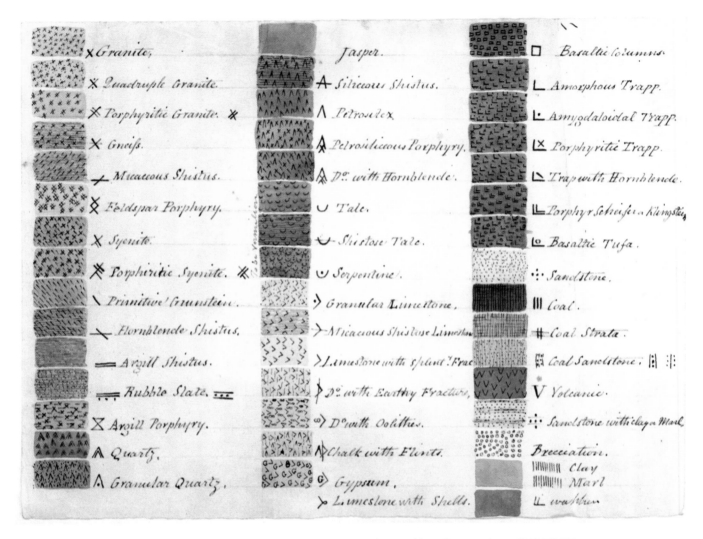

Figure 7. Watt's multi-colored table of the 46 different rock types he recognized on his Italian map (BPL 3219/7/56/7).

sandstone, limestone, were easily, and often, confused when Wernerian ideas were transferred to non-Saxon areas, as in Scotland (Torrens, 2003b, p. 147–148).

This problem of discriminating "repetitious lithologies" was *the* critical question and this was not solved until the likes of William Smith in England and Alexandre Brongniart and Georges Cuvier in France started to use fossils to identify, and to *separate*, such distinct, but lithologically similar, strata from 1795. These now allowed similar lithologies to be separated on the basis of the precisions that fossils provided. This was "a major geological breakthrough," with great significance for geological cartography (Torrens, 2003a, p. xxiii, 161–162; Torrens, 2003b, p. 139–141).

Exactly the same point is made by Vai (*in* Gadenz, 2004, p. 1) when discussing the development of geological cartography in Italy. He rightly noted that,

Wernerian geognostic maps usually represent petrological or mineralogical geography on a two-dimensional basis, and are not concerned with chronology ... having a geometric-tectonic more than a stratigraphic-chronologic meaning. Unlike conventional lithostratigraphic units, geognostic "*Formationen*" were petrographically-characterized elements or units used for describing a geometric model of Earth's crust which followed a static order of superposition that was thought to represent also a time sequence. Once the crustal system was dynamically changed, no tool for detecting ages and consequently the time sequence of the component units was available (Vai, *in* Gadenz, 2004, p. 1).

The tool used, at first, for separating these repetitious lithologies, and later for detecting their ages, was fossils. An additional, vital feature needed on any "modern geological" map must also be an accompanying cross section showing the strata in a third dimension (Vai, *in* Gadenz 2004, p. 1). This was provided on the first modern maps, but is missing on Watt's.

BATH AND WATT'S LAST DAYS

By 1804, Watt was dying of consumption, and so we learn nothing more of his Italian map. Another contemporary Anglo-American victim of the scourge of consumption, who also made important geological contributions, was Rev. Henry Steinhauer (1782–1818). He worked in palaeobotany (Torrens, 2005). It is painful to compare their experiences. Both came to Bath in hopes of a cure. A letter from T.C. Hope, dated 15 May 1804, addressed to Watt at the Post Office, Bath, explains how Hope had guessed that Watt was the reviewer of Breislak's book in the *Edinburgh Review* (no. 5 listed above). He ended,

> It will afford me sincere pleasure to learn that the air of Bath has exerted its full restorative powers to enable you to resume your experimental labours with renewed Vigour at the Foundry. Any Anti-Huttonian facts from that quarter will be truly formidable (BPL C2/14).

Between mid-May and early September 1804, Watt lived in Bath, at first at 42 Milsom Street and then at 7 Queens Parade. Bath had a high reputation as a place where the sick could receive good medical attention, and Watt now sought this, as soon did Steinhauer. The Royal Society's editor, Charles Francis Greville, wrote, on 9 August 1804, to tell Watt in Bath that 70 copies of specially printed offprints of his Royal Society paper (Watt, 1804a) "were at last printed" (BPL C2/14). Watt acknowledged receipt of these on 29 August (JWP C2/10), but it was then noted that this would not appear in the full *Transactions of the Royal Society* (Watt, 1804b) until the following Christmas, one Watt was never to see. Watt was able to distribute some offprints while he was in Bath, but the few still remaining in his possession were only sold in 2003 (Sotheby's, 2003, lot 168).

THE LINK WITH WILLIAM SMITH

On 11 August 1804, Samuel Galton Jr. (1753–1832), Lunar Society member, wrote to James Watt Sr. He noted,

> My affectionate Sympathy has often been excited on your account and that of Mrs Watt and the Health of your interesting Son, Mr Gregory Watt, has been the subject of my frequent and anxious Enquiry ... When I was at London I was present at the two sittings of the Royal Society when Mr Gregory Watt's account of his Experiments on Rowley Rag were read. They excited considerable interest and were justly considered as forming one of the most intelligent Papers that had been read there for a considerable time (BPL 3219/7/56/7).

As his postscript Galton added this fascinating note:

> There is a Mr Smith, land surveyor at Bath, who has a collection of the mineral[s] of this Country and professes to have paid particular attention to the strata in various Counties in this Kingdom. Perhaps Mr. Gregory Watt might wish to see him. He lives near Trim Bridge.

This was William Smith (1769–1839), English pioneer of ordering, identifying, separating, and mapping strata, against an accurately ordered list of strata, over wide areas of the whole of Britain using fossils. His offices were at Trim Bridge, Bath, between 1802 and 1804 (Torrens, 2003a, p. xxiv–xxv). This letter demonstrates how he was already not working in the complete isolation that some have claimed, like Laudan (1987, p. 168) who wrote how, "Smith can be pitied for his isolation." It is even possible that Watt could then have met Smith in Bath, as Smith was there between 30 August 1804 and 17 September (Smith's diaries, Oxford University Museum of Natural History [OUMNH]). But by 7 September, Watt's health had taken a turn for the worse, and he now moved to Sidmouth, in Devon, for the sea air. This means their period of overlap in Bath together would have been short. Gregory Watt died in Devon, after a final move to Exeter, on 16 October 1804, according to the mourning rings which his family prepared (Sotheby's, 2003, lots 166, 285). An account of him survives in manuscript, prepared later by James Watt Sr.'s biographer, James Patrick Muirhead (1813–1898). It has never been published (BPL C2/28). Of Watt's geological work it notes how:

> He spent some time in Paris, where he experienced the attentions of Mme de Stael and studied mineralogy under the Abbe Haüy and learned geology from Dolomieu. With this latter study he became enamoured and pursued it with enthusiasm, in a tour through the volcanic districts of Italy, studying deeply the vestiges of those extinct at Rome and the adjoining country, in the Euganean and Vincentini Districts and the remarkable phenomena which the living volcanoes of Vesuvius and the far famed Campi Phlegrae present.

Watt's early death was indeed a loss, perhaps a major loss, to European geology. He was buried in Exeter Cathedral on 22 October 1804 (*Exeter Flying Post*, 25 October 1804, p. 4, col. 3), where only a tiny tablet survives, in South Nave Aisle Bay 2, reading "G.W. Oct. 1804 ae[tat] 27." His father, in retirement, was devastated by his death. James Watt now placed all Gregory's books and papers in a trunk, which he put in his garret workshop at Heathfield, Birmingham, "where his father's glance might rest upon them ... The loss that was the greatest grief to him was that of [both] the children by his second wife" (Dickinson and Jenkins, 1927, p. 73–74; Dickinson, 1970, p. 4). It was the sale of the last of these papers, with books and objects later auctioned in 2003 (Sotheby's, 2003), that brought all of the Gregory Watt papers into the public domain to follow earlier major acquisitions of Watt papers by Birmingham Public Library.

A MYSTERY WATT GEOLOGICAL COLLECTION

An old geological collection survives in Guernsey Museum. This was presented in 1889 by the widow, then of Guernsey, Charlotte Brabazon, née de Saumarez (1818–1896), of Gilbert Hamilton (1803–1882). He had been a managing partner in the Soho Works from ca. 1845 (Dickinson and Jenkins, 1927, p. 278). Hamilton's grandmother, Kathleen MacGregor, was the sister of Anne, James Watt Sr.'s second wife. This collection is supposed to have been started by James Watt Sr. and to have then been given to Gilbert Hamilton by James Watt Jr. (Howell, 1989, p. 16). But in view of the possible Gregory Watt connection, this collection deserves investigation. Certainly a brother of Gilbert Hamilton was soon named Gregory Watt Hamilton (1812–1832) to emphasize the links.

ACKNOWLEDGMENTS

I thank Mary Atkinson (Sidmouth), Glen Caldwell (London, Ontario), John Doskey (Clayton, California), the late David Gibson-Watt and Julian Gibson-Watt (Doldowlod), Martin Guntau (Rostock), Alan Howell (St. Peter Port), Jean Jones (Edinburgh), Herbert Kaden (Freiberg), Kai Kanz (Halle), David Oldroyd (Sydney), the late Peter Schmidt, Ken Taylor (Oklahoma), Lydie Touret (Paris), Ezio Vaccari (Varese), and Gian Battista Vai (Bologna) for their help and encouragement. Staff at the Birmingham Public Library have given me every assistance. Since I have used these collections when they were in both private and public hands and both before and after they were recently recataloged, some of the BPL reference numbers I have given will not be final or current ones. Funds to deliver this paper at Firenze were provided by the School of Earth Sciences, Keele University.

REFERENCES CITED

Acton, F. and F.E., 1969, Genealogia degli Acton: Napoli, private, 55 unnumbered p.
Acton, H.M.M., 1956, The Bourbons of Naples (1734–1825): London, Methuen and Co., 729 p.
Anonymous, 1814, Classificazione delle Rocce secondo i piu celebri autori: Milano, Tipografia Sonzogno e Comp.
Beddoes, T., 1791, Observations on the affinity between basaltes and granite: Philosophical Transactions of the Royal Society, v. 81, p. 48–70.
Boud, R.C., 1975, The early development of British geological maps: Imago Mundi, v. 27, p. 73–96.
Breislak, S., 1812, Introduction à la géologie: Paris, J. Klostermann fils.
Butcher, N., 1983, The advent of colour-printed geological maps in Britain: Proceedings of the Royal Institution of Great Britain, v. 55, p. 149–161.
Carozzi, A.V., 2000, Manuscripts and publications of H.-B. De Saussure on the origin of basalt (1772–1797): Carouge, Geneva, Editions Zoë, 769 p.
Carozzi, A.V., and Newman, J.K., 2003, Lectures on physical geography given in 1775 by Horace-Bénédict de Saussure: Carouge, Geneva, Editions Zoë, 527 p.
Challinor, J., 1953, The early progress of British geology—I: Annals of Science, v. 9, p. 124–153, doi: 10.1080/00033795300200093.
Cockburn, H.C., 1852, Life of Lord Jeffrey: Edinburgh, A. and C. Black, 2 volumes.
Daubeny, C.G.B., 1826, A description of active and extinct volcanos: London, W. Phillips.
d'Aubuisson de Voisons, J.F., 1803, Mémoires sur les basaltes de la Saxe: Paris, Courcier (translated by Patrick Neill into English and published in 1814: Edinburgh, A. Constable and Co.).
Davy, H., 1800, Researches, chemical and philosophical: chiefly concerning nitrous oxide: London, J. Johnson.
Davy, J., 1858, Fragmentary remains literary and scientific of Sir Humphry Davy: London, J. Churchill.
Dean, D., 1989, New light on William Maclure: Annals of Science, v. 46, p. 549–574.
Dean, D., 1992, James Hutton and the history of geology: Ithaca and London, Cornell University Press, 303 p.
De Beer, G.R., 1957, Gregory Watt's tour on the continent 1801: Annals of Science, v. 13, p. 127–136.
De Beer, G.R., 1960, The sciences were never at war: London, Thomas Nelson and Sons, 279 p.
Den Tex, E., 1996, Clinchers of the basalt controversy: Empirical and experimental evidence: Earth Sciences History, v. 15, p. 37–48.
Dickinson, H.W., 1970, The garret workshop of James Watt: London, Science Museum, 21 p.
Dickinson, H.W., and Jenkins, R., 1927, James Watt and the steam engine: Oxford, Clarendon Press, 415 p.
Doskey, J., 1988, The European journals of William Maclure: Philadelphia, Memoirs of the American Philosophical Society, v. 171, 815 p.
Ellenberger, F., 1984, Early French geological maps: Trends and purposes, in Dudich, E., ed., Contributions to the history of geological mapping: Budapest, Akadémiai Kiadó, p. 73–82.
Ellenberger, F., 1985, Recherches et réflexions sur la naissance de le cartographie géologique: Histoire et Nature, v. 22–23, p. 3–54.
Forbes, R.J., ed., 1970, Martinus van Marum: Life and work: Volume 2: Haarlem, H.D. Tjeenk Willink and Zoon, 401 p.
Fox, H.R.V., 1905, Further memoirs of the Whig party 1807–1821: London, John Murray.
Fritscher, B., 1991, Vulkanismusstreit und geochemie: Stuttgart, Franz Steiner Verlag, 346 p.
Gadenz, S., ed., 2004, Past, present and future of the Italian geological maps from ink to digital cartography: 4 DVDs and text, San Giovanni Valdarno, e-Geo.
Gottschalk, K.G., 1867, Verzeichniss derer, welche seit Eröffnung der Bergakademie und bis zum Schluss des ersten Säculam's auf ihr studiert haben, in Festschrift zum hundertjährigen jubiläum der Königl. Sächs. Bergakademie zu Freiberg: Dresden, C.C. Meinhold and Sohne, p. 221–295.
Hall, J., 1800, Experiments on whinstone and lava: [William Nicholson's] Journal of Natural Philosophy, v. 4, p. 8–18 and 56–65.
Hankin, C.C., 1860, Life of Mary Anne Schimmelpenninck, fourth edition: London, Longman, Green, Longman and Roberts.
Houghton, W., 1966, The Wellesley index to Victorian periodicals 1824–1900: Toronto, University of Toronto Press and London, Routledge and Kegan Paul, v. 1, 1194 p.
Howell, A., 1989, Guernsey Museum and Art Gallery: Natural History Collections: Journal of Biological Curation, v. 1, p. 1–16.
Hunt, L., and Jacob, M., 2001, The affective revolution in 1790s Britain: Eighteenth-Century Studies, v. 34, p. 491–521.
Imbruglia, G., 2000, Naples in the eighteenth century: The birth and death of a nation state: Cambridge, Cambridge University Press, 204 p.
Jameson, L., 1854, Biographical memoir of the late Professor Jameson: Edinburgh New Philosophical Journal, v. 57, p. 1–49.
Jenkins, I., and Sloan, K., 1996, Vases and volcanoes: London, British Museum Press, 320 p.
Jones, P.M., 1999, Living the Enlightenment and the French Revolution: James Watt, Matthew Boulton and their sons: Historical Journal (Cambridge, England), v. 42, p. 157–182.
Kazmer, M., 1999, An early Wernerian in Hungary, in Rozsa P., ed., Robert Townson's Travels in Hungary: Debrecen, Kossuth Egyetemi Kiadó, p. 55–58.
Laudan, R., 1987, From mineralogy to geology: Chicago, University of Chicago Press, 278 p.
Marvin, U.B., 1996, Ernst Florens Friedrich Chladni (1756–1827) and the origins of modern meteorite research: Meteoritics & Planetary Science, v. 31, p. 545–588.
Maton, W., 1797, Observations relative chiefly to the Natural History... of the Western Counties of England: Salisbury, J. Easton.
[Miller, J.], 1812, Rowley quarry: Gentleman's Magazine, part 2, p. 513.
Muirhead, J.P., 1859, The life of James Watt: New York, D. Appleton and Co.
Nazzaro, A., 1997, Il Vesuvio - Storia eruttiva e theorie vulcanologiche: Napoli, Liguori Editore, 374 p.

North, F.J., 1931, From Giraldus Cambrensis to the geological map: Transactions of the Cardiff Naturalists' Society, v. 64, p. 20–97.

Oxford Dictionary of National Biography, 2004, Matthew, H.C.B., and Harrison, B., eds., Oxford, Oxford University Press, 60 volumes.

Paris, J.A., 1831, The life of Sir Humphry Davy, two volumes: London, Henry Colburn and Richard Bentley.

Pinkerton, J., 1811, Petralogy: A treatise on rocks, two volumes: London, White, Cochrane and Co.

Porter, R., 1977, The making of geology: Earth science in Britain 1660–1815: Cambridge, Cambridge University Press, 288 p.

Robertson, J.L., 1907, The complete poetical works of Thomas Campbell: London, Henry Frowde.

Robertson, T., 1956, The presentation of geological information in maps: Advancement of Science, v. 13, p. 31–41.

Robinson, E., and McKie, D., 1970, Partners in science: Letters of James Watt and Joseph Black: London, Constable and Co. Ltd, 502 p.

Rupke, N.A., 1983, The great chain of history: Oxford, Clarendon Press, 322 p.

Schofield, R.E., 1963, The Lunar Society of Birmingham: Oxford, Clarendon Press, 491 p.

Silliman, B., 1844, Editorial note: American Journal of Geology, v. 47, p. 2.

Sotheby's, 2003, The James Watt sale: Art and Science, London, 20 March 2003: London, Sotheby's.

Stapleton, D.H., 1985, Accounts of European science, technology, and medicine written by American travelers abroad 1735–1860: Philadelphia, American Philosophical Society, 48 p.

[Thomson, W.], 1795a, Breve catalogo di alcuni prodotti ritrovati nell'ultima eruzione del Vesuvio, Giornale Letteraria di Napoli, v. 41, p. 51–55 (translated into English in 1797 in *Monthly Magazine*, v. 4, p. 91–92; *New York Magazine*, v. 2, p. 623–624).

[Thomson, W.], 1795b, Abbozzo di una classificazione de prodotti vulcanici, Firenze (and in Giornale Letteraria di Napoli, v. 41, p. 59–81) with a later edition of 1801 (of which Thomson's own copy survives in Edinburgh University Library, pressmark *L.30.4/1).

Torrens, H.S., 2002, The practice of British geology 1750–1850: Aldershot, Ashgate, 356 p.

Torrens, H.S., ed., 2003a, Memoirs of William Smith (1844) with an introduction: Bath, Royal Literary and Scientific Institution, 230 p.

Torrens, H.S., 2003b, William Smith (1769–1839) and the search for English raw materials: some parallels with Hugh Miller and with Scotland, *in* Borley, L., ed., Celebrating the life and times of Hugh Miller: Cromarty, Cromarty Arts Trust and the Elphinstone Institute of the University of Aberdeen, p. 137–155.

Torrens, H.S., 2005, The Moravian minister Rev. Henry Steinhauer (1782–1818); his work on fossil plants, their first "scientific" description and the planned Mineral Botany, *in* Bowden A.J., Burek C.V., and Wilding J., eds., History of Palaeobotany: Selected essays: Geological Society [London] Special Publication 241, p. 9–24.

Townson, R., 1797, Travels in Hungary: London, G.G. and J. Robinson.

[von Schleghel, C.W.F.], 1803, Deutsche Fremde in Paris, Europa: Eine Zeitschrift: v. 1, 1803, 2 Stück, p. 162–163.

Wagenbreth, O., 1967, Abraham Gottlob Werner's System der Geologie, Petrographie und Lagerstättenlehre: Freiberger Forschungshefte, C 223, Leipzig, VEB Deutscher Verlag für Grundstoffindustrie, p. 83–148.

Watt, G., 1804a, Observations on basalt and on the transition from the vitreous to the stony texture: offprint from the Philosophical Transactions (70 copies only).

Watt, G., 1804b, Observations on basalt and on the transition from the vitreous to the stony texture: Philosophical Transactions of the Royal Society, v. 1804, p. 279–313.

Woodward, H.B., 1907, The history of the Geological Society of London: London, Geological Society of London, 336 p.

Wyse Jackson, P.N., 2000, Science and engineering in Ireland in 1798: A time of revolution: Dublin, Royal Irish Academy, 81 p.

MANUSCRIPT ACCEPTED BY THE SOCIETY 17 JANUARY 2006

Geological Society of America
Special Paper 411
2006

Giovan Battista Brocchi's Rome: A pioneering study in urban geology

Renato Funiciello*
Dipartimento di Scienze Geologiche, Università degli Studi di Roma Tre, Roma, Italy

Claudio Caputo
Dipartimento di Scienze della Terra, Università degli Studi di Roma "La Sapienza," Roma, Italy

ABSTRACT

Rome was the largest and most important capital city of the ancient western world, but then, for about ten centuries—from the fall of the Roman Empire through the Middle Ages—it lost its demographic consistency and political influence, and with its decline, the advanced infrastructural systems for which the city has been noted were allowed to decay from a lack of upkeep and modernization. The "Eternal City" became a not-very-important village between the religious sites of Quirinal and the Vatican Hills, and one that paid little attention to its rich heritage of historical buildings and monuments.

Forty years after the first edition (1748) of the exceptional topographical map of Rome by Giovan Battista Nolli, Giovan Battista Brocchi, the noted geologist from Castelfranco Veneto, chose Rome as a basis for field surveys needed for his *Carta Fisica del suolo di Roma* (Physical map of the terrains of Rome), a map with explanatory notes that was published in 1820. The purpose of this later map was to illustrate the geology and geomorphology of Rome, and its issuance was timely because it preceded the growth of the city into the greatly expanded urban complex of today.

Brocchi's map was made by direct and systematic examination of the geological surfaces that underlay the more or less continuous superficial covering of man-made materials and structures. Today, this approach, together with the high quality lithological observations on the sedimentary and volcanic rocks and the close attention to the completeness and inter-relationship of the stratigraphic sections, allows Brocchi's map to be considered as the pioneering attempt to produce a geological map of an urban area.

Keywords: geological map, geognostic map, topographic map, Nolli, Cuvier, Brongniart, Smith, Lyell, Darwin, Breislak, Tibullus, Propertius.

INTRODUCTION

For the ancient world, Rome was the prototype modern city. Later, it nearly disappeared; the Roman heritage only slowly reemerging in the wake of the economic and social development of the western world during the tenth century. Man's activities, previously developing mainly in the countryside and with an agricultural focus, then became progressively concentrated in urban areas (Le Goff, 2004). Ancient Rome was rediscovered under a veneer of wild vegetation, where it lay half destroyed by time and neglect. During the Renaissance, artists re-exposed the city's classical origins in their architecture, paintings, and

---

*E-mail: funiciel@mail.uniroma3.it.

Funiciello, R., and Caputo, C., 2006, Giovan Battista Brocchi's Rome: A pioneering study in urban geology, *in* Vai, G.B., and Caldwell, W.G.E., The Origins of Geology in Italy: Geological Society of America Special Paper 411, p. 199–210, doi: 10.1130/2006.2411(12). For permission to copy, contact editing@geosociety.org. ©2006 Geological Society of America. All rights reserved.

sculptures. Beginning with the eighteenth century, the Vatican religious authorities also recognized the value of the city's classical roots, and they leant their weight to restorative measures when Pope Prospero Lambertini (Benedict XIV) directed a commission to revise existing records of Rome's topography and, in this context, to record with consistent accuracy the sites of ancient buildings and monuments. The map of the skilled geometrician, Giovan Battista Nolli, was used as a starting point for this important undertaking.

Nolli's topographic map came to have another important application in the hands of Giovan Battista Brocchi, who used it as a base map for the first "geognostic map" of Rome, which, following his own original methods, he produced in 1820. Although described somewhat vaguely as, "a map of the physical conditions of the terrains of Rome," this map contained adequate geological representation for it to be said that, in this way, Rome became one of the first cities in the world to have its own dedicated "geological" map. Leaving aside the imperfection of a minor mismatch between the legend and the map, or more precisely between virtual stratigraphic sequence and the true order of stratal superposition (Vai, 1995; Vai *in* Gadenz, 2004), Brocchi's map also plays a historic role in recording the transition from geognostic map to geological map in the modern sense—a significant step in the evolution of the geological map as a key element in urban geological studies today.

By focusing on the work of Nolli (1748) and Brocchi (1820), this paper seeks to trace the history of the geological map as a critical compilation in considerations of urban geology.

TOPOGRAPHIC SURVEYS OF URBAN AREAS IN THE SEVENTEENTH AND EIGHTEENTH CENTURIES

After having been *the* city of the ancient world, Rome, at the beginning of the eighteenth century, ranked little apart from the other major cities of Europe. In all of them, demographics and city-planning were becoming increasingly important issues, and, concurrently, there developed an intensified realization of the importance of preserving heritage and history. The concentration of goods and services within the cities, and, largely as a result, their renewed physical growth, began with the late Middle Ages and quickly became significant in Italy. These early signs of urbanization marked a historical turning point, insofar as they contrasted with emphasis on the rural-dwelling peasant population of the feudal system in the early Middle Ages.

Expansion of the cities generated a demand for a progressively more exact cartographical representation of their burgeoning boundaries and for a system of public record of the extent, value, and ownership of land for purposes of taxation. In short, urban expansion was paralleled by a demand for official, cadastral reference maps.

In Rome, furthermore, concern of the western world for the city's famous buildings and monuments and for preservation of the records and values of the classical world—a concern continuous from the time of the Renaissance—intensified to become a matter of primary importance. The attraction of the museum city with its concentration of past witness of the Roman empire generated a progressive demand for action from the time of Pope Pius VII (1800–1820) onward. Certainly, by the eighteenth century, the monuments of ancient Rome were no longer regarded as simply objects of an "impious culture," reported on the papal epigraphs on several monuments in Rome, and the need to recognize and respect them for the great range of values they represented had strengthened and widened. It was entirely logical that, in Rome, progressively more detailed and precise maps should be looked upon as a component of that effort because by then the city had a history of recording information in topographic documents and related maps—a history that dated from the time of the ancient city (for example, the topographic document *Forma urbis severiana*, the form of the city at Severus's time). Such early cartographic portrayals as "*Roma Antica*," by Pirro Lagorio (1561), and "*Roma*," by Stefano Du Pérac (1577), both published by Lafréty (Frutaz, 1962), were followed by the exceptional topographic map produced by Nolli in 1748.

Rome and Nolli's Map

Giovan Battista (Giambattista) Nolli (1692–1756) (Fig. 1), architect and engraver from Como, was an outstanding representative of the Lombard school of geometers and topographers, which, from the early eighteenth century, provided accurate maps to city and town administrations to permit them to assess the growth of their centers.

At the beginning of the Age of Enlightenment, the image of Rome was one rooted in the city's architectural distinctions. There was no shortage of perspective maps depicting the ancient buildings and monuments, but there was a total lack of maps drawn to scale as if looking down from the highest point of the city, or, like most modern maps, of the city in plan view, as if looking down from a point in the sky directly overhead. It was just at this time, too, that demand for such maps rose dramatically. There was a call for change from the perspective maps, compiled by various techniques, to maps compiled by a standard technique—one based on the "numerical" logic of the trigonometrically based topographic survey. What brought this change to reality was the introduction of the plane table in the early seventeenth century by the German mathematician Johannes Richter, also known as Pretorius, and what facilitated the change in Italy was the introduction of the handbook by Angelo M. Ceneri, *Uso dello strumento geometrico detto della tavoletta pretoriana*, which showed how, using the plane table, the topographic survey could be executed systematically and scientifically (Spondberg Pedley, 2005).

The change first took place in Milan when, between 1720 and 1722, Giovanni Filippini, a member of the acclaimed Lombard school of topographic cartography (*Catasto asburgico della Lombardia*), issued the first cadastral map of that great northern city. The map was a response to a convergence of needs—the need for an exact knowledge of the extent and position of real

Figure 1. Nolli's caricatural portrait in 1748.

used to construct that map, Nolli, together with fellow architect and engraver, Giovan Battista Piranesi, who became famous for developing a special means of portraying and glorifying the monuments of Rome (Frutaz, 1962), used the techniques paradigmatic of the Age of Enlightenment: they adopted the more accurate and reproducible numerical techniques based on trigonometric methods, the plane table, and a more refined, systematic use of the compass for architectural surveys. The result was a map noted for its many advances and innovations. These included (1) the degree of accuracy designed into the map by relying on large-scale surveys (1:100–1:300), then subsequently reproducing the results, initially on maps at a scale of 1:1000 and finally at a scale of 1:3000; (2) a planimetric portrayal of all the churches and buildings, with architectural details consistent with the scale of the map; (3) detailed representation of the "fabric" of the city; (4) detailed representation of open spaces, even to the point of marking the main types of trees; (5) portrayal of the topography and morphology of all the major elements, with an indication of differently sloping areas; and (6) the detailed portrayal of monuments.

Although sketched by 1736, the first draft of Nolli's map appeared in 1744 as an untitled and unpublished map of Rome (Frutaz, 1962). Benedict XIV (1740–1758) used this map as a launching point for a new organization of the fourteen districts of the city, as ordered earlier by Pope Sixtus V (1585–1590), who wanted to "make of Rome the model of all cities."

Pope Benedict's approach to the reorganization was to direct a commission, chaired by the papal chamberlain, Cardinal Alessandro Albani, to revise the topographic districts of the city and their associated administrations (Castagnoli et al., 1958), a task that, understandably, was greatly facilitated by Nolli's new map.

The synthesis of information that came to be accurately and precisely portrayed in outstanding detail on this new map made it unique. Today, it serves not only as a tribute to the mythological origins and history of the Eternal City, but also to the level of cartographic accomplishment that had been attained by the time it was produced. That it was still used to define the urban cadastre in the century after that in which it was compiled and published (Fagiolo, 2005) testifies to its durability.

In 1748, Nolli's great map was issued at a scale of 1:3000, this version engraved by Nolli himself and his son, Carlo, and published after the work of the Cardinal Albani Commission had ended. Divided into three sheets and depicting the city in a north-south orientation, the 1748 map contains much more detailed and precious information, including plans of the most important buildings. Drawn in iconographic and not perspective portrayal (Castagnoli et al., 1958), to this day the map serves as a source of various kinds of information, ranging, for example, from the detailed topographic configuration of the eighteenth-century city to the history of urban patrimony.

Also in 1748, Nolli published a smaller map of Rome, the cartouche of which contains the following dedicatory information: *Sig.r Cardinale Alessandro Albani E.mo e R.mo Principe ardisco d'offerire all'Em.za V.ra ristretta in questa picciola*

property, the need for an estimation of relevant values, and the need to provide appropriate warranties to both administrators and owners. With the issuance of the cadastral map of Milan, the figure known today as the "surveyor" came to be acknowledged as an indispensable player in a society becoming increasingly dependent on technology, and the plane table came to replace the surveyor's cross.

Nolli's work in Rome may be seen as fitting in with the major, European, urban cartographic experiences of the time, among which must be noted Ogilby and Morgan's 1676 map of London—the first in vertical projection, required for planned reconstruction of the city after The Great Fire of 1666—and a "plan Turgot" map of Paris (1737), a magnificent portrayal of the city, but one devoted more to commemorative purposes than to meeting increasing demand for accurate and objective urban cartographic documentation (Bevilacqua, 2005b).

In Rome, the last important attempt to map the city prior to the survey that led to Nolli's map of 1744 was that recorded by the celebrated map of Giovan Battista Falda in 1676 (Frutaz, 1962; Bevilacqua, 2005a). But relative to the techniques

Pianta la Topografia di Roma.... Roma il dì primo del 1748 (The Cardinal Alessandro Albani, His Eminence, Right Reverend Prince, I dare to offer His Eminence Rome's topography in this reduced small map ... Rome, the first day of 1748). Further information specified by Nolli is printed at the bottom-center of the map, where the "title" appears: *La Topografia di Roma di Giobatta Nolli dalla maggiore in questa minore Tavola dal medesimo ridotta* (Rome's topography by Giobatta Nolli, reduced from the greater to the smaller plate by the creator himself). The title of the map is also found on the side wall of the pedestal of Trajan's Column (Frutaz, 1962), and the following phrase is written on a small architectural feature close to it: *Scala di 2500 palmi romani d'architettura* (Scale of 2,500 Roman architectonic palms). On the left side of this feature, the following words appear: *Piranesi e Nolli incisero* (Piranesi and Nolli engraved). The small map of 1748 was in fact drawn up by Giovan Battista Nolli, but engraved by his son, Carlo, and by Giovan Battista Piranesi.

Brocchi and Nolli's Map

Giovan Battista (Giambattista) Brocchi stated in his 1820 publication (p. 208) that he used the 1773 reprinted version of Nolli's map as a cartographic base for his *Carta fisica del suolo di Roma* (Physical map of Rome's terrain). In reality, however, this statement is an oversimplification. The working map actually used by Brocchi (Fig. 2) was recently found in manuscript form in the Servizio Geologico Nazionale (Italian Geological Survey), and this revealed that Brocchi used a new version of Nolli's map, reduced by Nolli, but then revised by Ignazio Benedetti. In this working map, Nolli's original cartouche was replaced by one in which Ignazio Benedetti offered the work to Cardinal Giovanni Carlo Boschi, with the following rather elaborate specification: "*essendo divenuta rara la Pianta Topografica di Roma publicata già in un sol foglio dal celebre Giovanbattista Nolli nel 1748, ho intrapreso, e condotta a termine questa nuova edizione per soddisfare alle ricerche, che se ne fanno, ed ho aggiunto alla medesima la Pianta delle Fabbriche da quel tempo inalzate.... Roma il dì primo del 1773* (as Rome's topographic map, already issued in 1748 in one sheet by the famous Giovan Battista Nolli, became rare, I undertook the preparation of this new edition in order to meet demand, and I added to it the Plan of Edifices built from that time [i.e., 1748] onward. Rome, the first day of 1773). All of this makes sense when it is recalled that Nolli died in 1756.

More specifically, in the 1773 version, the index of the *Fabbriche più ragguardevoli* (most important edifices) contains three new items, added in layout and writing distinct from those of the earlier version. Moreover, in the indication of the map's type on the side wall of the pedestal of Trajan's Column, the following words are inserted: "*e da Ignazio Benedetti incise*" (and engraved by Ignazio Benedetti), and on the stone building block close to the foot of the pedestal, the reference to Nolli and Piranesi (see above) is deleted.

Despite its smaller scale, the 1773 map—the topographic working map that Brocchi used—shows essentially all of the outstanding features of Nolli's great map of 1748. Of particular note, the city's fabric inside the Aurelian Walls is drawn in remarkable detail, except for the northern part where only country houses and vineyards are shown. Furthermore, a three-dimensional variation in the steepness of slope of the hills upon which Rome is built is portrayed with seemingly plastic gradualism by using a pattern of fine hatching. Such technique used to expose the relative prominence of the city's positive features does double duty by exposing also the negative depressions, such as the troughs running from the Palatine (even more so from the Esquiline and Coelian hills) to the Coliseum "valley"—the depression that was once the center of the Neronian Domus Aurea. The same may be said of the area between the Palatine and the Aventine (the ancient Vallis Murcia, later site of the Circus Maximus), on the eastern side of which the slope toward the Circus Maximus exhibits an apparent but not geomorphologically meaningful break in its alignment with the main part of the present Via del Circo Massimo.

It is doubtless that Brocchi derived great benefit from the wealth of detail offered by Nolli's map, especially in the site-specific investigations he conducted himself.

When Brocchi's map was completed in draft, it was engraved by Pietro Ruga in 1820 (Fig. 3) and released as the *Carta fisica del suolo di Roma ne' primi tempi della fondazione di questa città* (Physical map of Rome's terrain during the early period, after the foundation of the city had been established).

GIOVAN BATTISTA BROCCHI AS NATURALIST AND GEOLOGIST

Evolutionary Changes in His Scientific Approach

Giovan Battista Brocchi (1772–1826) (Fig. 4) was born in Castelfranco Veneto and went to Rome for the first time in 1791, long after Nolli's death in 1756. Brocchi's eclectic way of thinking was already manifest in his youth, and he maintained it into adulthood. He began to study law at the University of Padua, but then shortly thereafter he turned to ancient history and dedicated the first part of his life in Rome to the pursuit of Egyptology. Following upon that, he became devoted to natural history, which he approached with a kind of Renaissance thought process, radically changing his approach from an initial, highly practical and utilitarian one directed toward an effective exploitation of natural resources to a much more theoretical one directed toward the fundamental understanding and appreciation of the beauty and wonder of the natural process and product. This change in thinking can be dated to ~1796, being first evident, perhaps, in his *Trattato delle piante odorifere e di bella vista da coltivare ne' i giardini* (Treatise on sweet-smelling and good-looking plants to cultivate in the gardens) of that date. In the early years of the nineteenth century, a blend of Baconian-Newtonian-Lockian scientific thinking with Platonic-Leibnizian nonscientific thinking is evident in Brocchi's writings, as it is also in the thought processes of a number of his contemporaries. His interest in, and quality of, his studies in natural history progressed, evident in his

Figure 2. The Nolli map, the first Roman modern topographic map, was the tool used by Brocchi in preparing the *Carta Fisica del suolo di Roma* (1820). Working document recovered within the Italian Geological Survey historical library in Rome.

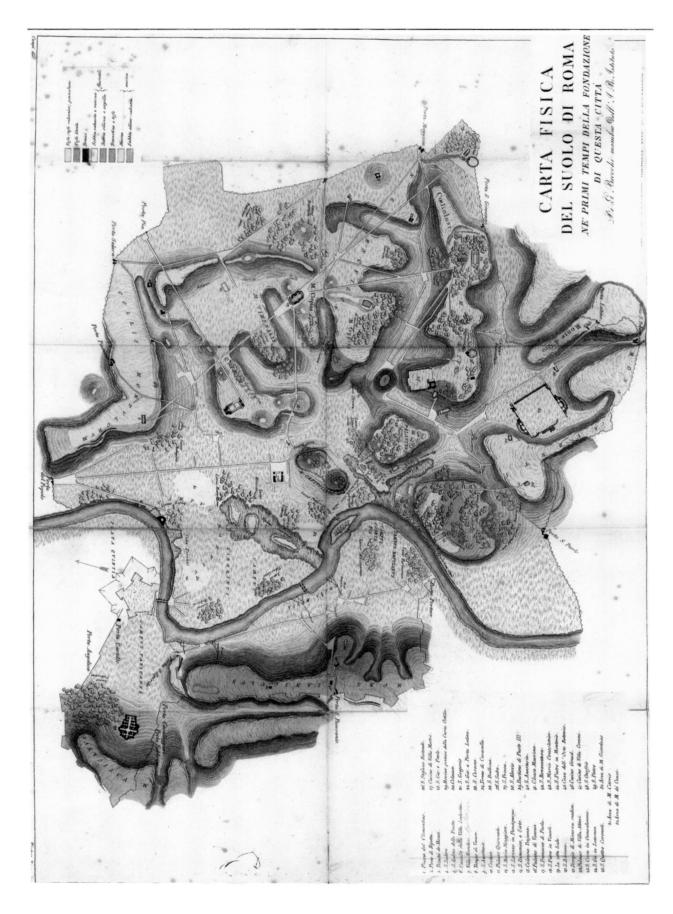

Figure 3. *Carta fisica del suolo di Roma* (Brocchi, 1820).

Brocchi and the Earliest Geological Maps

In France in 1811, Cuvier and Brongniart published their *Essai sur la Géographie mineralogique des Environs de Paris,* which immediately became the reference of note in the European literature on natural history. The volume contains a splendid hand-colored map, together with a description of the rocks and fossil biotas of the Paris basin, and fourteen cross sections to illustrate the geology of Paris itself, including such well-known districts as Montmartre, Saint Germain des Près, and Vaugirard. A complete and updated version was published by Cuvier and Brogniart in 1822.

At about the same time, William Smith (1769–1839), engineering surveyor and meticulous analyst of the Mesozoic fossiliferous rocks he encountered in his early work in southwestern England, had formulated his principle by which successive "formations" could be distinguished and correlated by the fossil remains they contained. Denied access to the leading geologists of his day by means of fellowship in the Geological Society of London and ill at ease in providing written accounts of his investigations—both a consequence of his lowly birth and poor education—he developed the technique of illustrating the stratigraphical geology of different region by means of geological maps. Fairly early in his career he exhibited a number of these in the Agricultural Society of Bath, an organization into which he was welcomed and in which his geological discoveries were not only recognized but championed. Later in his career, Smith went on to complete his celebrated map of the southern part of the United Kingdom: *A Delineation of the Strata of England and Wales with Part of Scotland* (Smith, 1815). This large (3m x 2m) and impressively detailed compilation, recently hailed as *The Map that Changed the World* (Winchester, 2001), is now accommodated in Burlington House, Piccadilly, London, the home of the Geological Society.

Thus, the turn of the eighteenth century and the early years of the nineteenth century may be looked upon as a time when the geological map came to the fore as a means of visually explaining the stratigraphical and structural relationships among different rocks. The differences between the various cartographical expositions of geology dating to this time were numerous and great. The extent of the area covered by the maps, their scales, what they were designed primarily to expose, and who their prime users were expected to be, were by no means the same. Despite the fact that Smith, in Britain, was working on a regional and ultimately national scale, Cuvier and Brongniart in France on a regional scale, and Brocchi in Italy on an urban scale, these naturalists pursued parallel paths toward common goals—those of persuading the contemporary community of scholars and artisans of the distinctiveness and importance of the emerging science of geology, of its theoretical and, more particularly, its practical applications and values, and of the powerful role that may be played by a graphical exposition of geology in reaching those goals (Table 1).

Figure 4. Brocchi's portrait.

Trattato mineralogico e chimico sulle miniere del Dipartimento di Mella (Mineralogical and chemical treatise on the iron mines of the Mella Department). At this time too, the advanced nature of Brocchi's thinking about natural processes became clear (Brocchi, 1808), most evident perhaps in the steadfast adherence in his works to the principle of the universality of nature's laws.

Brocchi made a name for himself as an Italian geologist of international stature, who became recognized particularly in the field of paleontology for his *Conchiologia Fossile Subapennina* (Sub-Apennine fossil conchology) (Brocchi, 1814). In his paleontological endeavors, he emerged as a forerunner to Lamarck and Charles Darwin, although he did not adhere to the evolutionary interpretations of either of these individuals, but instead had his own distinct views on the transmutation of species (Berti, 1987). Charles Lyell knew of Brocchi's work and drew Darwin's attention to the essential elements of it. Darwin in turn, although unable to read Italian himself, gave careful consideration to Brocchi's thinking and his data. The evolutionary views of the two remained far apart, however, the fundamental divergence caught in their respective appreciation of purpose and progress in evolutionary theory. Brocchi's stated position was that the destiny of species was "planned," Darwin's that it was "unforeseeable" in the sense of being "blind" and thus unpredictable.

TABLE 1. HISTORICISTS (DARWINIAN*) VS. DETERMINISTS (NEWTONIAN°)

Darwinian vision	Newtonian vision
The more you observe, the more the observed object becomes complex	The more you observe, the more the system becomes simple
Priority to contingency and complexity of historical factors	Priority to the knowledge of starting conditions
Reluctance to draw general laws	The search of general laws
Contempt for sketched and caricatural visions of Nature	Predictive (chaos and quantum mechanics)
	Central role for the search after ideal systems (ideal gases, harmonic oscillations

Note: From Harte (2002).
*From *Hutton to Ippolito* until models of geodynamics.
°From *Newton to Laplace to Kelvin* until uncertainty principle.

More specifically, Cuvier and Brongniart, in France, and Smith, in Britain, showed how it was possible to use contemporary principles and methods to represent bidimensionally, with geometric accuracy and precision, the lithostratigraphic characteristics of regional rock successions, while Brocchi, in Italy, when defining the "first cartographic sketch of the city of Rome," proposed widely accepted methods for deciphering the natural foundation of a densely populated area, recognized original geognostic units, and began correctly to refine interpretation of the city's Quaternary deposits.

With regard to Brocchi's more localized efforts, it must be noted that, in 1801, Scipione Breislak published his two-volume, more than 600-page *Voyage physique et lithologique dans la Campanie* (Physical and lithological journey in Campania), which follows upon an accompanying memoir titled *Constitution physique de Rome* (Physical structure of Rome). Part of the latter work describes geological and geomorphological features of Rome. Indeed, Breislak's study was the first to attempt a synthesis of the geological features of the Roman area following upon the observations made earlier by Leopold von Buch in his *Sur la constitution physique de la plaine de Rome* (On the physical structure of the Roman plain). Von Buch had made some interesting observations: for example, he had challenged the attribution to alluvial processes of some of the prevailingly volcanic deposits traceable to the Alban Hills, a matter that Breislak, soundly experienced in the Naples volcanic region, had no difficulty in settling by recognizing the primary origin of these and other volcanic products. The only open question remaining after Breislak's study was that of the inferred remnants on the Aventine of an ancient volcanic crater that could have been the feeder for the volcanic rocks located in the other Roman hills.

As part of his study, Breislak contributed a map titled *Plan physique de la Ville de Rome* (Physical map of the city of Rome), which the young Brocchi seems to have taken as further fuel for his fire to undertake a cartographical revision of the works of his distinguished predecessors—a fire that had been ignited by his development of an increasingly critical mind and innovative methods, to say nothing of having perceived his predecessors work to have been too definitive and hasty in assumption and conclusion. Some insight into the innovative methods that Brocchi brought to bear on his investigations of the Roman foundation is apparent in the following remarks (freely translated):

> Modern Rome rests on a foundation of fluvial deposits. They are nowhere exposed, however, and may be seen only by breaking deeply into the ground or otherwise extracting samples from beneath the surface. This latter method I mostly used, availing myself of a solid iron drill, which could be extended as required. Using this instrument, and excepting where I encountered invisible obstacles that were in vain to move, I could come to know the native terrains wherever convenient to do so (Brocchi, 1820, p. 86).

Needless to say, systematic application of this technique generated much original data on Rome's ancient geological environment.

Brocchi's Reconstruction of the Original Ground Surface of Rome

Brocchi decided to take advantage of the accuracy and reliability of Nolli's topographic map for his first survey of Rome. Preserved in the offices of the Italian Geological Survey, the original copy of Brocchi's first sketch of the geological map of Rome, based on the 1773 version of Nolli's map, speaks to Brocchi's dependency on the base map. By accepting Nolli's confirmed altitudes, he used it to establish the geomorphological basis for his *Carta geognostica della città* (Geognostic map of the city), in which he showed the relief boundaries of the tuff on the left side of the Tiber River (the Seven Hills), the articulated complex of the Ianiculum-Vatican hill system, and the correct configuration of the alluvial plain with its artificially raised parts (Monte Citorio, Monte Savello, Monte Testaccio, Monte Giordano) and depressions (Palus Caprae, Lacus Curtius). To define these elements, Brocchi wisely combined the topographic information available from Nolli's map with observations on known exposures and with data gathered with the aid of his ingenious hand drill. Brocchi also made handwritten annotations directly on Nolli's map about the limits of different geological units—continental, marine, and volcanic in origin—recognized in the restricted, but highly significant, historical center of the city (Fig. 2). Today, this unpublished sketch map serves as a historical geological "proto-map" of the city, just as it also served Brocchi in his day as a proto-map he conveniently could use to better organize the observational data he collected during his long periods in the Eternal City. The proto-map, moreover, could be construed as Brocchi's initial answer to Breislak, insofar as it demonstrated that the latter's nineteen-year-old physical map of the city had already been outdated (Fig. 3).

Brocchi's reconstruction of the original surface of Rome correlates well with the digital elevation model (DEM) provided by the Italian Geological Survey of the historical center, obtained by Amanti et al. (1995) through the processing of data gathered by direct underground exploration and through a recently compiled database (in Funiciello et al., 1995) (Fig. 5). In short, the basic modernity of Brocchi's work has lasted for nearly two centuries. Leaving aside the technological artifice he relied upon, his investigative methods remain central to the study of any urban area today. Particularly impressive in this respect are his detailed assessments of the notable thicknesses of the backfill layer, which he determined by direct observational reference to the basal structures of some celebrated Roman monuments (the imperial columns of Marcus Aurelius and Trajan, for example), which are now themselves completely buried in that backfill.

The Explanatory Notes to Brocchi's Map

The explanatory notes attached to Brocchi's (1820) *Carta fisica del suolo di Roma*—the first notes of their kind accompanying such a map—were designed by him to enhance the "explanation of the geognostic map of this City." They reveal that Brocchi saw natural outcrops and viewed urban landscapes that

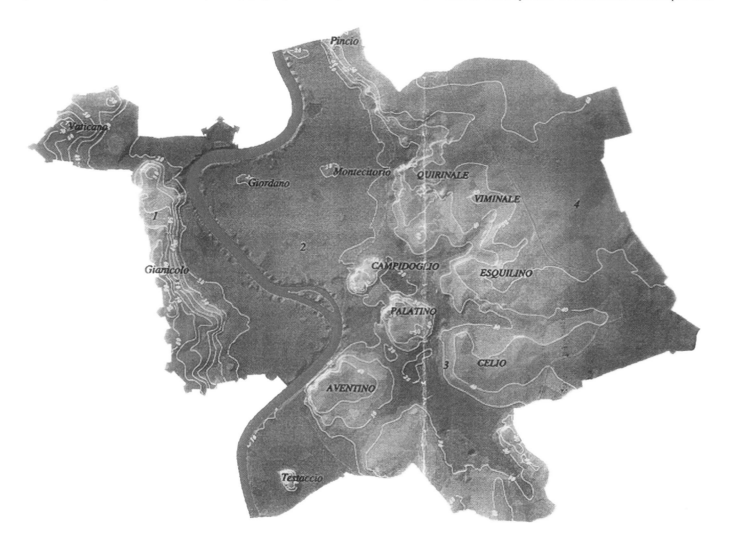

Figure 5. Digital model of the real surface beneath the human cover of the Rome historical center. The computer-generated document fits with Brocchi's 200-year-old map. From *Modello digitale della superficie reale del Centro Storico di Roma* (Amanti et al., 1995).

have long since been lost to the city's burgeoning growth. Today, these notes form a valuable component of Rome's geological and geomorphological heritage, which still permit conclusions to be reliably drawn on documented historical evidence rather than inferentially drawn on the restrictive evidence of the drill sample or speculatively drawn on even less.

Brocchi's fifteen geological cross sections of the Roman area constitute a further valuable supplement to the map, and in many respects, they serve a similar function in graphical form as the notes do in verbal form. In particular, the sections portray what was to be seen geologically and geomorphologically before the huge growth of the city that followed the Italian unification. Some of the cross sections carry hand-written footnotes of their own, like those by Alessandro Portis (Figs. 6A, 6B). The cross sections are variable in the detail that they show and in the extent to which they continue to provide a satisfactory interpretation of sequence and structure. Thus, the interesting section of the Via della Consolazione remains inadequately investigated and explained even in the present day, whereas, in contrast, all of the sections relevant to the Roman hills (especially the Seven Hills) on the left of the Tiber, together with those on the right of the river (Monte Mario, Vatican, Ianiculum, and Monteverde), exhibit remarkable detail and remain durable interpretations.

The combination of maps, sections, and notes left by Giovan Battista Brocchi give rise to an informational inheritance of a unique kind and value. Although the graphical renderings may be of more instant appeal to the geological researcher, it is the notes that offer the greatest real insight into Brocchi's innovative ideas and depth of understanding. The following examples are provided to make this point.

Brocchi was the first to recognize flow structures in some pyroclastic rocks, allowing him to interpret them in a thoroughly original and modern way as volcaniclastic sediments deposited in a fluid medium, their emplacement controlled by the preexisting physiography of the district in which they are found. Using such rocks and his convincing interpretation of their depositional mechanism, he was able, moreover, to tackle Breislak's (1801) problem of the origin and mode of accumulation of these explosive volcanic products and refute his hypothesis to explain them. His criticism of Breislak, freely translated, carries more than a touch of irony:

Mister Breislak, intending to use his talent to support an opposing view and being entirely persuaded that active belching from volcanic vents actually took place within the precincts of Rome [more specifically on the Aventine], having advanced some physical observations believed to corroborate his assumption, then immerses himself in mythology (Brocchi, 1820, p. 186).

In an entirely different vein, Brocchi's notes draw attention to the extraordinary mammalian (more specifically proboscidean) fauna recovered from fluvio-lacustrine Pleistocene deposits of the Galerian cycle found in different parts of Rome (Tor di Quinto-Vigna Giardini, Monteverde, Colli della Farnesina, Sedia del Diavolo) and also mixed with alluvial deposits in volcanic terrains (Montesacro). In similar fashion, the fresh-water molluscan fauna of Pleistocene (Galerian) and Holocene continental deposits and the much better-known marine molluscan fauna of Monte Mario and Vatican Hill are well described in Brocchi's explanatory notes.

Brocchi's comments also reveal his perceptive interpretation of the role that Holocene waters and their deposits played in the early historic period in modifying the configuration of the land surface on which the city was built. He suggested correlating surficial alluvial deposits with historically recorded floods of the Tiber and with hydrometric measurements made inside Rome itself, the last completed in additional handwritten notes by Alessandro Portis (1853–1931). Portis was professor of geology in the University La Sapienza of Rome; and author of a new contribution on the geology of the Rome area in 1893. Brocchi also pointed out in his notes the effect that water had had on the earliest stages of human settlement, observing the importance of the Velabrum River bordering and protecting the prehistoric settlements on the slopes of the Palatine and Capitol hills, and how waters of the Tiber drainage net had shaped the left side of the river and brought about isolation of the small tuffaceous Seven Hills. He was not the first, however, to report historically on some of these matters. Propertius (1984) had already described the ford for sailing boats of the Vallis Murcia-Circus Maximus in the following terms:

Qua Velabra suo stagnabant flumine, quaque
Nauta per urbanas velificabat aquas (Propertii Elegiarum, IV, IX, 5)
(where the Velabra River was stagnant, the seamen sailed through the city waters).

This claim was supported by Tibullus (1988) who wrote,

et qua Velabri regio patet, ire solebat
Exiguus pulsa per vada linter aqua (Tibulli Carminum, II, 5, 33)
(and across the Velabra area, small-oared boats were floating in shallow water).

It may be that Propertius was referring to the Velabrum Maior or Vallis Murcia (the present valley of the Circus Maximum), which was occasionally flooded by the Tiber. Its oversaturated clay deposit covering the imperial remains at the confluence of the Forum Boarium and the Tiber also testify to the episodic flooding.

Lastly, and turning from surficial waters to groundwater, Brocchi (1820) reported on the different springs of the Forum, their supposed locations, and any anomalous properties they had (for example, the Aquae Lautolae, which was weakly hydrosulphuric, maybe warm). Manifestations of the causes of any chemico-physical abnormalities were also sought and identified, such

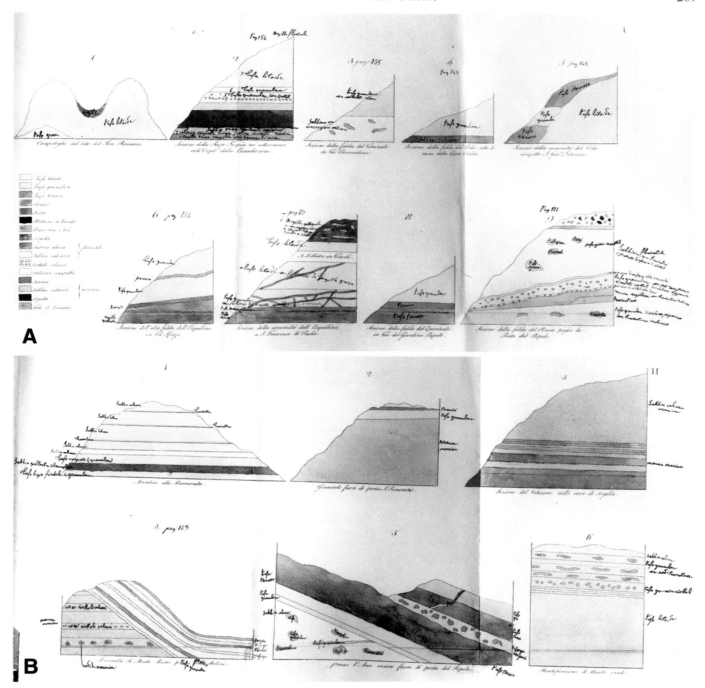

Figure 6. (A) Geological sections of Rome (Brocchi, 1820), with handwritten notes by A. Portis (1890). (B) More geological sections of Rome (Brocchi, 1820), with handwritten notes by A. Portis (1890).

as, for example, a correlation between limey springwater and the clear, thin-fracture, carbonate mineralization of the tuffs of the Tarpeius Mons and some travertine levels between the Tabularium and the Mamertine Prison. Different springs identified by Brocchi, for example, Castores, Iuturna, and Lupercal, were to be fed by the same aquifer that feeds the small, swampy basins common in the flat ground among the Roman hills.

SUMMARY AND CONCLUSIONS

The history of geological exploration within the city of Rome led to Giovan Battista Brocchi, a geologist from Castelfranco Veneto, producing the first geognostic or proto-geological map of an urban area in 1820—that is, in the later stages of the period when geological maps were being pioneered in other

European countries on larger regional and national scales. The value of Brocchi's map as both a historical document and a basic guide to urban expansion has diminished little in the nearly 200 yr of its existence.

Brocchi's map was based on carefully executed field work, completed not in a gentle, open, readily surveyed tract of land like the Campagna Romana, but within the relatively restricted bounds of a city that had already seen some 2000 yr of construction and historical development. Recognizing this, Brocchi decided to base his map on the original, natural, topographical surface of the city, and he used a primitive hand drill to probe beneath the archaeological-historical foundations of the city, much of which, in a decayed and ruined state, had been transformed into a leavening backfill for depressions in that original surface. The accuracy of Brocchi's map was greatly enhanced by his use, as a cartographical base, of Giovan Battista Nolli's remarkable 1748 topographical map of Rome. This map—north-orientated (thereby permitting azimuthal plots to be made on it)—was produced in different versions and on different scales. Brocchi used the 1773 version.

In the course of compiling the map, Brocchi brought his geological originality and knowledge to bear on a number of problems of Roman geology. For example, by recognizing flow structures in some of the pyroclastic rocks of the Seven Hills area, he was able to attribute their deposition to a fluid medium and thereby resolve a controversial issue. He also successfully resolved the lithostratigraphical classification of the complex and peculiar Plio-Pleistocene sedimentary and volcanic deposits of the city; and by making original observations and correlations between the channel ways and alluvial deposits of Holocene surface waters, and the configuration and extent of underground aquifers that fed important springs, he was able to supplement his geological and geomorphological portraits of the city with one that was fundamentally hydrological.

Brocchi's map is the synthetic product of a technologically advanced style of surveying, the availability of a sound topographical base map, and the perceptive observations and imaginative interpretations of a skillful broad-based geologist. Supplemented as it is by illustrative cross sections and reinforced by valuable explanatory notes, it stands as an exceptional pioneering contribution to what is recognized today as the important, developing field of urban geology.

ACKNOWLEDGMENTS

The authors thank Giovambattista Vai, Philippe Taquet, and Ellis Yochelson for their critical reviews of the manuscript.

REFERENCES CITED

Amanti, M., Cara, P., and Pecci, M., 1995, Modello digitale della superficie reale del centro Storico di Roma, in La geologia di Roma: Il Centro Storico: Memorie Carta Geologica d'Italia, Poligrafico dello Stato, v. 50, p. 309–319.

Berti, G., 1987, Aspetti della formazione scientifica e intellettuale di Giambattista Brocchi, in L'opera scientifica di Giambattista Brocchi (1772–1826): Atti del Convegno, Bassano del Grappa, 9–10 Nov. 1985, p. 13–40.

Bevilacqua, M., ed., 2005a, Nolli, Vasi, Piranesi: Immagine di Roma Antica e moderna: Rappresentare e conoscere la Metropoli dei Lumi: Roma, Artemide Editrice, 118 p.

Bevilacqua, M., 2005b, Percorsi e Incontri nella città del Settecento, in Bevilacqua M., ed., Nolli, Vasi, Piranesi: Immagine di Roma Antica e moderna: Rappresentare e conoscere la Metropoli dei Lumi: Roma, Artemide Editrice, p. 19–29.

Breislak, S., 1801, Voyages physiques et lithologiques dans la Campanie, suivi d'une mémoire sur la "Constitution physique de Rome": Paris Dentu Editrice, v. 1, 320 p., v. 2, 300 p.

Brocchi, G., 1808, Trattato mineralogico e chimico sulle miniere di ferro del Dipartimento del Mella, 2 volumes, Brescia, Bettoni.

Brocchi, G., 1814, Conchiologia fossile subapennina con osservazioni geologiche sugli Appennini e sul suolo adiacente: Milano, Dalla Stamperia Reale, v. 1, 56 + lxxx + 240 p.

Brocchi, G., 1820, Dello stato fisico del suolo di Roma—Memoria per servire d'illustrazione alla carta geognostica di questa città: Roma, Nella Stamperia De Romanis, 282 p. (with two color plates of cross sections and a color geognostic map).

Castagnoli, F., Cecchelli, C., Giovannoni, G., and Zocca, M., 1958, Topografia e urbanistica di Roma, in Storia di Roma: Bologna, Istituto di Studi Romani, Licinio Cappelli, v. 22, 795 p.

Cuvier, G., and Brongniart, A., 1811, Essai sur la géographie minéralogique des environs de Paris, avec une carte géognostique et des coupes de terrain: Paris, Baudouin, Imprimeur de l'Institut Impérial de France, viii + 278 p., 2 plates and a colored map.

Cuvier, G., and Brongniart, A., 1822, Description géologique des environs de Paris: Paris, II édition amplifiée.

Fagiolo, M., 2005, L'Immagine di Roma moderna, in Bevilacqua, M., ed., Nolli, Vasi, Piranesi: Immagine di Roma Antica e moderna: Rappresentare e conoscere la Metropoli dei Lumi: Roma, Artemide Editrice, p. 11–17.

Falda, G.B., 1676, Pianta di Roma, in Bevilacqua, M., ed., 2005, Nolli, Vasi, Piranesi: Immagine di Roma Antica e moderna: Rappresentare e conoscere la Metropoli dei Lumi: Roma, Artemide Editrice, 118 p..

Frutaz, A.P., 1962, Le piante di Roma: Istituto di Studi Romani, Roma, Stabilimento Arti Grafiche Luigi Salomone, v. 1, 350 p.; v. 2, plates 1–322 (from the III century A.D. to the year 1625); v. 3, plates 323–684 (from year 1630 to year 1960).

Funiciello, R., editor, 1995, La Geologia di Roma—Il centro Storico: Mem. Serv. Geol. It. no. 50, 550 p.

Gadenz, S., editor, 2004, Past, present and future of the Italian geological maps from ink to digital cartography, 4 DVDs and text: San Giovanni Valdarno, e-Geo.

Le Goff, J., 2004, Il Cielo sceso in terra: Le radici medievali dell'Europa: Bari, Laterza, 344 p.

Harte, J., 2002, Toward a synthesis of the Newtonian and Darwinian worldviews: Physics Today, no. 10, http://www.aip.org/pt/vol-55/iss-10/p29.html.

Nolli, G.B., 1748, Nuova Pianta di Roma data in luce da Giambattista Nolli l'Anno M DCC XLVIII.

Propertius, S., 1984, Elegie: (Latin text with Italian translation), Bologna Zanichelli Editore, 277 p.

Smith, W., 1815, A delineation of the strata of England and Wales with part of Scotland: London, J. Carey.

Spondberg Pedley, M., 2005, Scienza e cartografia: Roma nell'Europa dei Lumi, in Bevilacqua, M., ed., Nolli, Vasi, Piranesi: Immagine di Roma Antica e moderna: Rappresentare e conoscere la Metropoli dei Lumi: Roma, Artemide Editrice, p. 37–47.

Tibullus, A., 1988, Elegie (Latin text with Italian translation): Milano, Garzanti, 332 p.

Vai, G.B., 1995, L'opera e le pubblicazioni geologiche di Scarabelli, in Pacciarelli, M., and Vai G.B., eds., La collezione Scarabelli, 1: Geologia: Bologna, Grafis Edizioni, Musei Civici di Imola, Catalogo delle raccolte, p. 48–104.

Winchester, S., 2001, The map that changed the world: New York, HarperCollins Publishers, 330 p.

MANUSCRIPT ACCEPTED BY THE SOCIETY 17 JANUARY 2006

Printed in the USA

Leopoldo Pilla (1805–1848): A young combatant who lived for geology and died for his country

Bruno D'Argenio*

Department of Earth Sciences, University Frederick II, Naples, and Institute for Coastal Marine Environment, National Research Council, Calata Porta di Massa, Porto di Napoli, 80133 Naples, Italy

ABSTRACT

Leopoldo Pilla was born in Venafro, Italy, in 1805. As a young student, he moved to Naples, the capital city of the Bourbon's reign, where he pursued medical and other scientific studies. He began to express volcanological and other geological interests early on and tried ardently for scientific success as a geologist and professor. Though assailed by serious personal problems, mostly of a psychosomatic nature, he always managed to overcome the difficulties that led to his deep depression.

Pilla published many original papers in the most important journals of Europe. In 1842, he was called to Pisa as a professor of geology at the local university. He was on the verge of producing his most mature and widely recognized scientific work when, enthusiastically involved in the political turmoil of his times, he was killed by a grenade while marching along with his students in the battle of Curtatone in the spring of 1848. Italy not only lost a promising young professor, but also a great scientist in the most creative period of his career.

Keywords: volcanology, Vesuvius, volcanic monitoring, mineralogy, geology, uplift tectonics, heat flow, stratigraphy, Etrurian terrain, stratigraphic correlation, earthquake, history of geology, Steno, Murchison, Lyell.

INTRODUCTION

The name of Leopoldo Pilla (1805–1848) (Fig. 1) rarely appears in the old Italian treatises on geology and is not even briefly mentioned in modern university textbooks. To find a trace of this man outside of a few dedicated publications (Bassani, 1905; Ippolito, 1948; Scherillo, 1992; D'Argenio, 1996; Monsagrati, 1996), it is necessary to refer to the century-old, but ever-distinguished *History of Geology and Palaeontology*, by Karl Zittel (1901), in which Pilla is briefly mentioned for his volcanological studies. Accurate readings of Pilla's works are found in d'Archiac (1862). Other Italian references to Pilla may be found in Ciancio (1995), Marabini (1995), and Corsi (2001, 2003), in which there are comments on, and an appreciation of, the scientific influence Pilla had on his colleagues and students during the last part of his life, when he was professor of geology in Pisa. Corsi (2001) deals with Pilla's theories, geological upbringing, his personal and academic position in Pisa, and the reason why, until the 1890s, his name was rarely mentioned after his heroic death in 1848.

If one were to measure the importance and impact of Pilla's work and teachings in the last century and a half by these meager expressions of interest, one would have to conclude that the man was no more remarkable than many who were actively engaged in the earth sciences in the first half of the nineteenth century. And yet, Leopoldo Pilla deserves to be remembered in the second centennial of his birth, together with Giuseppe De Lorenzo (1871–1957), as one of the most outstanding southern Italian

---

*E-mail: b.dargenio@iamc.cnr.it.

D'Argenio, B., 2006, Leopoldo Pilla (1805–1848): A young combatant who lived for geology and died for his country, *in* Vai, G.B., and Caldwell, W.G.E., The Origins of Geology in Italy: Geological Society of America Special Paper 411, p. 211–223, doi: 10.1130/2006.2411(13). For permission to copy, contact editing@geosociety.org. ©2006 Geological Society of America. All rights reserved.

Figure 1. Portrait of Leopoldo Pilla (Venafro, 20 October 1805–Curtatone, 29 May 1848)

geologists. With De Lorenzo, he shared a profound and unique geological vision of the world that he was able to express in the effective, elegant, and flawless prose of his very personalized language. To Pilla, his influence on his students and peers was second only to that of his own work during the later years of his short career, which finds expression in the *Trattato di Geologia* (Pilla, 1847, 1847–1851). Pilla's *Trattato di Geologia* is indebted to Jean Baptiste d'Omalius d'Halloy (1783–1875) and other geological handbooks of his time. This work was completed with publication of a second volume three years after his tragic death at the age of 43 yr during the 1848 battle of Curtatone—one of the emblematic conflicts of the Italian Risorgimento. The Curtatone battle, as historians would suggest, was a slaughter. The poor Italian students and their young teachers never saw their Austrian enemy; they were simply bombed out of the river bank they were assigned to keep.

The present paper is not intended to provide a detailed analysis of Pilla's entire scientific production, but rather to be an assessment of some of the key aspects of his research and how these were affected by the cultural climate of the times. The most impressive aspect of Pilla's papers is the modern-day regional vision and insight (avant-garde in his time) that complemented his ability in observing various geological structures from distant locations and extracting valid general principles about the geological mechanisms that formed them.

At the same time, it is obvious from his recently published diary (Discenza, 1996) that Pilla was determined to reach an ever-more-profound understanding of these mechanisms. This determination was driven by a basic compulsion coupled with an enormous enthusiasm, which never failed to overpower feelings of defeat and frustration that followed upon hardships or obstacles he encountered along the way, both in Naples and in Pisa. This compulsion and enthusiasm also allowed Pilla to overcome frequent times of deep depression clearly referred to in the short dedication of his *Trattato di Geologia* to his father.

PILLA'S EARLY YEARS

Pilla commenced to write his diary in 1830. In a short autobiographical introduction, the 25-yr-old Leopoldo announced his fields of endeavor as mineralogy and geology. Along with this cultural program in mind, which despite various adversities he would master successfully, there was an eagerness for research, at times naïve in its scope, and an ambition to excel among the experts of the time. His desire to "create" was sufficiently strong to dispel any illusion that he might be able to do so simply by reading the works of others and proceeding to write without any personal "experience" or observations.

At the age of fifteen, Leopoldo Pilla arrived in Naples from his birthplace of Venafro to become a physician like his father, Nicola, from whom he had inherited a strong interest in the natural sciences (Pilla, 1795, 1823). As a physician, Leopoldo was trained also as a surgeon and worked as such in the Naples Military Hospital.

What is known about the 10 yr that preceded the start of his diary is mainly what Leopoldo Pilla himself narrated (Discenza, 1996): the hardships of living in the capital with meager financial resources; the meetings with scientists which brought about not only his educational growth but also the academic recognition he struggled to achieve; the strong desire to improve his knowledge of geology through travel and fieldwork; the conviction that a modern naturalist should be competent in physics and chemistry in order to understand the phenomena he observed and in mathematics in order to describe them. Recognizing the need for an interdisciplinary background led him to propose a permanent committee to observe the volcanic processes of Vesuvius—a precursor to the foundation of the Osservatorio Vesuviano in 1845.

Although a student of Matteo Tondi (Fig. 2) at the University of Naples, Pilla was never permitted to become Tondi's successor. The reasons for this were diverse, but among them is certainly Pilla's "liberal" outlook, which probably offended the prevailing socially correct views of the times. Tondi was succeeded by Arcangelo Scacchi (Fig. 3), a schoolfellow of Pilla, who devoted himself to mineralogy, whereas Pilla had wider

Figure 2. Matteo Tondi (1762–1855), Pilla's respected teacher, was among the founders of the Real Museo Mineralogico in Naples.

Figure 3. Angelo Scacchi (1810–1893), who was director of the Naples Royal Museum of Mineralogy after Tondi, and schoolfellow of L. Pilla. Scacchi and Pilla greatly advanced mineralogical and volcanological studies in the Naples kingdom. While Scacchi was more inclined to mineralogy, Pilla had wider interests in rocks and geological processes, among which included his predilection for volcanology.

interests in rocks and geological processes, with a predilection for volcanology. Apparently, the Naples University preferred a narrow specialist in mineralogy over a generalist like Pilla.

As Scherillo (1992) states in his remembrance of Pilla, his career was greatly aided by excellent verbal skills acquired at the Basilio Puoti school, which he attended for several years. In Naples, he also associated with Teodoro Monticelli and Gaetano Tenore, two distinguished mineralogists, and with Nicola Covelli from whom he learned the value of direct observation and hands-on experience when seeking valid scientific results. From Covelli he also learned the passion for theories and speculation in natural sciences. Why Scacchi was preferred to Pilla is a complex issue, dealing with social status and even, perhaps, the psychological qualities or defects displayed by the two friends-rivals.

From 1830, Leopoldo Pilla repeatedly sent his written scientific contributions to the most prominent periodical publishing agencies of foreign academic institutions (*Societé Géologique de France, Académie des Sciences Paris, Neues Jahrbuch für Mineralogie*, and so on), as is indicated by the list of papers appended to the end of this article. He involved himself with all of the great naturalists of his time and with many eminent scholars visited Naples and the surrounding Flegraean Fields and Vesuvius (even then an exceptional natural laboratory) to witness active volcanic phenomena. Pilla was, therefore, well known particularly as a volcanologist (he had gathered a rich collection of volcanic rocks), and his early work, published in France, Germany, and Switzerland, was almost exclusively dedicated to Vesuvius and other Italian volcanoes. His investigations led him to propose that volcanoes were the result of accumulation of repeated lava flows, even though he later partly modified his views in favor of the von Buch theory of crater upheaval (Corsi 2001).

Of great importance to his scientific contacts was not only his correspondence but also his personal encounters with Italian geologists during the scientific meetings frequently held in various Italian cities, where he was able to present himself and his work. In six of the meetings that took place between 1839 (the first meeting of Italian scientists in Pisa) and 1845 (the seventh meeting of Italian scientists in Naples; Fig. 4), he had the opportunity to read his works on volcanology and on stratigraphical and regional geology, as well as on south Italian ore bodies (Discenza, 1996; Pancaldi, 1983; Scherillo, 1992). Among the

Figure 4. The Royal Mineralogic Museum of Naples, established in 1801. (A) This engraving, presently at the San Martino Museum in Naples, shows the inauguration of the 7th Congress of Italian Scientists on 20 September 1845. On the right side, next to the column, Arcangelo Scacchi, director of the museum, and Carlo Buonaparte (from Scherillo, 1966).

foreign scientists who had either written or personal contacts with Pilla were the famous Leopold von Buch and Elie de Beaumont. Also, François Arago and Alexander von Humboldt personally intervened with the Grand Duke of Tuscany to favor his nomination as professor of geology at the University of Pisa, perhaps the most advanced school of geology in Italy at the time, with teachers such as Paolo Savi (1798–1871) and Giuseppe Meneghini (1811–1889), (Corsi, 2001, 2003; Scherillo, 1992).

THE SCIENTIFIC AMBIENCE

Between 1830, when Pilla started to write his *Notizie storiche della mia vita quotidiana,* and his death in 1848, European geologists were gradually overcoming their sense of insecurity about the age and evolution of Earth. In fact, those years of the nineteenth century witnessed the death of George Cuvier (1832), the publication of the first volume of *Principles of Geology* by Charles Lyell (1830), and Charles Darwin's departure on the *Beagle* for his voyage of discovery around the world (1831–1836), which laid the groundwork for his much-contested theory in the *Origin of Species* (1859). Cuvier, founder of modern comparative anatomy, was the most influential exponent of catastrophism in geology, which he applied in his studies of the sedimentary succession in the Paris basin of France. Starting off with a study of the large assortment of vertebrate fossils found in the younger strata of this area, he theorized that a succession of catastrophic floods had caused the phenomena that today would be described as mass extinctions, with repeated "migrations" from adjacent regions repopulating the ancient lands from which the fauna had been annihilated. Among some of Cuvier's followers, however, such migrations were regarded as acts of creation. Pilla was a firm believer in such acts. Consequently, he matured during a time of heated debate, prompted mainly by the increasing acceptance and consolidation in Britain of Lyell's Huttonian gradualism and its opposition to the catastrophism advocated by the French school led by Cuvier. Actually, Lyell's gradualism and uniformism met with little success for over 50 yr. In Italy, only the Pisa school of geology (and especially Giovanni Capellini, 1833–1922) accepted Lyell's ideas, which were considered as a further development of Constant Prévost's views (Corsi, 2001).

Figure 4, continued. (B) The Naples Mineralogy Museum as it is now.

Over a period of about 50 years, Lyell, who had visited Naples and its surroundings several times, published twelve editions of his treatise (1831–1875), consistently integrating it with new data and findings. Lyell had given his most important achievement a meaningful title: *Principles of Geology or the Modern Changes of the Earth and its Inhabitants Considered as Illustrative of Geology*. Following the title page, Lyell quoted Playfair's famous words:

> Amid all the revolutions of the globe the economy of Nature has been uniform and her laws are the only things that have resisted the general movement. The rivers and the rocks, the seas and the continents, have been changed in all their parts; but the laws which direct those changes, and the rules to which they are subject have remained invariably the same (and are still those of today).

Lyell's notion of geology was thus represented by a kind of viscous view of sequential events gradually shaping Earth, with uniformitarianism and gradualism being the mainstreams of his ideas on how change was wrought in the geological realm and his concepts, which agreed well with the liberal culture of Victorian England, obstinately permeated geological culture worldwide also within the twentieth century.

Leopoldo Pilla attempted to mitigate the strong ideological influence of gradualism with a pragmatic analysis of known geological data and his own field observations (Pilla, 1840), which he discussed without prejudice (see also Corsi, 2001). Traces of gradualism are, therefore, less evident in his papers and in his *Trattato di Geologia*, in which, if anything (vol. II), there is a tendency to consider the intensity of geological phenomena as declining continuously with time, as proposed by one of the founding fathers of stratigraphical geology, William D. Conybeare (Zittel, 1901). This observation suggests an attempt to reconcile the perceived gradualism of "present causes" in recent times with the geological revolutions caused by the greater intensity of the "ancient causes" of the past.

PILLA'S TRAVELS

Pilla's conception of Earth's natural history is better understood with some knowledge of his numerous travels, during which he carefully examined the geomorphology and geology of the regions he visited. From the very beginning, Pilla sensed the importance of systematically comparing the different settings he observed. In doing so, he argued as a regional geologist along the lines of a knowledge based on the comparison of different areas of the planet (see also Vai, this volume, chapter 7), as it came into vogue only subsequently. Pilla was so aware of the importance of this kind of comparative anatomy of Earth that he added the following subtitle to his *Trattato di Geologia* (Treatise of Geology; Fig. 5): "Specially intended for a comparison between the physical structures of Northern and Southern Europe." Pilla tried to dialogue with northern European geologists and even planned a long journey to France, Belgium, Germany, and England; he wanted to underline the differences between the constitution of the crust in those regions and in the southern ones. Yet, Savi, a zoologist who also taught geology in Pisa before Pilla, was more "regionalist" than Pilla ever was. Savi maintained that those of Pilla's were sweeping generalizations and that too little was understood of the geology of Tuscany to warrant hasty comparisons with the north of Europe. Moreover, Savi deeply resented the rapid incursions eminent foreign geologists were making in Tuscany, which led to even more unwarranted generalizations. In particular, Savi disliked Pilla's attempts at getting approval from France and England for the views his younger colleague was putting forward, views that contrasted with his own work.

Pilla traveled not only in Campania and in his native, rocky Molise, but also in other regions of southern Italy, such as Puglia and Calabria. It was natural that his attention to direct observation of the geological characteristics of southern Italy and his eagerness to unveil the meaning of the ancient processes recorded in rocks (*ex libro lapidum historia mundi*) followed in the steps of his father, Nicola, who, already in 1795, had recognized the volcanic origins of the Roccamonfina, north of Naples, and had been the author of two small volumes on Campanian volcanism in 1823 (Pilla, 1795, 1832).

Pilla's first important journey out of Italy was to Vienna. In his capacity as a surgeon at the Naples Military Hospital, he was sent to Vienna with a medical delegation in 1831 to study the cholera epidemic. Readers can learn about the vicissitudes of that journey from his autobiographic pages (Discenza, 1996), in which he recounts the ailments, some of them very serious, that hindered the course of his trip and cut it short. That journey, however, was of tremendous importance to his studies, not only because he was able to visit the "giant extinct volcanoes in the Latium region" (around Rome), but also because in the Apenninic stretch between Florence and Bologna, he was able to closely observe extensive outcrops of Tertiary sandstones to which he dedicated much of his comparative studies. Two years later, Pilla undertook another exciting journey, this time to Abruzzo, which inspired many reflections and scientific works. On a quest for peat and lignite, he was also able to make important observations on the Mesozoic and Tertiary stratigraphical successions of the region.

Between 1834 and 1837, Pilla visited the Neapolitan and Pontine volcanic archipelagoes and the Aeolian Islands. From this experience, he extracted important information for writings on his elements of volcanology. Finally, in January of 1840, because of his skill, he was sent to the Campania and Puglia regions by the Ministry of Internal Affairs to identify the outcrops of some ornamental stones from the Gargano limestone area. On this occasion he discovered the famous and much-discussed Paleocene rocks of Punta delle Pietre Nere, by Lesina Lake, on the Adriatic coast. These rocks are now considered as basaltic dykes intruding Late Triassic limestones, but earlier were described as amphibolites. This was Pilla's last field work in the south of Italy because in 1842

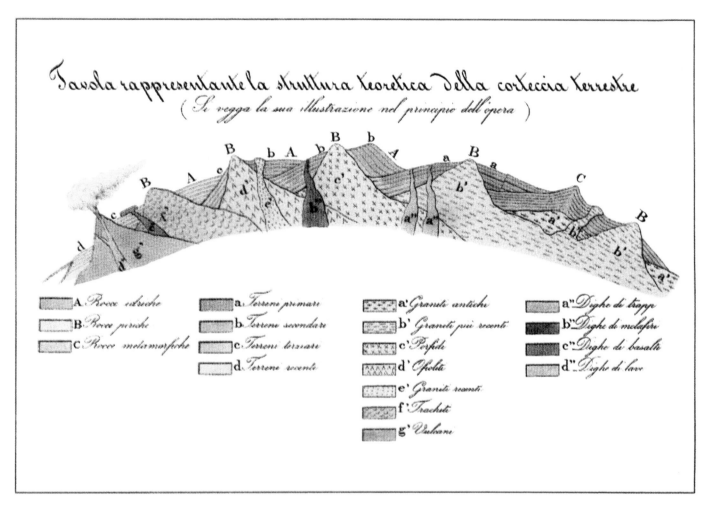

Figure 5. This sketch, which faces the frontispiece of the *Trattato di Geologia* (vol. 1, 1847), represents the structure of Earth's crust. Here, Pilla offered an extremely concise clarification of the fundamental aspects of geology in his time through what he defined as the genealogy of rocks and the chronology of rocks. In the genealogy, he outlined the main groups of rocks: hydrogenetic (or sedimentary), pyrogenetic (or igneous), and metamorphic. In the chronology, he described the relationships among the strata and, through them, he indicated the succession of events that produced these rocks. An interesting aspect worth noting is Pilla's modern use of color gradations to illustrate different types of rocks, their mutual relationships, and their subsequent tectonic deformation.

he relocated to Pisa and began work in Tuscany, a region that he traveled widely and on which, from 1843 to 1847, he was able to write several papers before his death.

Thus, in retrospect, Pilla's scientific production falls into two phases: volcanology and mineralogy, and regional geology (see also Fig. 7). His volcanological studies were his first venture. Before the age of 30, he had actively addressed scientific issues and gone on site to "look" for himself, thus following the advice of one of his teachers, Nicola Covelli. He tenaciously and systematically described the activity of Vesuvius and of the Campi Flegrei, later shedding light on other active and inactive volcanoes in southern Italy to which he dedicated ~20 publications. In these, he came to some very significant conclusions for his time, among them the close link between volcanic activity and marine waters (Pilla, 1847); which is still considered.

Pilla, moreover, endeavored objectively to reconcile, using field examples, the von Buch theory of *uplifting craters* with gradualistic principles, making a genuine effort to demonstrate the validity of both. Pilla himself applied the von Buch ideas to the Roccamonfina volcano, north of Naples, which had supposedly uplifted, splitting the opposing blocks of Monte Massico and Monte Camino.

His most groundbreaking scientific contributions, however, concerned comparative regional geology and, in particular, his discussions on lower Tertiary terrigenous deposits, which he observed and described south and north of the Alps. In Italy, he took the stratigraphical discrimination of these deposits a step further by distinguishing them from the Mesozoic strata with which they were sometimes confused and labeling them "Terreno Etrurio" (Etrurian terrain). He then placed this division in an intermediate position between the Cretaceous strata with rudists and Eocene strata with gypsum, sulfur, and lignite (Pilla, 1846), which today are dated as late Miocene in age.

Another of Pilla's meritorious accomplishments was his studies of Apenninic deformations, which he attributed to a succession of events and illustrated by schematic geological sections. He also investigated the dynamics of deformation and endogenous heat on which he formulated both observations and speculations (Pilla, 1842).

All of these ideas and general scientific results of Pilla's were eventually consolidated in his *Trattato di Geologia* (1847). Reasonably enough, for Pilla this work represented a fulfillment of his ambitions. He took special care to endow both volumes with ample information, bibliographic references, and personal accounts and observations (Fig. 6).

A PREMATURE DEATH

The prolific scientific production of Pilla between 1830 and 1847 (see appendix and Fig. 7) and the significance of his findings serve only to underscore the tragedy of his premature death.

"Too soon" are said to have been his own last words, when he fell in battle at Curtatone (Ippolito, 1948). Though cut short of reaching full maturity, his ability as a consolidator and synthesizer was already evident from his last papers and his *Trattato di Geologia*. The methodological approach of his work, the vastness of his interests and solidity of his logics on general problems, his skill in regional synthesis, combined with his original comparative approach, forces the conclusions that, had he lived, he would have continued his brilliant and steadfast ascent among the distinguished pioneers of geology.

Yet, if we look at the pages of his diary, the crucial difficulties he encountered to overcome his psychosomatic ailments, and the tenacious resolve with which he sought progress (see, for instance, his list of propositions in "Observations on the year 1834" where he embarked on a relentless confrontation between " my current knowledge" and "the knowledge I am lacking"), then we cannot but admire, respect, and give credit to Leopoldo Pilla for having been such a courageous man, not only on the field of battle, but also in his studies and in fighting against his personal struggles with himself.

In the words of Scherillo (1992), in his beautiful memorial note on Pilla, "140 years now after his glorious death, we recognize Pilla—the University professor fallen in battle in his fight for independence—as one of the most dignified figures of the Italian Risorgimento, the scientist who joined his devotion to knowledge to a longing for liberty, the geologist who combined the study of his country with love for that country and who was willing to make the ultimate sacrifice for that love."

More critically, Pilla firstly adhered to a very moderate federal view of the future of Italy, under the aegis of the Pope; only very late did he express pro-unity views, but always with a moderate stint. His death, as witnesses recorded, was sudden, having been struck by shrapnel that cut one of his arms and opened his bowels: he probably did not even have time to understand what was happening to him. To better understand the real contribution of this man to the science of geology, it would be useful to assess the real value of some of his ideas, like the "Etrurian terrain," a proposition Pilla's colleagues at Pisa simply buried with the awkward Neapolitan professor in the *damnatio memoriae*.

APPENDIX

Pilla's publications (1830–1847)

1. *Cenno biografico su Nicola Covelli,* Napoli, 1830, 43 p.
2. *Cenno Storico sui progressi della Orittognosia e della Geognosia in Italia I. Orittognosia,* "Il Progresso delle Scienze delle Lettere e delle Arti," Napoli, v. II, p. 37–814, 1832.
3. *Cenno Storico sui progressi della Orittognosia e della Geognosia in Italia II. Geognosia,* "Il Progresso delle Scienze delle Lettere e delle Arti," Napoli, v. III, p. 165–234, 1832.
4. *(Modo con cui Broussais considera il cholera morbus)* "L'Esculapio, Giornale delle Scienze Fisico Mediche," Napoli, agosto 1832.
5. *Narrazione di una gita al Vesuvio fatta nel dì 26 Gennaio 1832,* "Il Progresso delle Scienze delle Lettere e delle Arti," vol. I, fasc. II, 1832, p. 232–240.
6. *Nota su l'ultima eruzione del Vesuvio* (27 maggio–5 giugno 1833), "Il Progresso delle Scienze, Lettere e delle Arti," Napoli, v. IV, 1833, p. 317.
7. *Cenni storici sui progressi della Orittografia e della Geognosia in Italia III,* "Il Progresso delle Scienze delle Lettere e delle Arti," Napoli, v. V, IX, 1833, p. 5–41.
8. *Osservazioni Geognostiche su la parte settentrionale ed orientale della Campania,* "Annali Civili del Regno delle Due Sicilie," Napoli, v. III, II, 1833.
9. *Osservazioni geognostiche su la parte settentrionale e meridionale della Campania,* "Annali Civili del Regno delle Due Sicilie," Napoli, v. III, VI, 1833, p. 117–147.
10. *Osservazioni intorno a' principali cangiamenti e fenomeni avvenuti nel Vesuvio nel corso dell'anno 1832,* "Annali Civili del Regno delle Due Sicilie," Napoli, v. I, II, 1833, p. 155–160.
11. *Sopra una singolare formazione di calcare lacustre giacente in alto e nel grembo degli Appennini delle Mainardi nella Provincia di Terra di Lavoro,* "Annali Civili del Regno delle Due Sicilie," Napoli, v. II, IV, 1833, p. 101–107.
12. *Sur l'éruption du Vésuve en Juillet et Août 1832,* "Bibl. Univ.," Losanna, 1833, p. 351–356.
13. *Ricerche geologiche sul carbon fossile dell'Abruzzo Ultra,* "Annali Civili del Regno delle Due Sicilie," Napoli, v. III, V, 1833.
14. *VI Escursione al Vesuvio fatta nel dì 9 dicembre* 1832), "Lo Spettatore del Vesuvio e de' Campi Flegrei," Napoli, fasc. II, no. 1-2, 1833, p. 3–4, (with Cassola, P.).

QUADRO DI CONFRONTO

DEI TERRENI STRATIFICATI DEL SETTENTRIONE E DEL MEZZOGIORNO DI EUROPA.

NB. Le caselle senza indicazione dinotano che le serie parallele sono presso a poco le stesse.

Serie generali	*Serie particolari* — Zona settentrionale	*Serie particolari* — Zona meridionale	*Serie incerte della zona meridionale*
Terreno moderno			
Terreno alluviale antico			
Terreno terziario	Terreno pliocenico recente		
	Terreno pliocenico antico		
	Terreno miocenico		
	Terreno eocenico	Non ancora bene distinto — Terreni a gesso, zolfo, ligniti d' Italia?	
Terreno, etrurio	Manca, o è poco distinto	Superiore — Terreno nummulitico inferiore — Macigno	
Terreno cretaceo	Creta di Maestricht	Manca	
	Creta bianca	Manca quasi generalmente, e non si scorge che in qualche raro sito	
	Creta tufacea	Creta nummulitico-ippuritica	
	Gault	Manca, o non è stato riconosciuto	
	Arenaria verde inferiore	Strati superiori del terreno neocomiano	
	Terreno wealdiano	Terreno neocomiano?	
Terreno giurassico	Terreno oolitico superiore	Queste divisioni non si possono bene riconoscere; ma elle formano molti strati, di cui fanno parte gli scisti ammonitiferi della Spezia	
	Terreno oolitico medio		
	Terreno oolitico inferiore	Calcare rosso ammonitifero	
Terreno liassico	Marne liassiche	Il lias non si mostra sempre bene distinto dal terreno giurassico. Ancora le divisioni del lias non si possono riconoscere. Il calcare nero del lago di Como e di alcuni luoghi d'Italia mostra appartenere a questo terreno	
	Calcare liassico		
	Arenaria liassica		
Terreno triassico	Marne iridate	Manca quasi generalmente. Non si ravvisa che nel Vicentino ed in Lombardia, principalmente il calcare conchiglifero.	*Serie inferiori di lias*: Arenarie rosse
	Calcare conchiglifero		
	Arenaria screziata		
Terreno permiano	Calcare permiano	Manca generalmente, o non si può riconoscere.	Anageniti quarzose
	Scisto ramifero		
	Arenaria rossa		
Terreno carbonifero	Terreno litantracifero	Manca per lo più. Comparisce in qualche parte della Spagna e della Sardegna.	Scisti di varie sorte, in gran parte cristallini.
	Calcare carbonifero	Manca — Calcare di Bleiberg in Carintia?	
Terreno devoniano	Terreno devoniano	Comparisce ben distinto nelle Asturie in Spagna, ed in Sardegna.	
Terreno silurio	Terreno silurio	Poco esteso. Comparisce nella Estremadura, in Sardegna, e nel Bosforo.	

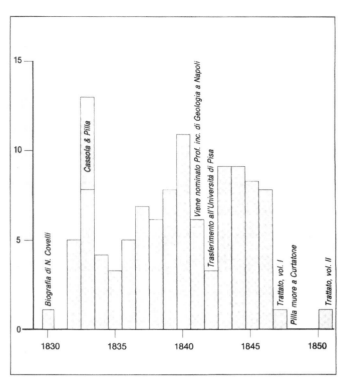

Figure 7. Pilla's scientific productivity as expressed by his publications (number of papers per year). From 1832 to 1846, he wrote more than one hundred papers, with three peaks of intensive production: the first was coincident with his research on the activity of Vesuvius, and the second was most probably a reflection of his effort to demonstrate his scientific value when he tried to be nominated as professor of geology in Naples University. (This attempt failed; he received only the temporary position of *professore incaricato*). The third peak of productivity coincides with his scientific activity as professor of geology in the University of Pisa, where he could spend time more freely in studies and in carrying out his research (from D'Argenio, 1996).

Figure 6. "Stratified rocks of northern and southern Europe" in the comparative table of the second volume of Pilla's *Trattato di Geologia* (1851). The stratigraphical nomenclature suffers from the uncertainties of the time: the first International Geological Congress was only held 30 yr later in Paris and the second in Bologna in 1882, when the first formalization of stratigraphical nomenclature was made. Note the position of Pilla's *Terreno Etrurio* between the Mesozoic and Tertiary, which (notwithstanding the age-related inaccuracies) is situated between the "Creta nummulitico-ippuritica" (the nummulites associated with rudists could be orbitoids) and the Eocene "Terreni a gesso, zolfo e ligniti" (gypsum-evaporites, sulfur and lignite), which today we recognize as being of Miocenic age.

15. *VII Escursione al Vesuvio fatta nel giorno 23 dicembre in occasione dell'eruzione avvenuta in questo mese,* "Lo spettatore del Vesuvio e de' Campi Flegrei," Napoli, fasc. II, no. 1–2, 1833, p. 8–21 (with Cassola, P.).
16. *VIII Escursione al Vesuvio, fatta nel dì 12 gennaio (1833), ed ascensione sul cono interno,* "Lo spettatore del Vesuvio e de' Campi Flegrei," Napoli, fasc. II, no. 1–2, 1833, p. 23–57 (with Cassola, P.).
17. *Misura della elevazione della Punta più alta del cono interno (del Vesuvio) presa prima e dopo l'eruzione di Dicembre 1832,* "Lo spettatore del Vesuvio e de' Campi Flegrei," Napoli, v. II, no. 1–2, 1833, p. 57–58 (with Cassola, P.).
18. *Stato dell'atmosfera durante la eruzione (dicembre 1832),* "Lo spettatore del Vesuvio e de' Campi Flegrei," Napoli, v. II, no. 1–2, 1833, p. 22–23 (with Cassola, P.).
19. *Bullettino geologico del Vesuvio e de' Campi Flegrei,* destinato a far seguito allo Spettatore del Vesuvio, no. 1, "Progresso delle Scienze, delle Lettere e delle Arti," Napoli, v. VIII, 1834, p. 126–156.
20. *Bullettino geologico del Vesuvio e de' Campi Flegrei,* destinato a far seguito allo Spettatore del Vesuvio, no. 2, "Progresso delle Scienze, delle Lettere e delle Arti," Napoli, v. IX, 1834, p. 126–149.
21. *Catalogue de collection de minéraux et de laves Vésuve à vendre,* s.n.t., (1834), 4 p.
22. *Exposé des phénomènes observés dans le cratère du Vésuve pendant l'éruption de 1833,* "Bullettin de la Société Géologique de France," Paris, v. V, 1834, p. 324.
23. *Osservazioni geognostiche che possonsi fare lungo la strada che da Napoli conduce a Vienna attraversando le province romane, toscane, venete, la Carinzia, la Siria, l'Austria,* Napoli, 1834, 92 p.
24. *(Ausbrueche des Vesuv im Anfange Aprils 1835),* "Neues Jahrbuch fuer Min. etc.," Stuttgart, 1835, p. 454–455.
25. *Bullettino geologico del Vesuvio e de' Campi Flegrei,* destinato a far seguito allo Spettatore del Vesuvio, no. 3, "Progresso delle Scienze, delle Lettere e delle Arti," Napoli, v. X, 1835, p. 262–295.
26. *Note sur un soulèvement remarquable du grès houiller par le granite, dans les environs de Gérace dans la Calabre Ultérieure,* "Bulletin de la Société Géologique de France, Première Série," Paris, v. VII, 1835, p. 306, 2 pl.
27. *Fait remarquable de soulèvement produit par le granite dans la Province de la premiere Calabre Ulterieure,* "Bulletin de la Société Géologique de France," Paris, v. VII, 1836, p. 306–307.
28. *Matteo Tondi. Necrologia,* Il Progresso delle Scienze delle Lettere e delle Arti," Napoli, v. XV, 1836, p. 37–74.
29. *Sur l'âge des grès de Gérace,* "Bulletin de la Société Géologique de France," Parigi, v. VII, 1836, p. 306.
30. *Sur la sortie de véritables flammes par la bouche du Vésuve,* "Bulletin de la Société Géologique de France," Parigi, v. VII, 1836, p. 262.
31. *Sur une mine de sel gemme de la Calabre,* "Bulletin de la Société Géologique de France," Parigi, v. VIII, 1836, p. 199, 1 tav.
32. *Brief an die franzoesiche Académie des Sciences.* (ueber Fossilien im Epomeotuff), "Froriep's Not. Geb. Nat. Heilk.," Erfurt, v. XXVI, 1837, p. 54.
33. *Bullettino geologico del Vesuvio e de' Campi Flegrei,* destinato a far seguito allo Spettatore del Vesuvio, no. 4, "Progresso delle Scienze, delle Lettere e delle Arti," Napoli, v. XVI, 1837, p. 223–253, 1 pl.
34. *Catalogo di una collezione di rocce della Calabria disposte secondo l'ordine della loro posizione relativa,* "Annali Civili del Regno delle Due Sicilie," Napoli, v. XIII, XXV, 1837, p. 52–59.
35. *Lettres découvertes dans les Appennines,* "Bulletin de la Société Géologique de France," Parigi, v. VIII, 1837, p. 99–201.
36. *Lettre sur la géologie de la Calabre et sur un tuf argileux fossilifère de la Somma,* "Bulletin de la Société Géologique de France," Paris, v. VIII, 1837, p. 198–199, 1 pl.
37. *Observations tendant à prouver que le cône du Vésuve a été primitivement formé par soulèvement,* "Comptes Rendus de l'Académie des Sciences," Paris, v. IV, 1837, p. 527.
38. *Parallelo tra i tre vulcani ardenti dell'Italia,* "Atti Accademia Gioenia di Scienze Naturali," Catania, serie prima, v. XII, 1837, p. 129–135.
39. *Prospetto di un insegnamento di geologia in Napoli,* "Il Progresso delle Scienze, delle Lettere e delle Arti," Napoli, v. XVIII, 1837, p. 129-135.
40. *Relazioni de' tremuoti che afflissero la città di Sangermano ed il Monastero di Montecassino nella primavera del corrente anno 1837",* "Annali civili del Regno delle Due Sicilie," Napoli, v. XIV, XXVIII, 1837, p. 91–105.
41. *Sur l'âge carbonifère des grès de Gérace,* "Bulletin de la Société Géologique de France," Paris, v. VIII, 1837, p. 198.
42. *Bullettino geologico del Vesuvio e de' Campi Flegrei,* destinato a far seguito allo Spettatore del Vesuvio, no. 5, "Progresso delle Scienze, delle Lettere e delle Arti," 1838.
43. *Catalogo ragionato delle rocce del Volture,* "Museo Mineralogico. Università di Napoli," Napoli, 1838.
44. *Notizie geologiche sopra il vulcano estinto di Roccamonfina,* "Il Lucifero," Napoli, v. I, no. 36.
45. *Notizie geologiche sopra il vulcano estinto di Roccamonfina,* "Il Lucifero," Napoli, v. I, no. 37.
46. *Osservazioni fisiche sopra il vulcano di Stromboli. Articolo primo,* "Il Lucifero," Napoli, v. I, no. 8, 1838, p. 30.
47. *Osservazioni fisiche sopra il vulcano di Stromboli. Articolo secondo,* "Il Lucifero," Napoli, v. I, no. 14, 1838, p. 54–55.
48. *Osservazioni fisiche sopra il vulcano di Stromboli. Articolo terzo,* "Il Lucifero," Napoli, v. I, no. 23, 1838, p. 89–90.
49. *Osservazioni fisiche sopra il vulcano di Stromboli. Articolo quarto,* "Il Lucifero," Napoli, v. I, no. 27, 1838, p. 106.

50. *Relazione su' principali fenomeni avvenuti nella eruzione del Vesuvio del mese di agosto 1838,* "Il Lucifero," Napoli, v. I, no. 29, 1838, p. 114.
51. *Sulla questione del Serapeo toccata dal cav. Tenore,* "Il Progresso delle Scienze, delle Lettere e delle Arti," Napoli, vol. XIX, 1838, p. 242–245.
52. *Tertiäre Gebirge in Calabrien,* "Neues Jahrbuch fuer Min. etc.," Stuttgart, Bd. 3m, 1838.
53. *Ventitreesima gita al Vesuvio nella notte del 13 al 14 settembre 1838,* "Il progresso delle Scienze, delle Lettere e delle Arti," Napoli, vol. XIX, 1838, p. 242–245.
54. Articolo negli "Atti della Reale Accademia delle Scienze di Torino," Torino, 1839.
55. *Ausbruch des Vesuvs Anfangs Januar 1839,* "Neues Jahrbuch fuer Min. etc.," Stuttgart, 1839.
56. *Notizie geologiche sopra il vulcano estinto di Roccamonfina,* "Il Lucifero". Napoli, v. II, no. 46, 1839, p. 142.
57. *Sur l'éruption du Vésuve de Janvier 1839. Lettre à M. Elie de Beaumont,* "Comptes Rendus de l'Académie des Sciences," Paris, v. VIII, I semèstre 1839, p. 250–253.
58. *Relazione de' fenomeni avvenuti nel Vesuvio ne' i primi del corrente anno 1839,* "Il Progresso delle Scienze, delle Letterature e delle Arti," Napoli, v. XXXII, 1839, p. 28–40.
59. *Discorso accademico intorno ai principali progressi della geologia ed allo stato presente di questa scienza recitato nella sala dell'Accademia Pontaniana il dì 29 aprile 1839,* Napoli, Tipografia Platina, 1840, 35 p.
60. *Illustrazione a due spaccati geologici degli Appennini presi alle due estremità, meridionale e settentrionale del Regno di Napoli,* nel volume "Atti della I Riunione degli Scienziati italiani. Pisa, 1840.
61. *Observation sur le groupe montagneux de la Roccamonfina,* "Annales des Mines," Paris, série 3, v. XVIII, 1840, p. 127–144.
62. *Observations sur le groupe volcanique de Roccamonfina en Campanie. Lettre à M. Elie de Beaumont,* "Comptes Rendus de l'Académie des Sciences," Paris, v. X, 1840, p. 766–770.
63. *Presentazione di due sezioni geologiche del Regno di Napoli,* nel volume "Atti della I Riunione degli Scienziati Italiani, Pisa 1839," Pisa, 1840.
64. *Primo rapporto a S. E. il Ministro Segretario di Stato degli Affari Interni* (sulla struttura geologica del Gargano in data 8 gennaio 1840), "Giornale degli Atti della Società Economica di Capitanata," Foggia, v. V, 1839–1840, p. 93–96.
65. *Rapporto diretto all'Intendente di Capitanata sopra la struttura geologica del monte Gargano,* "Giornale degli Atti della Società Economica di Capitanata," Foggia, v. V, 1839–1840, p. 109–111.
66. *Relazione diretta all'Intendente di Capitanata sopra la struttura geologica del monte Gargano,* "Giornale degli Atti della Società Economica di Capitanata," Foggia, v. V, 1839–1840, p. 96–100.
67. *Secondo rapporto a S. E. il Ministro Segretario di Stato degli Affari Interni* (sulla struttura geologica del Gargano in data 20 gennaio 1840), "Giornale degli Atti della Società Economica di Capitanata," Foggia, v. V, 1839–1840, p. 96–100.
68. *Studi di geologia, ovvero conoscenze della scienza della Terra. Parte prima,* Napoli, 1840, 136 p.
69. *Sulla giacitura de' marmi ed alabastri del monte Gargano,* "Annali Civili del Regno delle Due Sicilie," Napoli, v. XXII, 43, 1840, p. 14–18.
70. *Applicazione della teorica dei crateri di sollevamento al vulcano di Roccamonfina,* nel volume "Atti della III Riunione degli Scienziati italiani," Firenze, 1841, p. 169–171.
71. *Conoscenze di mineralogia necessarie per lo studio della geologia,* Napoli, 1841, 71 p.
72. *Observation relatives au Vésuve. Extrait d'une lettre a M. E. de Beaumont,* "Comptes Rendus de l'Académie des Sciences," Paris, v. XII, 1841, p. 997–1000.
73. *Observation sur le groupe volcanique de Roccamonfina en Campanie. Lettre à M. Elie de Beaumont,* "Giornale di mineralogia e geologia," edited by Leonhard, v. II, 1841.
74. *Sur une chaleur extraordinaire ressentie à Naples le 16, 17 e 18 juillet 1840,* "Comptes Rendus de l'Académie des Sciences," v. XIII, II semèstre, 1841, p. 448–449.
75. *Ueber die vulkanische Gruppe von Roccamonfina,* "Neues Jahrbuch fuer Min. etc.," Stuttgart, 1841, p. 162–175, 1 tav.
76. *Application de la théorie des cratères de soulèvement au volcan de Roccamonfina,* "Bulletin de la Société Géologique de France," Paris, v. XIII, p. 402–403.
77. *Discorso proemiale recitato nell'apertura della Cattedra di Mineralogia e Geologia della I. R. Università di Pisa il 15 Novembre 1842,* Pisa, 1842.
78. *E dissertatione Nicolai Stenonis de solido intra solidum naturaliter contento.Eexcerpta, in quibus doctrinas geologicas quae hodie sunt in honore facile est reperire. Curante L. Pilla*: Florentiae, Ex Typographia Galilaeana, 1842.
79. *Notizie geologiche sopra il carbon fossile trovato in Maremma,* 1843.
80. *Sopra la produzione delle fiamme ne' Vulcani, e sopra le conseguenze che se ne possono tirare,* 1843.
81. *Sopra la temperatura di un pozzo aperto a Monte Massi vicino Grosseto,* "Miscellanee di Chimica, Fisica e Storia Naturale di Pisa," Pisa, v. I, no. 7 and 8, 1843.
82. *Sopra la temperie del pozzo di Monte Massi in Toscana,* "Il Lucifero," Napoli, v. VI, no. 15 and16, 1843.
83. *Spaccato dell'Appennino napoletano,* nel volume "Atti della IV Riunione degli Scienziati italiani. Padova 1842," Padova, 1843, p. 393–396.
84. *Sur la production des flammes dans les volcans, et sur les conséquences qu'on peut en tirer,* "Comptes Rendus de l'Académie des Sciences," Paris, v. XVII, no. 17, 1843, p.889–895.

85. *Sur la production des flammes dans les volcans, et sur les conséquences qu'on peut en tirer,* "Bibl. Université de Genève," v. 48, 1843.
86. *Sur la temperature d'un puit ouvert à Monte Massi près de Grosseto en Toscane. Lettre à M. de Beaumont,* "Comptes Rendus de l'Académie des Sciences," v. XVI, no. 23 1843.
87. *Aggiunte al discorso sopra la produzione delle fiamme ne' Vulcani* "Il Cimento," Pisa, settembre-ottobre 1844, p. 380–396.
88. *Application de la théorie des cratères de soulèvement au Volcan de Roccamonfina dans la Campanie,* mémoire traduit de l'italien par L. Frappoli, "Mémoires de la Société Géologique de France," série 2, v. I, 1844, p. 163–179, 3 pl.
89. *Application de la tHéorie des cratères de soulèvement au volcan de Rocca Monfina, dans la Campanie,* Parigi, P. Bertrand Librarie, 1844.
90. *Della epidoside, nuova specie di roccia distinta in Toscana tra la famiglia dei Gabbri,* "Il Cimento," Pisa, maggio-giugno 1844.
91. *Intorno alla separazione della Calabria meridionale dalla Penisola Italiana nel periodo terziario subappennino,* "Annuario Geografico Italiano," Bologna, Ranuzzi, v. I, 1844.
92. *Lettera alla egregia donna Giuseppa Nobile Guachi sopra il golfo della Spezia,* "Indicatore Pisano," no. 25 and 25, 1844.
93. *Sopra la produzione delle fiamme ne' Vulcani, e sopra le conseguenze che se ne possono tirare,* in "Atti della V Riunione degli Scienziati Italiani. Lucca 1843," Lucca, 1844, p. 255–258 and p. 293–318, 2 pl.
94. *Sopra la produzione delle fiamme ne' Vulcani, e sopra le conseguenze che se ne possono tirare,* "Museo di Scienze e Letteratura," Napoli, v. 27, marzo 1844.
95. *(Sull'età del calcare secondario nell'Appennino napoletano),* "Atti della V Riunione degli Scienziati italiani. Lucca 1843," Lucca, 1844, p. 242.
96. *Sur les phénomènes volcaniques de l'Italie méridionale,* "Mémoires de la Société Géologique de France," Parigi, série 2, v. I, 1844, p. 179.
97. *(Sul Tempio di Serapide),* nel volume "Atti della V Riunione degli Scienziati italiani. Lucca 1843," Lucca, 1844, p. 261.
98. *Aggiunte al discorso sopra la produzione di fiamme nei vulcani,* "Nuovi Annali di Scienze Naturali," Bologna, seconda serie, v., III, 1845, p. 161–178.
99. *Nouveau mèmoire sur le terrain etrurien,* "Comptes Rendus de l'Académie des Sciences de Paris," tome XXI, 20 octobre 1845.
100. *Ricerche intorno alla vera posizione geologica del macigno in Italia e nel Mezzogiorno di Europa,* "Il Cimento," Pisa, gennaio-febbraio 1845.
101. *Ricerche intorno alla vera posizione geologica del macigno in Italia e nel Mezzogiorno di Europa,* "Comptes Rendus de l'Académie des Sciences de Paris," v. XX, 13 janvier 1845.
102. *Saggio comparativo de' terreni che compongono il suolo d'Italia,* nel volume "Atti della VI Riunione degli Scienziati italiani. Milano 1844," Milano, 1845, p. 544 and 567–574.
103. *Saggio comparativo de' terreni che compongono il suolo d'Italia,* "Annali dell'Università Toscane," Pisa, v. I, second part, p. 205–339, 1 pl.
104. *Sur la production des flammes dans les volcans, et sur les conséquences qu'on peut en tirer,* "Bulletin de la Société Géologique de France," Paris, série 2, v. II, 1844–1845, p. 595–599.
105. *Sur le filons pyrotechnique et cuprifères de Campiglia en Toscana,* "Comptes Rendu de l'Académie des Sciences de Paris," Parigi, v. XX, 17 mars 1845.
106. *Sur quelques minéraux recueillis en Vésuve et à la Roccamonfina,* "Comptes Rendu de l'Académie des Sciences de Paris," v. XXI, aout 1845.
107. *Alcune osservazioni circa la dottrina delle cause geologiche attuali esposte dal sig. Lyell nei suoi ultimi principi di geologia,* "Il Cimento," v. 47, 1846.
108. *Breve cenno sulla ricchezza minerale della Toscana,* Pisa, Vannucchi, 1846.
109. *Descrizione de' caratteri del terreno Etrurio,* "Il Cimento," Pisa, Maggio-Giugno 1846.
110. *Distinzione del terreno Etrurio tra' piani secondari del mezzogiorno di Europa,* Pisa, Vannucchi, 1846, 107 p., 3 pl.
111. *Istoria del tremuoto che ha devastato i paese della costa Toscana il dì 14 agosto 1846,* Pisa, Vannucchi, 1846.
112. *Lettre à M. Arago sur le tremblement de terre qui vient de ravager la Toscane,* "Annales de chemie. et de physique.," v. XVIII, novembre 1846.
113. *Poche parole sul tremuoto che ha desolato i paesi della costa toscana,* Pisa, 1846.
114. *Sur le vraie position du Macigno en Italie et dans le Midi d'Europe,* "Mémoire de la Société Géologique de France," Paris, série 2, v. 1–2, 1846, p. 149.
115. *Notice sur le calcaire ammonitifère de l'Italie,* "Bulletin de la Société Géologique de France," 1847. *Trattato di geologia. Diretto specialmente a fare un confronto tra la struttura fisica del settentrione e del mezzogiorno di Europa.* Parte prima, Pisa, Tipografia Vannucchi, 1847, 549 p.
116. *Discorso recitato nella Università di Pisa nella occasione del conferimento di laurea in Scienze naturali,* Tip: Nicola Fabbrini, Firenze, 1847.
117. *Lettera al geologo Murchison sul terreno Etrurio,* "Il Cimento," Pisa, 1847.
118. *Osservazioni sulla pietra lenticolare di Cascina,* "Il Cimento," Pisa, 1847

119. *Riflessioni sulla pietra lenticolare di Cascina nelle colline pisane,* "Corrispondenze Scientifiche," Roma, v. I, 1847.

Posthumous

1. *Trattato di geologia. Diretto specialmente a fare un confronto tra la struttura fisica del settentrione e del mezzogiorno di Europa.* Parte seconda, Pisa, Tipografia Vannucchi, 1847–1851, 616 p.
2. *Rapporto a S. E. il Tenente Generale marchese Nunziante, Presidente della Società Sebezia, intorno al terreno carbonifero ed al carbon fossile della Provincia di Teramo,* in Carelli Vincenzo, *Esplorazioni disposte dal R. Governo per la ricerca di nuove miniere negli Abruzzi e nel contado di Molise,* "Annali Civili Del Regno delle Due Sicilie," Napoli, v. LXV, CXXIX, gennaio-febbraio 1859. (The report is dated 1833.)
3. *Lettere inedite al Prof. Antonio Orsini,* Cassino, Tip: L. Ciuffi, 1890, 36 p.
4. *Alcuni passi di Dante interprete Leopoldo Pilla,* Tipografia Francesco Mariotti, Firenze, 1911.
5. *Un manoscritto inedito di Leopoldo Pilla sul malgoverno borbonico,* edited by Michel, E., Officine grafiche G. Chiappini, Livorno, 1912.

ACKNOWLEDGMENTS

I'd like to thank G.B. Vai for his invitation to write this brief memorial on Leopoldo Pilla that expresses my admiration for this forerunner of the contemporary "runaway brains," and I hope to be forgiven for this unpretentious essay being a dilettante in science history. I am very thankful to P. Corsi, A.G. Fischer, and G.B. Vai, and to the anonymous reviewers for their helpful implementations and suggestions. I also wish to thank Patricia Sclafani who helped with the English translation and Lucia Toro for having patiently typed the text. This paper partly derives from a previous work edited by M. Discenza in 1996. Also, the list of Pilla papers is from Discenza (1996; with permission).

REFERENCES CITED

Bassani, F., 1905, In memoria di L. Pilla: Napoli, Rendiconti Accademia Scienze Fisiche e Matematiche, s. 3, v. 11, p. 477–492 (cum bibl.).

Ciancio, L., 1995, La geologia italiana dell'800 fra storia naturale e specializzazione disciplinare: la normale anomalia di Giuseppe Scarabelli, *in* Pacciarelli, M., and Vai, G.B., eds., Musei Civici di Imola: La Collezione Scarabelli: Bologna, Grafis Edizioni, Casalecchio di Reno, p. 25–48.

Corsi, P., 2001, La scuola geologica Pisana, *in* Storia dell'Università di Pisa: Pisa, Giardini Editore, v. 2, no. 3, p. 889–927.

Corsi, P., 2003, The Italian Geological Survey: The early history of a divided community (La Carta Geologica d'Italia: agli inizi di un lungo contenzioso), *in* Vai, G.B., and Cavazza, W., eds., Four centuries of the word geology: Ulisse Aldrovandi 1603 in Bologna: Bologna, Minerva Edizioni, p. 271–299.

D'Archiac, A., 1862, Corse de Paléontologie stratigraphique: Précis de l'histoire de la Paléontologie stratigraphique: Paris, Savy, 491 p.

D'Argenio, B., 1996, Gli anni della maturità di Leopoldo Pilla geologo, *in* Discenza, M., ed., Leopoldo Pilla: Notizie storiche della mia vita quotidiana, a cominciare dal primo gennaro 1830 in poi: Venafro, Edizioni Vitmar, p. xix–xxx.

Discenza, M., cur., 1996, Leopoldo Pilla: Notizie storiche della mia vita quotidiana, a cominciare dal primo gennaro 1830 in poi: Venafro, Edizioni Vitmar, xlvii + 655 p.

Ippolito, G., 1948, Leopoldo Pilla (1805–1848): Bollettino Società Naturalisti in Napoli, v. 57, p. 24–35.

Marabini, S., 1995, L'esplorazione degli inediti geologici di Scarabelli: appunti per una biografia scientifica, *in* Pacciarelli, M., and Vai, G.B., eds., Musei Civici di Imola: La Collezione Scarabelli: Bologna, Grafis Edizioni, Casalecchio di Reno, p. 105–147.

Monsagrati, G., 1996, Leopoldo Pilla: Scienziato per vocazione, eroe per scelta, *in* Discenza M., ed., Leopoldo Pilla: Notizie storiche della mia vita quotidiana, a cominciare dal primo gennaro 1830 in poi: Venafro, Edizioni Vitmar, p. xxxi–xlvii.

Pancaldi, G., ed., 1983, I congressi degli scienziati italiani nell'età del positivismo: Bologna, Clueb.

Pilla, N., 1795, Saggio litologico dei vulcani estinti di Roccamonfina, di Sessa e di Teano: Napoli, Orsino, xiii + 74 p.

Pilla, N., 1823, Geologia vulcanica della Campania: Napoli, v. 1, xix + 125 p.; v. 2, 161 p. (published by the author).

Pilla, L., 1842, Discorso proemiale recitato nell'apertura della Cattedra di Mineralogia e Geologia della I.R.: Pisa, Università di Pisa il 15 Novembre 1842.

Pilla, L., 1846, Distinzione del terreno Etrurio tra' piani secondari del mezzogiorno di Europa: Pisa, Vannucchi, 107 p., 3 pl.

Pilla, L., 1847, 1851, Trattato di Geologia: Pisa, Tipografia Vannucchi, v. 1 (1847), xiv + 550 p., v. 2 (1851), 616 p.

Scherillo, A., 1966, La Storia del Real Museo Mineralogico di Napoli nella storia napoletana: Atti Accademia Pontaniana, Napoli, new series, v. 15, p. 3–47.

Scherillo, A., 1988, 1992, Ricordo di Leopoldo Pilla: Memorie Società Geologica Italiana, v. 41, p. 161–168.

Vai, G.B., 2006, this volume, Isostasy in Luigi Ferdinando Marsili's manuscripts, *in* Vai, G.B., and Caldwell, W.G.E., The origins of geology in Italy: Geological Society of America Special Paper 411, doi: 10.1130/2006.2411(07).

Zittel (von), K., 1901, History of geology and paleontology to the end of the nineteenth century: London, W. Scott, (reprinted in 1962 by Wheldon and Wesley, Ltd, and Hafner Publlication Co., New York), xv + 562 p.

MANUSCRIPT ACCEPTED BY THE SOCIETY 17 JANUARY 2006